# Handbook of Plant Classification and Biodiversity

# Handbook of Plant Classification and Biodiversity

Editor: Matthew Taylor

R CALLISTO
REFERENCE

www.callistoreference.com

**Callisto Reference,**
118-35 Queens Blvd., Suite 400,
Forest Hills, NY 11375, USA

Visit us on the World Wide Web at:
www.callistoreference.com

ISBN: 978-1-64116-522-8 (Hardback)

**Cataloging-in-Publication Data**

Handbook of plant classification and biodiversity / edited by Matthew Taylor.
    p. cm.
Includes bibliographical references and index.
ISBN 978-1-64116-522-8
1. Plants--Classification. 2. Plant diversity. 3. Biodiversity. I. Taylor, Matthew.
QK95 .H36 2022
581.012--dc23

# Table of Contents

# Preface

This book has been a concerted effort by a group of academicians, researchers and scientists, who have contributed their research works for the realization of the book. This book has materialized in the wake of emerging advancements and innovations in this field. Therefore, the need of the hour was to compile all the required researches and disseminate the knowledge to a broad spectrum of people comprising of students, researchers and specialists of the field.

Plant classification is the practice of placing known plants into categories or groups in order to establish a relationship. It follows a system of rules that standardizes the results and the group's successive categories into a hierarchy. The classification of plants provides an organized system for the cataloging and naming of the future specimen. It ideally reflects scientific ideas about inter-relationships between plants. Biodiversity refers to the variety and variability of life on earth. It acts as a measure of variation at the ecosystem, genetics and species level. The distribution of biodiversity of plants varies greatly across the globe, as well as within the region. Taxonomic diversity of plants is measured at the level of species diversity. This book includes some of the vital pieces of work being conducted across the world, on various topics related to plant classification and biodiversity. Its aim is to present researches that have transformed this field of study and aided its advancement. It is an essential guide for both academicians and those who wish to pursue this discipline further.

At the end of the preface, I would like to thank the authors for their brilliant chapters and the publisher for guiding us all-through the making of the book till its final stage. Also, I would like to thank my family for providing the support and encouragement throughout my academic career and research projects.

**Editor**

# ELECTROPHORETIC PATTERN OF SEED PROTEINS IN *TRIFOLIUM* L. AND ITS TAXONOMIC IMPLICATIONS

N.M. George[1], A. Ghareeb, N.M. Fawzi[2] and S. Saad

*Department of Botany, Faculty of Science, Zagazig University, Egypt*

*Keywords*: Trifolium; Numerical analysis; SDS-PAGE; Seed protein.

## Abstract

The taxonomic delimitations of 61 taxa of the genus *Trifolium* L. belonging to presently accepted five sections, namely *Lotoidea, Mistyllus, Vesicaria, Chronosemium* and *Trifolium* are evaluated, based on numerical analysis of their electrophoretic seed protein profiles. The dendrogram, resulted from the hierarchical cluster analysis of SDS-PAGE profiles of seed proteins conform, with some restrictions, to the present splitting of the genus *Trifolium* into the sections but not into the subsections and series.

## Introduction

The genus *Trifolium* L. (Clover) is one of the important genera of Papilionoideae of the Leguminosae with agricultural value. It contains 237 species and represented in all continents (Zohary, 1972b). The Mediterranean region and its adjacent countries are one of the main centres of distribution of *Trifolium* species, and also the centre of domestication and breeding of the cultivated species (Zohary and Heller, 1984).

Several taxonomic treatments were made by botanists to divide the genus into natural groups. Linnaeus (1753) divided the genus into five groups, some of which were later accepted as sections. Seringe (1825) proposed the genus with seven sections. Presl (1832) splitted the genus into nine new genera and all of these genera are retained today as sections. Lojacono (1883) distinguished two subgenera within the genus and divided the first subgenus into 11 sections and the second one into only two sections. Boissier (1873) reduced the number of sections to seven. Hossain (1961) divided the genus into eight subgenera. Another approach was adopted by Zohary and Heller (1984), who recognized eight sections for the genus. The first and largest section is tentatively divided into nine subsections and 13 series. Based on morphological characters alone, it is difficult to distinguish the subordinate taxa of the genus *Trifolium* from one another because they have overlapping variations in terms of the major delimiting morphological and biological characters.

The importance of electrophoretic evidence in plant systematics has been discussed in detail by mamy workers (Boulter and Derbyshire, 1971; Gottlieb, 1977; Ghareeb *et al.*, 1999; Kamel, 2005). Electrophoretic profiles of seed proteins have been used in different systematic studies (Badr *et al.*, 2000; Zecevic *et al.*, 2000). In Leguminosae many studies have been carried out based on the electrophoresis of seed proteins (Hussein and George, 2002; Hussein *et al.*, 2005). Electrophoretic patterns of total seed proteins as revealed by polyacrylamide gel electrophoresis (PAGE) with sodium dodecyl sulphate (SDS) have been successfully used to resolve the taxonomic and evolutionary problems of some plant species (Ladizinsky and Hymowitz, 1979; Potokina *et al.*, 2000; Ghafoor and Arshad, 2008; Ayten *et al.*, 2009). Badr (1995) and Nikolic *et al.* (2010) studied the electrophoretic seed profiles of some taxa of the genus *Trifolium*. Recently the phylogeny of the genus *Trifolium* was studied based on DNA sequencing (Ellison *et al.*, 2006).

[1]Corresponding author. Email: nmgtadrous@yahoo.com
[2]Flora and Phytotaxonomy Research Department, Horticultural Research Institute, Agriculture Research Center, Cairo-Egypt.

In the present study, the taxonomic delimitations of 61 taxa of *Trifolium* are re-assessed based on the data resulted from SDS-PAGE profiles of their seed proteins.

## Materials and Methods

In the present study, 61 taxa of *Trifolium* have been investigated. Sources of the seeds directly used for protein extraction are given in Table 1. To extract the seed proteins, 0.5 g of mature seeds ground to meal using a mortar and pestle. The meals were homogenized with 0.5 ml of Tris-HCl buffer containing 2% SDS and 10% sucrose at pH 6.8 for overnight at 4°C. The slurry was centrifuged at 9000 rpm for 6 min. The supernatant (protein extract) was taken for loading on 12.5% polyacrylamide gel. Protein samples (20 µl) including loading dye were loaded in the stacking gel. Electrophoresis was carried out under non-reducing conditions in 12.5% polyacrylamide gel. The assay was carried out by an electric supply of 15 mA for 30 min, and then raised to 25 mA for 5-6 h, using a protein marker with low molecular weights. Gels were then stained in Coomassie brilliant blue for 16 h at room temperature, distained and photographed. The bands produced by each sample were counted. The similarity coefficient between the species based on comparisons of their SDS-PAGE profiles was calculated by Jaccard's coefficient using the SPSS program (version 10.1).

The data obtained from the seed protein banding patterns, each species, were subjected to the numerical analysis. The presence or absence of each of the bands (coded as 1 and 0 respectively) was treated as a binary character in a data matrix. The OUTs (Operational Taxonomic Units), produced from the analysis of SDS-PAGE profiles of seed proteins, collected from the investigated taxa of *Trifolium*, resulted in a dendrogram and it was compared with the current taxonomic treatments of the genus *Trifolium*.

## Results and Discussion

The banding patterns of *Trifolium* taxa are shown in Figure 1. The seed protein profiles of examined taxa illustrated that bands in between marker weight 116KDs and 55KDs are homogenous in comparison to bands in between 50KDs and 14KDs. The relationships among the taxa of *Trifolium* are presented in Figure 2. The dendrogram resulted from the hierarchical cluster analysis of SDS-PAGE profiles of seed proteins of 61 *Trifolium* taxa conform, with some restrictions, to the splitting of this genus into sections, but not with the sub-sectional arrangement under the section *Lotoidea* and section *Trifolium* considered by Zohary and Heller (1984).

The dendrogram shows that the investigated taxa of *Trifolium* are split into two major clusters. The first major cluster includes 20 taxa belonging to section *Trifolium* and the second major cluster includes 41 taxa belonging to four sections, viz., *Lotoidea*, *Mistyllus*, *Vesicaria* and *Chronosemium*. Within the first major cluster, the taxa are divided into two clusters. The first one included *T. alexandrinum*, *T. caudatum* and *T. canescens* in which *T. alexandrinum* was delimited leaving *T. caudatum* and *T. canescens* as a group. In the second cluster, the taxa are divided into two groups. The first group includes *T. arvense*, *T. bocconei*, *T. cherleri* and *T. incarnatum*. The second group includes 12 taxa of section *Trifolium*. *Trifolium ligusticum* represents the subsection *Phleoidea*. The similarity between the taxa belonging to section *Trifolium* ranged from 36.4% to 100%. Zohary and Heller (1984) showed that section *Trifolium* ranks second in the number of species, after section *Lotoidea*, which is consistent with the results of this study. It is heterogeneous in appearance but have several distinctive proteins banding pattern after SDS-PAGE. But their agreement in splitting of section *Trifolium* into 17 small and natural clusters by Zohary (1971, 1972a, b), regarded as subsections does not conform to the results of this study (Table 1, Fig. 2). The grouping of *T. caudatum* and *T. canescens*, as well as, the high similarity

**Table 1. Sections, subsections and series based on Zohary and Heller (1984) and sources of *Trifolium* samples.**

| Section | Subsection | Series | *Trifolium* | Source | Serial no. |
|---|---|---|---|---|---|
| *Lotoidea* | *Loxospermum* | | 8. *T. decorum* Chiov. | ICLA | 9437 |
| | | | 17. *T. multinerve* A. Rich. | ICLA | 13321 |
| | *Ochreata* | | 19. *T. polystachyum* Fresen. | ICLA | 6298 |
| | | | 28. *T. simense* Fresen. | ICLA | 324903 |
| | *Lotoidea* | Lotoidea | 2. *T. amabile* H.BK | RPIS | 262412 |
| | | | 5. *T. burchellianum* Ser. | RPIS | 369911 |
| | | | 6. *T. burchellianum* ssp.*johanstonii* Gillett (Oliv.) | ICLA IPK | 10179 53179 |
| | | | 7. *T. cernum* Brot. | | |
| | | | 10. *T. hybridum* L. | RPIS | 184555 |
| | | | 12. *T. masaiense* Gillett. | ICLA | 896 |
| | | | 14. *T. michelianum* Savi. | IPK | 79181 |
| | | | 15. *T. michelianum* var. *balansae* (Boiss.) Azn. | IPK | 145176 |
| | | | 18. *T. nigrescans* ssp. *nigrescens* Viv. | IPK | 117179 |
| | | | 22. *T. repens* L. | RPIS | 282378 |
| | | | 23. *T. repens* var. *giganteum* Larg.-Foss. | RPIS | 324903 |
| | | | 24. *T. occidental* Coombe. | IPK | 254191 |
| | | | 25. *T. semipilosum* var. *semipilosum* Fresen. | ICLA | 905 |
| | | | 26. *T. semipilosum* var. *glabrescens* Gillett | ICLA | 6235 |
| | | | 30. *T. thalii* Vill. | RPIS | 308090 |
| | *Platystylium* | Platystylium | 1. *T. africanum* Ser. | RPIS | 369885 |
| | | | 3. *T. ambiguum* M. Bieb. | RPIS | 440689 |
| | | | 4. *T. bilineatum* Fresen. | ICLA | 8355 |
| | | | 11. *T. isthmocarpum* Brot. | IPK | 7719 |
| | | | 16. *T. montanum* L. | RPIS | 234914 |
| | | | 20. *T. ruppellianum* var. *ruppellianum* Fresen. | ICLA | 9229 |
| | | | 21. *T. ruppellianum* var. *lanceolatum* Fresen. | ICLA | 6260 |
| | | | 29. *T. tembense* Fresen. | ICLA | 8501 |
| | | Micrantheum | 9. *T. glomeratum* L. | IPK | 136180 |
| | | | 28. *T. suffocatum* L. | IPK | 71179 |
| *Mistyllus* | *Calycospatha* | | 13. *T. mattirolianum* Chivo. | ICLA | 8444 |
| | | | 31. *T. quartinianum* A. Rich. | ICLA | 9428 |
| | | | 32. *T. spumosum* L. | IPK | 67183 |
| | | | 33. *T. teudneri* Schweinf. | ICLA | 9720 |
| | | | 34. *T. xerocephalum* Fenzl. | IPK | |
| *Vesicaria* | | | 35. *T. fragiferum* L. | RPIS | 13322 |
| | | | 36. *T. physodes* Stev. *ex.* M.B. | RPIS | 243229 |
| | | | 37. *T. lumens* Stev. *ex.* M.B. | IPK | 181189 |
| | | | 38. *T. resupinatum* L. | ICLA | 9224 |
| | | | 39. *T. tomentosum* L. | IPK | 138180 |
| *Chronose-mium* | | Agraria | 40. *T. campestre* Schreb. | IPK | 98180 |
| | | Filiformia | 41. *T. dubium* Sibth. | IPK | 234186 |
| *Trifolium* | *Intermedia* | | 54. *T. heldreichianum* (Gib. Belli) Hausskn. | RPIS | 419289 |
| | | | 60. *T. Medium* var. *medium* L. | RPIS | 259988 |
| | | | 61. *T. medium* var. *sarosiense* (Hajsl.) Savul. | RPIS | 179191 |

**Table 1 Contd.**

| Section | Subsection | Series | Trifolium | Source | Serial no. |
|---|---|---|---|---|---|
| | Alpestria | | 43. *T. alpestre* L. | RPIS | 210191 |
| | Stellata | | 56. *T. incarnatum* L. | ILCA | 7018 |
| | Trichoptera | | 48. *T. bocconei* Savi. | IPK | 81187 |
| | Phleoidea | | 59. *T. ligusticum* Balb. *ex.* Loisel. | IPK | 137189 |
| | Lappacea | | 51. *T. cherleri* L. | IPK | 135182 |
| | | | 55. *T. hirtum* All. | IPK | 213175 |
| | | | 57. *T. lappaceum* L. | IPK | 140182 |
| | Arvensia | | 47. *T. arvense* L. | IPK | 40186 |
| | Angustifolia | | 44. *T. angustifolium* L. | IPK | 419304 |
| | | | 47. *T. purpureum* Loisel.var. *desvauxii* (Boiss). | IPK | 143182 |
| | | | 53. *T. dichroanthum* Boiss. | IPK | 130179 |
| | Alexandrina | | 42. *T. alexandrinum* L. | ILCA | 6810 |
| | | | 46. *T. apertum* Bobrov. | IPK | 44182 |
| | Urceolata | | 58. *T. leucanthum* M. Bieb. | IPK | 131177 |
| | Clypeata | | 52. *T. clypeatum* L. | IPK | 129192 |

ILCA = International Livestock Center for Africa at Addis Ababa, Ethiopia; RPIS = Regional Plant Introduction Station, Pullman, Washington, USA; IPK = Institut fur Pflanzengenetik und Kulturpfanzenforschung, Germany.

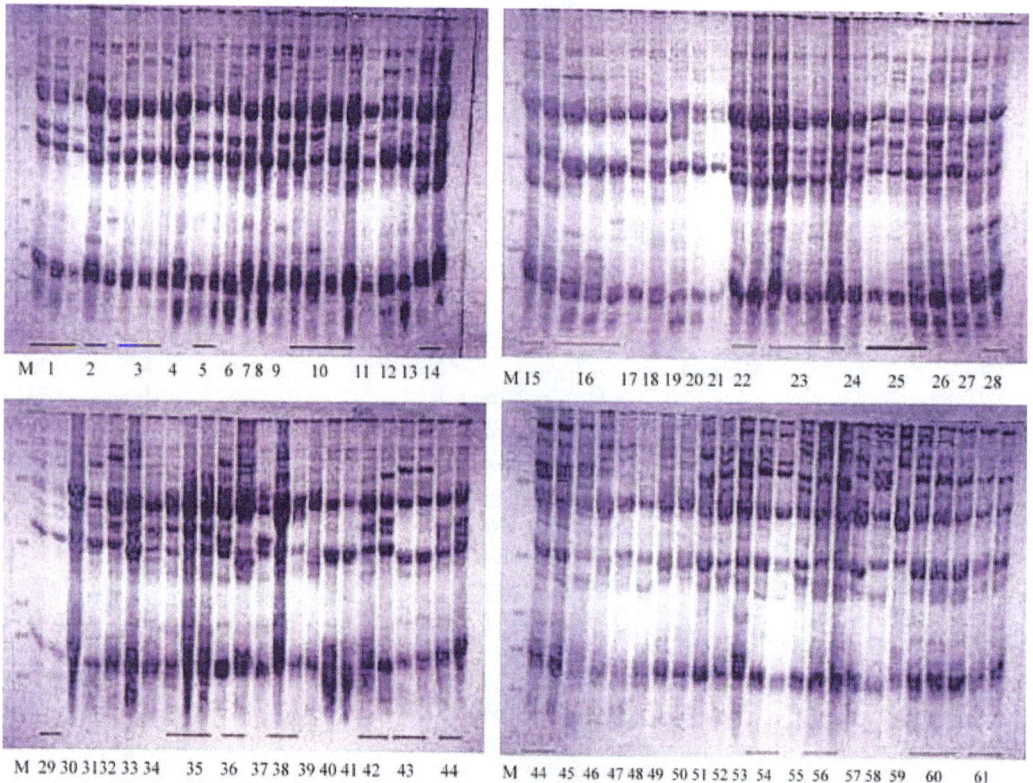

Fig. 1. Electropherogrames produced by SDS-PAGE analysis of seed proteins of 61 *Trifolium* taxa, under non-reducing conditions, numbered as in Table 1. M = Marker protein standards.

(95.7%) between them support their position in subsection *Ochroleuca*. On the other hand, *T. alexandrinum* show low similarity (36.4%) with *T. apertum*, although the obtained results, in the present work, referred that both the two species still delimited under the same section *Trifolium*, it may be claimed that the inclusion of them in the same subsection *Alexandrina* is inconsistent and needs further investigation. Three species *T. cherleri, T. hirtum and T. lappaceum* representing subsection *Lappacea* are distant from one another. This result implies that it may be better to treat them under separate subsections. Although *T. angustifolium, T. purpureum* var. *desvauxii* and *T. dichroanthum* belonging to subsection *Angustifolia, T. angustifolium* and *T. purpureum* var. *desvauxii* grouped together but *T. dichroanthum* grouped with *T. clypeatum* showing similarity (87.0%). Among the three taxa *T. heldreichianum, T. medium* var. *medium* and *T. medium* var. *sarosiense* comprising the subsection *Intermedia*, the two varieties of *T. medium* shows no difference with each other with a similarity of 100% and *T. heldreichianum* differs from these two varieties with a similarity of 52.05%. The present data show that the taxonomic delimitations in section *Trifolium* requires reconsideration and the number of its subsections as proposed by Zohary (1971, 1972a, b), should be reduced.

The second major cluster comprising of four sections (*Lotoidea, Mistyllus, Vesicaria* and *Chronosemium*) is divided into two large clusters. One includes 30 taxa belonging to the section *Lotoidea* and other includes 11 taxa belonging to the sections *Mistyllus, Vesicaria* and *Chronosemium*. Within the large cluster of section *Lotoidea* the taxa combine variously and form six similarity groups as described below. The pairs of taxa *T. tembense* and *T. thalii, T. mattirolianum* and *T. polystachyum*, and *T. isthmocarpum* and *T. masaiense* are consequently segregated as separate groups. The remaining taxa of the section are separated into three groups. The first group is formed by *T. africanum, T. amabile, T. ambiguum, T. bilineatum, T. burchellianum, T. burchellianum* var. *johanstonii, T. cernum, T. decorum, T. glomeratum* and *T. hybridum*. The second group comprised *T. nigrescans* ssp. *nigrescen*s and *T. suffocatum* and the third group is formed by the remaining 12 taxa of the section *Lotoidea*. These groupings of taxa also show that the members included in the subsections *Loxospermum, Ochreata, Lotoidea, Platystylium* and *Calycospatha* or that included in the series *Lotoidea, Platystylium* and *Micrantheum* by Zohary and Heller (1984) do not belong to these subsections or series (Table 1, Fig. 2). The similarity between the taxa belonging to this section ranged from 34.3% to 100%. Among these taxa, *T. ruppellianum* var. *lianruppeelum* and *T. repens* shows no difference respectively with *T. ruppellianum* var. *lanceolatum* and *T. repens* var. *giganteum*, rather a similarity of 100%. *Trifolium semipilosum* and *T. semipilosum* var. *glabrescens* presented the same similarity (100%). *Trifolim michelianum* differs from *T. michelianum* var. *balansae* with a similarity of 96.4%. These results show that the nine subsections and 13 series recognized in section *Lotoidea* by Zohary and Heller (1984) based on morphological characters should be reconsidered. Their view to consider this section as the most primitive group of the genus should be justified by its robust phylogeny. George and Hussein (2002) separated tribe *Ononidea* based on chromosome study, as well as the seed proteins analysis of 10 taxa of tribe *Trifolieae*. Badr (1995) illustrated that, on the basis of seed protein electrophoresis, section *Lotoidea* appears as a heterogenous group in which species relationship requires reconsideration.

The large cluster formed by 11 taxa following the section *Lotoidea* is segregated into three groups, one of which including *T. quartinianum, T. spumosum, T. teudneri* and *T. xerocephalum* is consistent with the section *Mistyllus* recognized by Zohary and Heller (1984). The unique structure of the symmetrically vesicular calyx and the persistent corolla, the manifestly bracteolate flowers and 2-4 seeded pod dehiscing suturally, sharply delimits this section from the others (Zohary and Heller, 1984). The other two groups that include the taxa of sections *Vesicaria* and *Chronosemium* and share the similarities between 53.8% and 76.2% do not completely conform to these sections, as recognized by Zohary and Heller (1984). The two taxa *T. resupinatum* and *T.*

*tomentosum* belonging to the section *Vesicaria* group with the taxa of section *Chronosemium* (Table 1, Fig. 2) which is inconsistent with Zohary and Heller (1984).

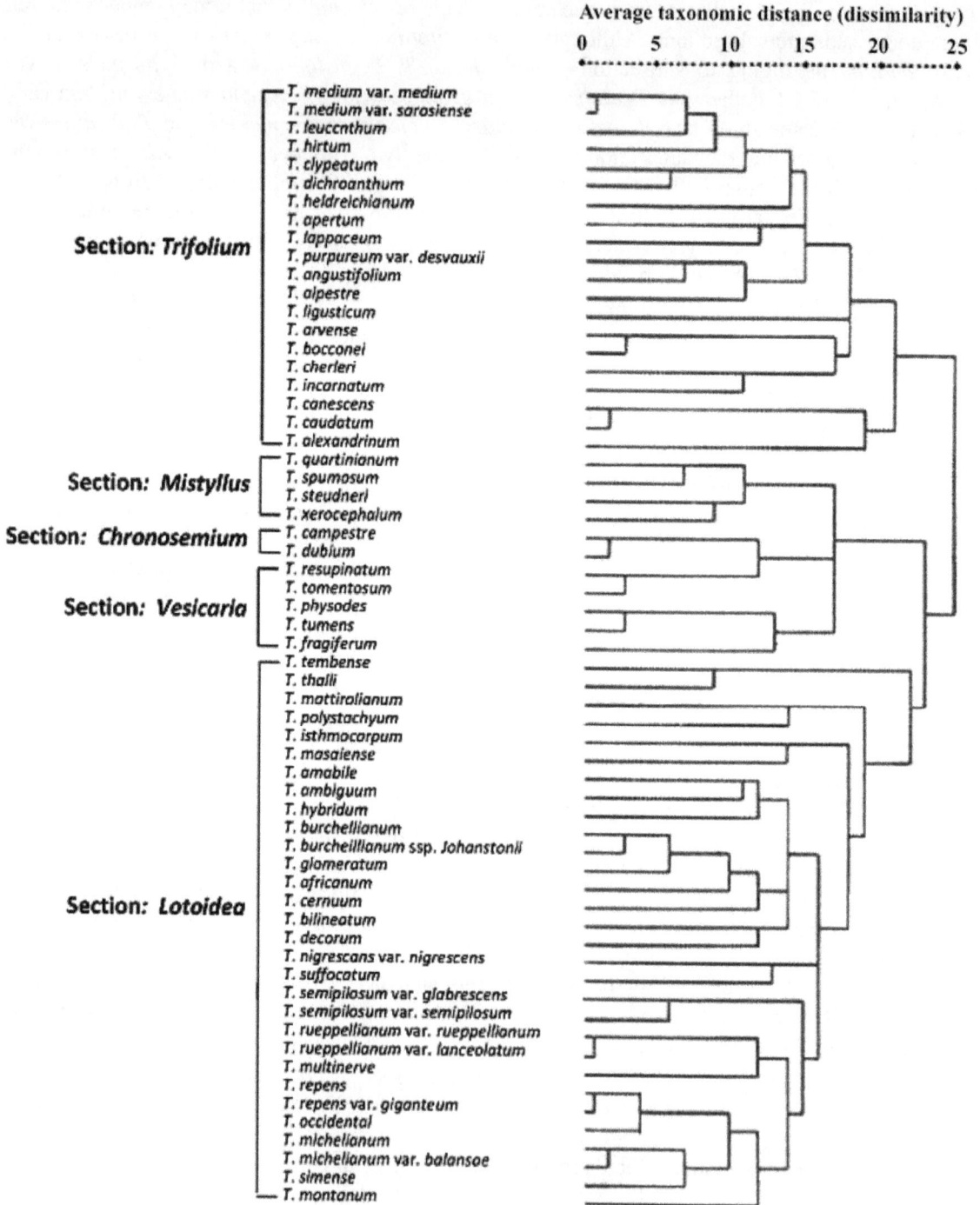

Fig. 2. Dendrogram illustrating the average taxonomic distance (dissimilarity) between the *Trifolium* taxa studied, based on the SDS-PAGE of seed protein characters under non-reducing conditions; numbered as in Table 1.

# References

Ayten, C., Leyla, A. and Zeki, A. 2009. Biosystematics studies among *Ebenus* L. species based on morphological, RAPD-PCR and seed protein analysis in Turkey. Pak. J. Bot. **41**(5): 2477-2486.

Badr, A. 1995. Electrophoretic studied of seed proteins in relation to chromosomal criteria and the relationships of some taxa of *Trifolium*. Taxon **44**: 183-191.

Badr, A., El-Shazly, H.H. and Abou El-Enain, M.M. 2000. Seed protein diversity and its implications on the relationships in the genus *Lathyrus* L. (Fabaceae). Proceedings of the 1st International Conference Biological Sciences, 7-8 May, 2000, Tanta University, pp. 333-346.

Boissier, E. 1873. *Trifolium. In:* Flora Orientalis. Vol. **2**: 110-156, Genevae et Basileae.

Boulter, D. and Derbyshire, E. 1971. Taxonomic aspects of the structure of Legume proteins. *In*: Harborne, J.B., Boulter, D. and Turner, B.L. (Eds), Chemotaxonomy of the Leguminosae. London, pp. 285-305.

Ellison, N.W., Liston, A., Steiner, J.J., Williams, M.W. and Taylor, N.L. 2006. Molecular phylogenetics of the Clover genus (*Trifolium* - Leguminosae). Mol. Phylogenet. Evol. **39**: 688-705.

George, N.M. and Hussein, H. 2002. Taxonomic significance of chromosome number and seed protein electrophoretic analysis in some taxa of tribe Trifolieae (Papilionoideae). Egy. J. Biotechnol. **11**: 323-329.

Ghafoor, A. and Arshad, M. 2008. Seed protein profiling of *Pisum sativum* L., germplasm using sodium dodecyl sulphate polyacrylamide gel electrophoresis (SDS-PAGE) for investigation of biodiversity. Pak. J. Bot. **40**(6): 2315-2321.

Ghareeb, A., Khalifa, S.F. and Nael, F. 1999. Molecular systematic of some *Cassia* species. Cytologia **64**: 11-16.

Gottlieb, L.J. 1977. Electrophoretic evidence and plant systematics. Ann. Miss. Bot. Gard. **64**: 161-180.

Hossain, M. 1961. A revision of *Trifolium* in Nearer East. Notes Roy. Bot. Gard. Edinburgh **23**: 387-481.

Hussein, H. and George, N.M. 2002. Taxonomic importance of floral morphology, chromosome number and seed protein electrophoretic patterns in some species of tribe Vicieae (subfamily: Papilionoideae - Leguminosae). Egy. J. Biotechnol. **11**: 106-123.

Hussein, H., George, N.M. and El-Dimerdash, M.M. 2005. Taxonomic importance of seed protein electrophoretic patterns in some taxa of the subfamily Mimosoideae-Leguminosae. Assiut Univ. J. Bot. **34**(2): 101-130.

Kamel, E.A. 2005. Biochemical and molecular variations in the genus *Raphanus* L. based on SDS-PAGE seed proteins and isozymes patterns. Bull. Fac. Sci. Assut. Univ. **34**(1): 95-113.

Ladizinsky, G. and Hymowitz, T. 1979. Seed protein electrophoresis in taxonomic and evolutionary studies. Theor. Appl. Genet. **54**: 145-151.

Linnaeus, C. 1753. Species Plantarum. 2 vols. Stockholm. Facsimile ed. 1957, Royal Society, London.

Lojacono, M. 1883. Clavis specierum Trifoliorum. Nuova Gironale. Bot. Ilal. **15**: 225-278.

Nikolic, Z., Vasiljevic, S., Karagic, D., Vujakovic, M., Jovicic, D., Katic, S. and Surlan-Momirovic, G. 2010. Genetic diversity of red clover cultivars (*Trifolium pratense* L.) based on protein polymorphism. Genetika J. **42**(2): 249-258.

Potokina, E., Duncan, A., Vaughan, A., Eggi, E.E. and Tomooka, N. 2000. Population diversity of the *Vicia sativa* agg. (Fabaceae) in the flora of the former USSR deduced from RAPD and seed protein analysis. Genet. Resour. Crop Evol. **47**: 171-183.

Presl, C.B. 1832. *Trifolium. In:* Symbolae Botanica. Sumpt. Auctoris Vol. **1**: 44-50.

Seringe, N.C. 1825. *Trifolium. In:* de Candolle A.P., Prodromus Systematis Naturalis Regni Vegetabilis, Parisiis. Vol. **2**: 189-207.

Zecevic, B., Stevanovic, D., Mladenovic-Drinic, S., Konstantinov, K. and Jakovjevic, T. 2000. Proteins as genetic markers in the study of phylogenetic relations in pepper. Acta Hortic. **579**: 113-116.

Zohary, M. 1971. A revision of the species of *Trifolium* sect. *Trifolium* (Leguminosae). I. Introduction. Candollea **26**: 296-308.

Zohary, M. 1972a. Flora Palaestina 2: *Trifolium,* pp. 157-193. Plates 232-267. The Israel Academy of Sciences and Humanities, Jerusalem, Israel.

Zohary, M. 1972b. A revision of the species of *Trifolium* sect. *Trifolium* (Leguminosae). II. Taxonomic treatment. Candollea **27**(1): 99-158.

Zohary, M. and Heller, D. 1984. The Genus *Trifolium.* 1st edition. Ahva Printing Press, Jerusalem, pp. 4-5.

# 2

# *ONOPORDUM MYRIACANTHUM* SUBSP. *ARACHNOIDEUM* COMB. & STAT. NOV. (ASTERACEAE: CARDUEAE)

S. Mesut Pinar[1] and Lütfi Behçet[2]

*Yüzüncü Yıl University, Faculty of Science, Department of Biology, 65080, Van/Turkey*

*Keywords:* Nomenclature; Asteraceae; *Onopordum bracteatum; O. myriacanthum*; Taxonomy.

## Abstract

Turkish endemic taxon *Onopordum bracteatum* Boiss. & Heldr. var. *arachnoideum* Erik & Sümbül is transferred to *O. myriacanthum* Boiss. as *O. myriacanthum* subsp. *arachnoideum* (Erik & Sümbül) Pınar & Behçet comb. & stat. nov. It is characterized by the phyllaries with densely and persistently arachnoid hairs both inside and outside, and upper stem leaves are 2–8 cm far from capitulum. In addition, the pollen characteristics and achene features are presented. The conservation status of *O. myriacanthum* subsp. *arachnoideum* has been assessed according to IUCN criteria. A distribution map of *O. myriacanthum* subsp. *arachnoideum* and its related taxa is also presented.

## Introduction

The genus *Onopordum* L. (Asteraceae, Cardueae) is distributed in the Western and Central Asia, Europe, Northern Africa and the Canary Islands, comprising c. 60 taxa (Susanna and Garcia-Jacas, 2007). Danin (1975) reported 17 species of *Onopordum* for the *Flora of Turkey*. After the publication of this flora, 1 new taxon and 3 new records were added in the subsequent works (Davis *et al.*, 1988; Tuzlacı, 2000; Özhatay *et al.*, 2009; Pınar and Behçet, 2014). Currently, the genus *Onopordum* is represented by 21 taxa in Turkey, of which 7 are endemic. Within the scope of the PhD study of the first author, *O. bracteatum* var. *bracteatum*, *O. bracteatum* var. *arachnoideum* and *O. myriacanthum* specimens have been collected from different localities for revision of *Onopordum* in Turkey. In this study, we aimed to explore the similarity and variation among these taxa and to accurately determine their taxonomic status.

## Materials and Methods

During an expedition carried out in 2010–2011, the endemic taxon *O. bracteatum* Boiss. & Heldr. var. *arachnoideum* Erik & Sümbül was collected from the type locality, Kazancı (Ermenek) district in Central Anatolia. This taxon was first collected by Mecit Vural in 1978 and identified as *O. bracteatum*. Afterwards, it was collected by Hüseyin Sümbül in 1984 and described as a new variety (Erik and Sümbül, 1986). These collections are compared with the type photo of *O. bracteatum* and *O. myriacanthum*, which were obtained from G, K, BM herbaria, and samples deposited in EGE, GAZI, HUB, KNYA, AEF and ANK herbaria. After a careful examination it is concluded that this taxon considerably differs from *O. bracteatum s.l.* Based on the morphological, palynological and geographical evidences presented in this study, we propose the collected taxon as new combination *Onopordum myriacanthum* subsp. *arachnoideum*. The examined specimens of *O. bracteatum*, *O. myriacanthum* subsp. *myriacanthum* and *O. myriacanthum* subsp. *arachnoideum* from different localities are also cited.

---

[1]Corresponding author. E-mail: mesutpinar@hotmail.com
[2]Bingöl University, Science and Art Faculty, Department of Biology, 12000, Bingöl/Turkey.

Pollen and achenes of *O. bracteatum, O. myriacanthum* subsp. *myriacanthum* and *O. myriacanthum* subsp. *arachnoideum* were studied with both light microscope (LM) and scanning electron microscope (SEM). The pollen materials were obtained from either fresh or dried specimens. For LM studies, Wodehouse technique was followed for the preparation of the pollen slides (Wodehouse, 1935). For the SEM investigations, the achenes and pollen were mounted to aluminium stubs, coated with gold in a sputter-coater, and examined under LEO 440 SEM. The descriptive terminology of pollen was adopted from Erdtman (1969), Faegri and Iversen (1975) and Punt *et al.* (2007), and for the seed terminology Koul *et al.* (2000), Tantawy *et al.* (2004) and Hacıoğlu *et al.* (2012) were followed.

## Results and Discussion

**Onopordum myriacanthum** Boiss. subsp. **arachnoideum** (Erik & Sümbül) Pınar & Behçet, **comb. & stat. nov.**                                                                              (Figs 1-3).

*O. bracteatum* Boiss. & Heldr. var. *arachnoideum* Erik & Sümbül in Notes Roy. Bot. Gard. Edinburgh 44: 155 (1986).

*Type*: Turkey. [C4 Konya] Ermenek, around of Kazancı, 650-850 m, 21.6.1984, *H. Sümbül 3024* (Holotype: HUB!).

*Onopordum myriacanthum* subsp. *arachnoideum* differs from *O. bracteatum s.l.* in having phyllaries with densely and persistently arachnoid hair on both sides. In addition, plant is longer, upper stem leaves are 2-8 cm far from capitulum, with smaller and sparse arrangement in upper stem leaves, longer peduncle and presence of glands in corolla lobes. *O. myriacanthum* subsp. *arachnoideum* differs from the typical subsp. *myriacanthum,* by having densely arachnoid hairy phyllaries both inside and outside, more intensive inflorescence (not sparse) and larger capitulum and phyllaries. Comparison of morphological characters of the related taxa *O. myriacanthum* subsp. *arachnoideum, O. myriacanthum* subsp. *myriacanthum,* and *O. bracteatum* is shown in Table 1.

*Onopordum bracteatum* belongs to the Irano-Turanian element and is distributed mainly in Central and Southwestern Anatolia, between 150-1500 m. *O. myriacanthum* subsp. *arachnoideum* belongs to the Irano-Turanian element and is a local endemic to south of Central Anatolia, confined to Karaman province. It grows in rocky slopes and *Pinus brutia* glades, associated with *Picnomon acarna* L., *Euphorbia aleppica* L., *Centaurea solstitialis* L. subsp. *solstitialis, Quercus trojana* Webb, between 600-1200 m altitude. *O. myriacanthum* subsp. *myriacanthum* in comparison, belongs to the Mediterranean element and is distrubuted mainly in West and Southwestern Anatolia; grows on edge of the field, roadside, glades at 100-1000 m (Fig. 2).

*Specimens examined:*

**Onopordum myriacanthum** Boiss. subsp. **arachnoideum** (Erik & Sümbül) Pınar & Behçet, **comb. & stat. nov.**: Turkey. Ermenek, Kazancı, Özlüce village, 1150 m, 11.7.1978, *M. Vural 1094* (*Paratypes*: ANK!, KNYA!, GAZI!); Ermenek, Ermenek-Kazancı, *Pinus brutia* glade, 1000 m, 12.7.1989, *H. Sümbül 3408, J. Venter* (HUB!); Ermenek, Kazancı, around of Çavuşköy, slopes, 785 m, 20.7.2011, 36°33'294" N, 32°58'689" E, *M. Pınar 3205* (VANF!).

**O. myriacanthum** Boiss., Diagn. ser. 2(6): 114 (1859) subsp. **myriacanthum** – Syntypes: Greece. In regione media montis Malevo Laconiae, *Orphanides 55* (G photo!, BM photo!); in regione media Parnassi inter Rachova et Gourna, *Heldreich 3203* (G photo!, K photo!); Turkey. B1 İzmir: Bergama, Kozak paşa, around of çeşme, 21.7.1962, *K. Karamanoğlu* (AEF!); B1 Manisa: Turgutlu, 2 km west of Turgutlu, roadside, 100 m, 8.8.2011, 38°29'309" N, 27°40'007"

E, *M. Pınar* 3377 (VANF!); C2 Denizli: Çukurköy-Serinhisar (Kisilhisar), 10 km away from Çukurköy, edge of field, 985 m, 22.7.2011, 37°37' 098" N, 29°14'442" E, *M. Pınar* 3292 (VANF!); Çukurköy-Kızılhisar, 13.7.1947, *Davis* 13285 (ANK!); Çukurköy, Denizli river, 14.7.1947, *Davis* 13460 (ANK!) C2 Muğla: Fethiye, south slopes of Babadağ, around of Kozaağaç, glade, 750 m, 10.8.2011, 36°31'288" N, 29°09'284" E, *M. Pınar* 3420 (VANF!); Baba Dağ, 610 m, *Davis* 13666 (ANK!).

Fig. 1. Habit and capitulum of *Onopordum bracteatum* (a, b), *O. myriacanthum* subsp. *myriacanthum* (c, d) and *O. myriacanthum* subsp. *arachnoideum* (e, f).

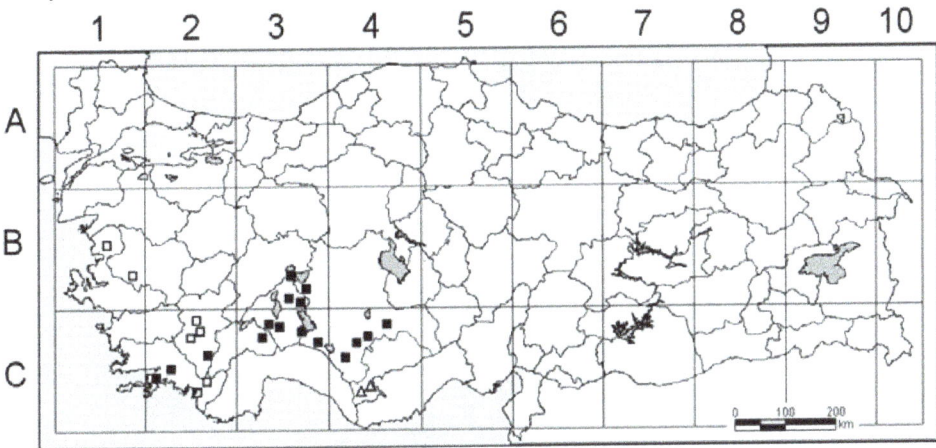

Fig. 2. Distribution map of *Onopordum bracteatum* (■), *O. myriacanthum* subsp. *myriacanthum* (□) and *O. myriacanthum* subsp. *arachnoideum* (Δ) in Turkey.

**Table 1. Comparison of the diagnostic characters of *Onopordum myriacanthum* subsp. *arachnoideum*, *O. myriacanthum* subsp. *myriacanthum* and *O. bracteatum*.**

| Characters | *O. myriacanthum* subsp. *arachnoideum* | *O. myriacanthum* subsp. *myriacanthum* | *O. bracteatum* |
|---|---|---|---|
| Plant height | 2.0–2.5 m | Up to 2 m | Up to 1.3 m |
| Stem | Branching from near to the base or from the middle | Branching from near to the base or from the middle | Branching from the middle |
| Stem branching | Dense | Sparse | Sparse |
| Upper stem leaves | | | |
| size | 2–5 × 1–2 mm | 2–5 × 1–2 cm | 2–10 × 1–5 cm |
| position | away from capitulum or below the capitulum | away from capitulum or below the capitulum | below the capitulum |
| number | 1–2 | 1–2 | many |
| arrangement | sparse | sparse | densely clustered |
| Peduncles | 2–8 cm long | 2–10 cm long | 0.5–3.0 cm long |
| Involucre | 3–4 × 6–10 cm (excl. spines) | 3–4 × 6–8 cm (excl. spines) | 3–5 × 6–11 cm (excl. spines) |
| Phyllaries indumentum | Densely and persistently arachnoid hairy | Glabrous | Glabrous |
| Outer phyllaries | 30–45 × 6–10 mm (spines 3–5 mm) | 20–35 × 5–8 mm (spines 2–5 mm) | 20–45 × 5–10 mm (spines 2–5 mm) |
| Median phyllaries | 35–50 × 6–10 mm (spines 5–8 mm) | 25–40 × 6–8 mm (spines 4–8 mm) | 25–60 × 6–10 mm (spines 4–10 mm) |
| Inner phyllaries | 30–45 × 2–4 mm (spines 3–5 mm) | 30–42 × 2–4 mm (spines 4–6 mm) | 30–50 × 2–5 mm (spines 2–5 mm) |
| Corolla | | | |
| length | 38–48 mm | 35–45 mm | 35–50 mm |
| lobes | glandular | glandular | eglandular |
| lobe lengths | 4 lobes equal, 10–12 mm 5th lobe longer, 13–15 mm | 4 lobes equal, 9–10 mm 5th lobe longer, 14–16 mm | 4 lobes equal, 9–12 mm 5th lobe longer, 13–16 mm |
| Corolla | | | |
| length | 38–48 mm | 35–45 mm | 35–50 mm |
| lobes | glandular | glandular | eglandular |
| lobe lengths | 4 lobes equal, 10–12 mm 5th lobe longer, 13–15 mm | 4 lobes equal, 9–10 mm 5th lobe longer, 14–16 mm | 4 lobes equal, 9–12 mm 5th lobe longer, 13–16 mm |

**O. bracteatum** Boiss. & Heldr. in Boiss., Diagn. ser. 1(10) : 91 (1849) – *Holotype:* Turkey. C3 Burdur: in saxosis prope Aglansoun (Ağlasun) ad radices montis Boudroun Pisidiae, *Heldreich* 1130 (G photo!); B3 Konya: Southwest of Akşehir lake (Yeniköy), around of lake, roadside, 1050 m, 7.8.1982, *M. Küçükodük* 170 (KNYA!); B3 Afyon: Sultandağı, west of Sultandağı, steppe, 1340-1370 m, 1.8.1993, *A. Dönmez, M. Ekici, Z. Aytaç* 6425 (GAZI!); B3 Isparta: Akşehir-Isparta, around of Bağıllı, 5 km before Gelendost, roadside, 980 m, 22.7.2011, 38°08'507" N, 31°03'726" E, *M. Pınar* 3271(VANF!); C2 Denizli: Acıpayam, Abbas, *Davis* 13470 (ANK!); C2 Muğla: Marmaris, Kozcakara Dağ, 150 m, 15.7.1960, *Khan et al* 64 (ANK!); Marmaris, Marmaris-Muğla, Çetibeli pass, edge of field, 550 m, 10.8.2011, 36°56'515" N, 28°15'577" E, *M. Pınar* 3416 (VANF!); C3 Isparta: Eğirdir, Barda mount, 1150-1250 m, 8.9.1982, *Y. Gemici, L. Bekat* 607 (EGE!); Ş. Karaağaç, between Kıyakdede and Göztepe mountains, 1200-1300 m, 23.7.1994, *B. Mutlu* 998 (HUB!); Eğridir, 10 km before Eğridir, edge of lake, 1200 m, 22.7.2011, 37°52'583" N, 30°54'282" E, *M. Pınar* 3276 (VANF!); C3 Burdur: Ağlasun, East of Ağlasun, steppe, 1130 m, 22.7.2011, 37°38'499" N, 30°31'271" E, *M. Pınar* 3286 (VANF!); C3 Konya: Seydişehir, north of Koyucak mount, oak gap, 1500 m, 27.7.1983, *H. Ocakverdi* 1652 (KNYA!); Beyşehir, Hoyran-Kurucaova, steppe, 1160 m, 22.7.2011, 37°35'305" N, 31°33'540" E, *M. Pınar* 3266 (VANF); Çumra-Bozkır, around of Dinek, slopes, 1100 m, 20.7.2011, 37°20'532" N, 32°36'418" E, *M. Pınar* 3181 (VANF!).

The pollen polar axis average is 57.52 µm and equatorial axis 59.19 µm, P/E ratio 0.97, exine thickness 9.69 µm, colpus length 32.62 µm and colpus width 19.69 µm. Pollen shape is oblate-spheroidal, the ornamentation microreticulate and sculpture is echinate in *O. myriacanthum* subsp. *arachnoideum*. The achenes are greyish white, average size is 5.58 × 2.96 mm. Usually achenes shape are obovate and transversely rugose. Sculpture ornamentation of achene surface finely and irregularly undulate. A comparative account of palynological and achene properties of *O. bracteatum*, *O. myriacanthum* subsp. *myriacanthum* and *O. myriacanthum* subsp. *arachnoideum* are shown in Tables 2 and 3.

**Table 2.** Comparison of pollen characters of *Onopordum bracteatum*, *O. myriacanthum* subsp. *myriacanthum* and *O. myriacanthum* subsp. *arachnoideum* (Max.: maximum, Min.: minimum, M: mean, SD: standard deviation).

| Characters | *Onopordum bracteatum* | *O. myriacanthum* subsp. *myriacanthum* | *O. myriacanthum* subsp. *arachnoideum* |
|---|---|---|---|
| | Min. - Max. (M ±SD) | Min. - Max. (M ±SD) | Min. - Max. (M ±SD) |
| Polar diameter (P) (µm) | 54.87–58.84 (57.14 ±0.99) | 53.98–57.85 (56.60 ±1.01) | 54.85–60.60 (57.52 ±1.53) |
| Equatorial diameter (E) (µm) | 55.36–60.48 (58.27 ±1.06) | 55.79–59.41 (57.83 ±0.93) | 56.07–62.41 (59.19 ±1.37) |
| P/E ratio | 0.96–0.99 (0.98±0.01) | 0.95–0.99 (0.97±0.01) | 0.94–0.99 (0.97±0.01) |
| Exine thickness (µm) | 8.91–9.70 (9.44 ±0.23) | 9.15–10.10 (9.67 ±0.24) | 9.23–10.09 (9.69 ±0.23) |
| Colpus length (µm) | 29.40–31.41 (30.79 ±0.48) | 33.67–36.43 (34.79 ±0.80) | 32.15–33.22 (32.62 ±0.27) |
| Colpus width (µm) | 17.38–19.82 (18.28 ±0.63) | 19.68–21.48 (20.63 ±0.37) | 18.60–20.44 (19.69 ±0.40) |

**Table 3.** Comparison of the achene characters of *Onopordum bracteatum*, *O. myriacanthum* subsp. *myriacanthum* and *O. myriacanthum* subsp. *arachnoideum*.

| Characters | *Onopordum bracteatum* | *O. myriacanthum* subsp. *myriacanthum* | *O. myriacanthum* subsp. *arachnoideum* |
|---|---|---|---|
| Length (mm) | 5.7–6.2 (5.91 ± 0.10) | 5.2–6.0 (5.60 ± 0.24) | 5.3 –6.1 (5.58 ± 0.21) |
| Width (mm) | 2.8–3.1 (2.96 ± 0.09) | 2.5–3.5 (3.04 ± 0.30) | 2.5 –3.5 (2.96 ± 0.32) |
| Shape | Oblong-obovate | Obovate | Obovate |
| Surface ornamentation | Transversely rugose | Transversely rugose | Transversely rugose |
| Colour | Cream | Greyish brown | Greyish white |

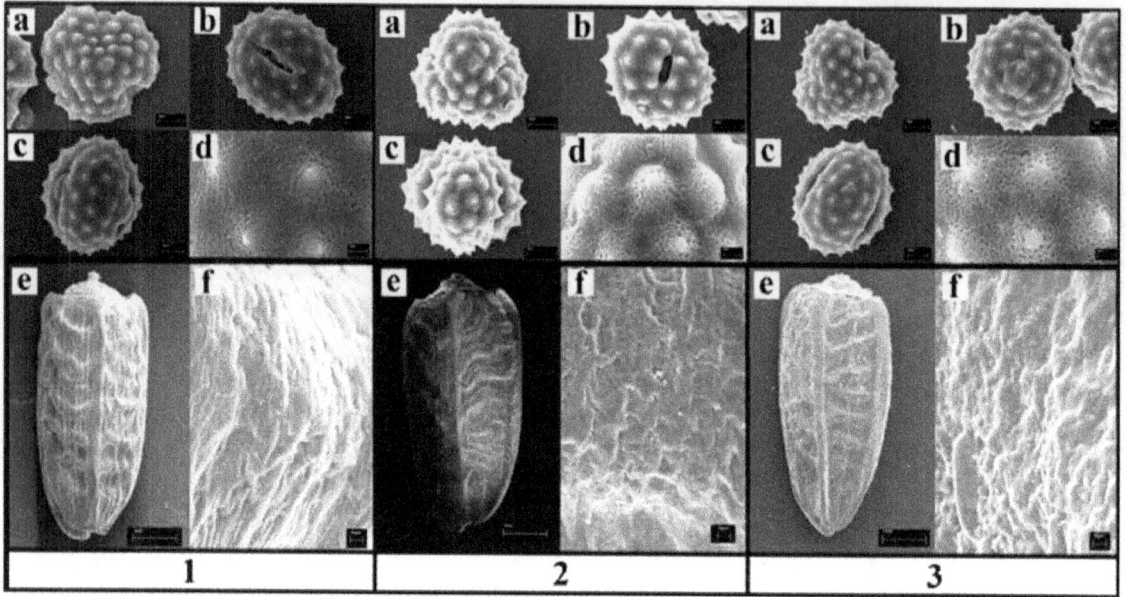

Fig. 3. SEM micrographs of pollen grains and achenes of *Onopordum bracteatum* (1), *O. myriacanthum* subsp. *myriacanthum* (2) and *O. myriacanthum* subsp. *arachnoideum* (3): a. polar view of pollen; b-c. equatorial view of pollen; d. ornamentation of pollen; e. gneral view of achenes; f. surface of achenes.

*Onopordum myriacanthum* subsp. *arachnoideum* is a local endemic taxon, with an estimated occupancy area of less than 10 km$^2$ [criterion B2ab(i)]. The population is endangered, with less than 100 individuals [criterion C2a(ii)]. Therefore, it should be classified as "Critically Endangered (CR)" based on the criteria of the IUCN Red List Categories (IUCN, 2011).

## Acknowledgements

The authors would like to thank Yüzüncü Yıl University (Project no: 2010-FBE-D0131) for its financial support. We are grateful to Dr. Laurent Gautier, head Curator of G (Geneve Herbaria) for the detailed type photographs of *O. bracteatum* and *O. myriacanthum*. We also thank the Curators of AEF, ANK, EGE, HUB, ISTE, ISTF, KNYA and GAZI herbaria for allowing us to study their *Onopordum* specimens.

## References

Danin, A. 1975. *Onopordum* L. *In:* Davis, P.H. (Ed.), Flora of Turkey and the East Aegean Islands. Vol. **5**, Edinburgh Univ. Press, Edinburgh, pp. 356-369.

Davis, P.H., Mill, R.R., Tan, K. 1988. Flora of Turkey and the East Aegean Islands (Supplement 1). Edinburgh Univ. Press, Edinburgh, pp. 317-550.

Erdtman, G. 1969. Handbook of Palynology. Hafner Publishing Co., New York, pp. 21-77.

Erik, S. and Sümbül, H. 1986. Three new taxa from Turkey. Notes Roy. Bot. Gard. Edinburgh **44**(1): 151-156.

Faegri, K. and Iversen, J. 1975. Textbook of Pollen Analysis. Fourth Edition, John Wiley and Sons, New York, pp. 283-284.

Hacıoğlu, B.T., Arslan, Y., Subaşı, İ. Katar, D., Bülbül, A.S. and Çeter, T. 2012. Achene morphology of Turkish *Carthamus* species. Australian J. Crop Sci. **6**(8): 1260-1264.

IUCN 2011. Guidelines for Using the IUCN Red List Categories and Criteria. Version 9.0. Prepared by the Standards and Petitions Subcommittee, Gland, Switzerland.

Koul, K.K., Ranjna, N. and Raina, S.N. 2000. Seed coat microsculpturing in *Brassica* and allied genera (Subtribes Brassicinae, Raphaninae, Moricandiinae). Ann. Bot. **86**: 385-397.

Özhatay, N., Kültür, Ş. and Aslan, S. 2009. Checklist of additional taxa to the Supp. Flora of Turkey IV. Turk. J. Bot. **33**: 191-226.

Pınar, S.M. and Behçet, L. 2014. *Onopordum hasankeyfense* (Asteraceae), a new species from south-eastern Turkey. Turk. J. Bot. **38**: 226-233.

Punt, W., Hoen, P.P, Blackmore, S., Nilsson, S. and Thomas, A. 2007. Glossary of Pollen and Spore Terminology. Review of Palaeobotany and Palynology **143**: 1-81.

Susanna, A. and Garcia-Jacas, N. 2007. Tribe *Cardueae* Cass. *In*: Kubitzki, K. (Ed.), Families and Genera of Vascular Plants, Vol. **VIII**, Flowering Plants, Eudicots, Asterales. Springer-Verlag, Berlin, pp. 123-146.

Tantawy, M.E., Khalifa, S.F., Hassan, S.A. and Al-Rabiai, G.T. 2004. Seed exomorphic characters of some Brassicaceae (LM and SEM Study). International J. Agri. Biol. **6**(5): 821-830.

Tuzlacı, E. 2000. *Onopordum* L. *In*: Güner, A., Ozhatay, N., Ekim, T. and Baser, K.H.C. (Eds), Flora of Turkey and the East Aegean Islands, (Supplement 2). Edinb. Univ. Press, Edinburgh, pp. 160-161.

Wodehouse, R.R. 1935. Polen Grains. McGraw-Hill, New York.

# 3

# TAXONOMY OF THE LEAFY VEGETABLES IN BANGLADESH

MAKSUDA KHATUN, MD. ABUL HASSAN, SHAIKH NAZRUL ISLAM[1]
AND M. OLIUR RAHMAN[2]

*Department of Botany, University of Dhaka, Dhaka 1000, Bangladesh*

*Keywords*: Leafy vegetables; New reports; Taxonomy; Bangladesh.

## Abstract

Thirty four exploration trips made throughout Bangladesh from 2000 to 2012 resulted in identification of 186 taxa used as leafy vegetables in the country, of which 173 taxa belong to angiosperms and 13 taxa to pteridophytes. Among the angiosperms, Magnoliopsida is represented by 153 taxa under 114 genera and 43 families, whereas Liliopsida is represented by 20 species under 15 genera and 8 families. Pteridophytes are symbolized by 13 species belonging to 10 genera and 10 families. Under each taxon updated nomenclature, vernacular names, habit, representative specimen and area of major consumption of the plant as a leafy vegetable have been provided. Out of 186 leafy vegetables identified in Bangladesh, 140 taxa are wild and 46 are cultivated. Among the cultivated ones 16 species are cultivated only as leafy vegetables and 30 are cultivated for other purposes but also used as leafy vegetables. A total of 61 species have been newly documented as leafy vegetables for Bangladesh.

## Introduction

Leafy vegetables are referred to leaves of any plants used as vegetables, sometimes accompanied by tender petioles and shoots. They constitute a major portion of our diet and play an important part in alleviating malnutrition. FAO (2012) has estimated that about 870 million people are chronically undernourished in the period 2010-12 representing 12.5% of the global population, or one in eight people. In order to arrest the undernourished situation, much attention has been paid on the exploitation and utilization of unusual plant materials for food (Kawatra *et al.*, 2001; Dini *et al.*, 2005). Leafy vegetables are important protective foods and highly beneficial for the maintenance of health and prevention of diseases as they contain valuable food ingredients. Usually they have no or very little poisonous alkaloids and do not cause any gastrological disturbance when they are consumed as food. The daily intake of at least 100 g of fresh leafy vegetables is recommended for the adult by nutrition experts (Reddy, 1999). It has been estimated that 100 g of tropical leafy vegetables can provide 60-140 mg of ascorbic acid, 100 mg of folic acid, 4-7 mg of iron and 200-400 mg of calcium (Saxena, 1999). Traditional leafy vegetables are said to be an invaluable substitute for meat and therefore form important part of daily diets of rural communities in particular.

Over the last decade many studies have shown that fresh leafy vegetables constitute important functional food components by contributing vitamins, iron, folic acid, minerals, biologically active compounds and photosynthetic pigments (Kmiecik *et al.*, 2001; Su *et al.*, 2002; Kimura and Rodriguez-Amaya, 2003). Traditional leafy vegetables have a proven nutritive value in terms of having more protein, minerals, carbohydrate and vitamins than several common   vegetables

---

[1]Institute of Nutrition and Food Science, University of Dhaka, Dhaka 1000, Bangladesh.
[2]Corresponding author. Email: dr_oliur@yahoo.com

(Sundriyal and Sundriyal, 2001; Fasuyi, 2006; Orech *et al.,* 2007). Leafy vegetables also contain antioxidants which offer protection against many chronic diseases including heart disease and certain types of cancer (Saxena, 1999).

In Bangladesh, people have a long heritage of taking leafy vegetables. However, very little attempt has been made to study the leafy vegetables of Bangladesh although they constitute a large proportion of the daily diet of the rural dweller of the country (Ali *et al.,* 1977; Sarker and Hossain 2009; Hassan, 2010). Despite the importance of leafy vegetables in the present day human lives, no systematic work has been carried out in Bangladesh to identify and document the plant species. In view of potential beneficial attributes of leafy vegetables, there is a need to explore, identify and document the leafy vegetables of the country. The objectives of the present study are therefore three-fold: i) to make an inventory, and identify the leafy vegetables, ii) to document the leafy vegetables including wild and cultivated ones, and iii) to carry out a detailed systematic study on the leafy vegetables of Bangladesh.

**Materials and Methods**

Thirty four field trips were conducted throughout Bangladesh during 2000-2012 to collect fresh plant materials and each field trip consisted of 4-8 days. The areas visited for collection of plant samples include: Bagerhat (Bagerhat Sadar, Mollarhat), Bandarban (Bandarban Sadar, Lama), Barisal (Barisal Sadar), Chittagong (Mirsharai, Sitakundu), Comilla (Comilla Sadar, Daudkandi), Cox's Bazar (Cox's Bazar Sadar, Teknaf), Dhaka (Nawabganj, Savar), Dinajpur (Birampur, Dinajpur Sadar, Phulbari), Faridpur (Faridpur Sadar, Madhukhali), Gazipur (Gazipur Sadar, Kaliganj, Kapasia, Tongi), Gopalganj (Gopalganj Sadar, Kashiani, Kotalipara, Tungipara), Habiganj (Chunarughat, Madhabpur), Jessore (Jessore Sadar, Jhikargacha, Keshobpur, Manirampur, Sharsha), Jhalokhathi (Jhalokathi Sadar, Rajapur), Jhenaidah (Jhenaidah Sadar, Kaliganj, Kotchandpur, Shailkupa), Khagrachari (Dighinala, Matiranga, Panchari), Khulna (Dighalia, Khulna Sadar, Phultala, Terokhada), Magura (Magura Sadar, Mohammadpur, Shalikha), Manikganj (Saturia, Singair), Moulvi Bazar (Barolekha, Kamalganj, Sreemangal), Munshiganj (Gazaria, Munshiganj Sadar), Mymensingh (Haluaghat, Muktagachha), Narail (Kalia, Lohagara, Narail Sadar), Narayanganj (Fatullah, Siddhirganj), Natore (Natore Sadar, Singra), Netrakona (Durgapur, Khaliajuri), Patuakhali (Kalapara, Patuakhali Sadar), Rajbari (Baliakandi, Pangsha, Rajbari Sadar), Rajshahi (Bagha, Bagmara, Godagari, Puthia), Rangamati (Kaptai, Rangamati Sadar), Sherpur (Jhenaigathi, Sherpur Sadar), Sunamganj (Jagannathpur, Sunamganj Sadar), Sylhet (Sylhet Sadar) and Tangail (Madhupur). Collection of fresh materials was made from local markets, village areas and forest lands. Prior to collection, assistance from local informants was taken regarding the use of plants as leafy vegetables. Local vegetable markets were also surveyed to record marketable items. The collected samples were processed following the standard herbarium technique (Hyland, 1972). Some samples were also kept in liquid preservative.

The collected specimens were critically studied and identified in the Dhaka University Salar Khan Herbarium (DUSH) by matching with the identified specimens housed at DUSH and DACB (Bangladesh National Herbarium), and with the help of standard literatures (Hooker, 1872-1897; Prain, 1903; Khan, 1972-1987; Dassanayake and Fosberg, 1980-1985; Khan and Halim, 1987; Khan and Rahman, 1989-2002). Nomenclatures have been updated using Siddiqui *et al.* (2007-2008), Ahmed *et al.* (2008-2009), and Rashid and Rahman (2011, 2012). The angiosperm families followed that of Cronquist (1981), while pteridophyte families are arranged in an alphabetical order. Under each family the genera and species have been arranged alphabetically. Updated nomenclature, vernacular names (Eng. = English, Beng. = Bangla), habit, representative specimen (only one cited because of page constraint) and area of major consumption have been furnished under each taxon. All voucher specimens have been deposited at DUSH in the Department of Botany, University of Dhaka, Bangladesh.

## Results and Discussion

A total of 186 leafy vegetable taxa have been identified in Bangladesh of which 173 belong to angiosperms, and 13 to pteridophytes. Among the angiospermic taxa Magnoliopsida is represented by 153 and Liliopsida is represented by 20 taxa (Table 1).

**Table 1. Number of taxa of leafy vegetables recorded in Bangladesh.**

| Taxa | Magnoliopsida | Liliopsida | Pteridophyta | Total |
|------|---------------|------------|--------------|-------|
| Families | 43 | 8 | 10 | 61 |
| Genera | 114 | 15 | 10 | 139 |
| Species | 153 | 20 | 13 | 186 |

The taxonomic enumeration of the leafy vegetables is briefly described below.

## MAGNOLIOPSIDA
### 1. Piperaceae C. A. Agardh (1825)

**1. Peperomia pellucida** (L.) H. B. K., Nov. Gen. Sp. 1: 64 (1815). Vernacular names: Pepper Elder (Eng.), *Luchi Pata* (Beng.), *Samol-hapang* (Garo).

A fleshy annual herb. *Representative specimen*: Netrakona: Boheratoli, 8.10.2001, M. Khatun 10. *Area of major consumption:* Netrakona district.

**2. Piper longum** L., Sp. Pl.: 29 (1753). Vernacular names: Long Pepper (Eng.), *Pipla-mul, Pipul, Pipul Morich* (Beng.).

A perennial herb. *Representative specimen*: Gopalganj: Tatulia, 3.3.2007, M. Khatun 428. *Area of major consumption:* Rajbari district.

### 2. Moraceae Link (1831)

**3. Ficus benghalensis** L., Sp. Pl.: 1059 (1753). Vernacular names: Banyan Tree (Eng.), *Bot, Botgachh* (Beng.), *Jalong* (Khasia).

A large tree. *Representative specimen*: Moulvi Bazar: Madhabpunji, Madhabkundu, 5.5.2003, M. Khatun 274. *Area of major consumption:* Moulvi Bazar district.

**4. F. carica** L., Sp. Pl. 2: 1059 (1753). Vernacular names: Common Fig, European Fig (Eng.), *Anjir, Dumur* (Beng.), *Soluya* (Khasia).

A large shrub or small tree. *Representative specimen*: Moulvi Bazar: Madhabkundu, 3.5.2003, M. Khatun 276. *Area of major consumption:* Moulvi Bazar district.

### 3. Nyctaginaceae A. L. de Jussieu (1789)

**5. Boerhavia diffusa** L., Sp. Pl.: 3 (1753). Vernacular names: Pigweed, Spreading Hog-weed (Eng.), *Punarnava* (Beng.)

A perennial, creeping or climbing herb. *Representative specimen*: Rajbari: Olangapur, 8.2.2004, M. Khatun 421. *Area of major consumption:* Rajbari district.

### 4. Aizoaceae Rudolphi (1830)

**6. Sesuvium portulacastrum** (L.) L., Syst. ed. 10: 1058 (1759). Vernacular names: Shoreline Sea Purslane (Eng.), *Nuna Shak* (Beng.), *Phru-bawn* (Rakhain).

A perennial herb. *Representative specimen*: No specimen was collected, but information was gathered from the local people. *Area of major consumption:* Patuakhali district.

## 5. Chenopodiaceae Ventenant (1799)

**7. Beta vulgaris** L., Sp. Pl. 1: 222 (1753). Vernacular names: Garden Beet, Common Beet (Eng.), *Beet, Palak* (Beng.).

A biennial herb. *Representative specimen*: Jhalokathi: Rajapur, 11.4.2010, M. Khatun 588. *Area of major consumption*: Dhaka district.

**8. Chenopodium album** L., Sp. Pl. 1: 219 (1753). Vernacular names: Pigweed (Eng.), *Batua Shak* (Beng.).

An annual herb. *Representative specimen*: Rajbari: Salmara, 8.2.2004, M. Khatun 402. *Area of major consumption*: Rajbari district.

**9. Spinacia oleracea** L., Sp. Pl. 1: 219 (1753). Vernacular names: White Goosefoot (Eng.), *Palong Shak* (Beng.), *Mui-yaa-bawn* (Rakhain).

A small, annual herb. *Representative specimen*: Rajshahi: Binodpur, 13.9.2001, M. Khatun 51. *Area of major consumption:* Dhaka district.

## 6. Amaranthaceae A. L. de Jussieu (1789)

**10. Achyranthes aspera** L., Sp. Pl. 1: 204 (1753).Vernacular names: Prickly Chaff-flower (Eng.), *Apang, Upatlengra* (Beng.), *Longra* (Santal).

A perennial, erect herb. *Representative specimen*: Dinajpur: Mukundupur, 16.6.2001, M. Khatun 76. *Area of major consumption:* Dinajpur, Mymensingh and Comilla districts.

**11. Aerva sanguinolenta** (L.) Bl., Bijdr.: 547 (1826). Vernacular names: *Nuriya, Lal Apang* (Beng.), *Nenga* (Chakma).

A perennial herb. *Representative specimen*: Moulvi Bazar: Adampur bit, 3.5.2003, M. Khatun 194. *Area of major consumption:* Sunamganj and Moulvi Bazar districts.

**12. Alternanthera bettzickiana** (Regel) Voss Nichols., III. Dict. Grad. 1: 59 (1884). Vernacular names: Joyweed (Eng.), *Nun-khuta Shak* (Beng.).

A perennial herb. *Representative specimen*: Patuakhali: Kolapara, 23.5.2008, M. Khatun 475. *Area of major consumption:* Patuakhali district.

**13. A. paronichyoides** St. Hil., Voy. Distr. Diamans Bresil. 2: 439 (1833). Vernacular names: Smooth Chaff-flower (Eng.), *Jhuli Khata* (Rakhain).

A perennial mat-forming herb. *Representative specimen*: Patuakhali: Kalachandpara, 23.5.2008, M. Khatun 476. *Area of major consumption:* Patuakhali district.

**14. A. philoxeroides** (Mart.) Griseb., Symb. Argent. in Abh. Ges. Wiss. Gott. 24: 36 (1879). Vernacular names: Alligator Weed (Eng.), *Malancha Shak* (Beng.), *Shergiti* (Santal).

A perennial, polymorphic herb. *Representative specimen*: Dinajpur: Noiyabad, 16.9.2001, M. Khatun 41. *Area of major consumption:* Dinajpur district.

**15. A. sessilis** (L.) R. Br. *ex* Roem. & Schult., Syst. 5: 554 (1819). Vernacular names: Sessile Joyweed (Eng.), *Chanchi, Sachishak* (Beng.), *Garundi* (Garo).

Annual or perennial herb. *Representative specimen*: Rajshahi: Binodpur, 13.9.2001, M. Khatun 37. *Area of major consumption*: Rajshahi district.

**16. Amaranthus blitum** L., Sp. Pl. ed. 1: 990 (1753). Vernacular names: Purple Amaranth (Eng.), *Natiyasag* (Beng.).

A tall, glabrous, succulent herb. *Representative specimen*: Narail: Lohagara, 15.9.2007, M. Khatun 458. *Area of major consumption:* Narail district.

**17. A. spinosus** L., Sp. Pl. ed. 1: 991 (1753). Vernacular names: Spiny Amaranth (Eng.), *Kantanotey, Kantadenga* (Beng.), *Katakailpha, Kuriakanta* (Tripura).

An annual herb. *Representative specimen*: Dhaka: Uttara, 14.1.2002, M. Khatun 168. *Area of major consumption*: Dhaka district.

**18. A. tricolor** L., Sp. Pl. ed. 1: 989 (1753). Vernacular names: Joseph's Coat (Eng.), *Lal Shak, Dengua* (Beng.), *Puspoo* (Santal).

An annual herb. *Representative specimen*: Jessore: Barobazar, 5.1.2004, M. Khatun 383. *Area of major consumption:* Mostly in urban areas.

**19. A. viridis** L., Sp. Pl. ed. 2: 1405 (1763). Vernacular names: Green Amaranth, (Eng.), *Notey, Notey Shak* (Beng.).

An annual herb. *Representative specimen*: Rajbari: Konagram, 8.2.2004, M. Khatun 416. *Area of major consumption:* Rajbari district.

**20. Celosia argentea** L., Sp. Pl. 1: 205 (1753). Vernacular names: Cock's Comb (Eng.), *Sada Moragphul* (Beng.), *Thinthinga* (Garo), *Thanthania* (Rakhain).

An erect, annual herb. *Representative specimen*: Netrakona: Bijoypur, 8.1.2000, M. Khatun 9. *Area of major consumption*: Patuakhali and Netrakona districts.

**21. C. cristata** L., Sp. Pl. 1: 235 (1753). Vernacular names: Crested Cock's Comb (Eng.), *Morog-ful* (Beng.), *Shibjota* (Garo).

A much branched herb. *Representative specimen*: Patuakhali: Kalachanpara, 23.5.2008, M. Khatun 480. *Area of major consumption:* Netrakona and Patuakhali districts.

**22. Digera muricata** (L.) Mart. in Nov., Acad. Caes. Leop. Carol. 13(1): 285 (1826). Vernacular names: False Amaranth (Eng.), *Boutibon Shak* (Beng.), *Latamouri* (Garo).

An annual herb. *Representative specimen*: Khagrachari: Golabari, 7.7.2003 M. Khatun 328. *Area of major consumption:* Netrakona and Khagrachari districts.

**23. Psilotrichum ferrugineum** (Roxb.) Moq. in DC., Prod. 13(2): 279 (1849). Vernacular names: *Rokto-sirinch* (Beng.), *Puti Shak* (Santal).

An annual herb. *Representative specimen*: Dinajpur: Noabad, 16.9.2001, M. Khatun 78. *Area of major consumption:* Dinajpur district.

## 7. Portulacaceae A. L. de Jussieu (1789)

**24. Portulaca oleracea** L., Sp. Pl.: 445 (1753). Vernacular names: Purslane (Eng.), *Bara Lunia, Kulfi, Lunia Shak* (Beng.), *Tee-jey-shey* (Marma).

An annual herb. *Area of major consumption*: Dhaka district. *Representative specimen*: Dhaka: Shambazar, 14.1.2002, M. Khatun 142.

**25. P. quadrifida** L., Mant. Pl. 1: 73 (1767). Vernacular names: Pot Purslane, Small-leaved Purslane (Eng.), *Chhota Lunia, Munia Shak* (Beng.).

A small, prostrate, annual herb. *Representative specimen*: Rajshahi: Shaheb Bazar, 13.9.2001, M. Khatun 54. *Area of major consumption:* Dhaka and Rajshahi districts.

### 8. Basellaceae Moquin-Tandon (1840)

**26. Basella alba** L., Sp. Pl. 1: 272 (1753). Vernacular names: Indian Spinach (Eng.), *Puishak* (Beng.).

A much branched, fleshy herb. *Representative specimen*: Dhaka: Diabari, 14.1.2002, M. Khatun 131. *Area of major consumption:* Dhaka district.

### 9. Molluginaceae Hutchinson (1926)

**27. Glinus oppositifolius** (L.) A. DC., Bull. Herb. Boiss. 2(1): 522 (1901). Vernacular names: *Gimashak* (Beng.), *Dima Tita* (Koach, Santal).

A diffusely branched, annual herb. *Representative specimen*: Dinajpur: Mukundopur, 16.9.2001, M. Khatun 85. *Area of major comsumption:* Dinajpur and Sherpur districts.

**28. Mollugo pentaphylla** L., Sp. Pl. 1: 89 (1753). Vernacular names: Mollugo (Eng.), *Khetpapra* (Beng.), *Tita Shak* (Garo).

An annual herb. *Representative specimen*: Tangail: Madhupur, Pirgacha, 18.4.2002, M. Khatun 176. *Area of major comsumption:* Tangail district.

### 10. Caryophyllaceae A. L. de Jussieu (1789)

**29. Polycarpon prostratum** (Forssk.) Aschers. & Schweinf., Oesterr. Bot. Zeitscher 39: 128 (1889). Vernacular names: *Ghimashak* (Beng.), *Beng-bong-jathong* (Koach).

A dichotomously branched herb. *Representative specimen*: Sherpur: Gajni, 31.10.2009, M. Khatun 520. *Area of major consumption*: Sherpur district.

**30. Stellaria wallichiana** Benth. *ex* Haines, Bull. Misc. Inf. Kew 1920: 66 (1920). Vernacular names: *Sada Fulki, Tara* (Beng.), *Murmuri Shak* (Koach).

A herb. *Representative specimen*: Moulvi Bazar: Barolekha, Madhabkundu, 4.5.2003, M. Khatun 272. *Area of major consumption:* Moulvi Bazar district.

### 11. Polygonaceae A. L. de Jussieu (1789)

**31. Ampelygonum chinense** (L.) Lindley, Bot. Reg. 24: 63 (1838). Vernacular names: Trailing Smartweed (Eng.), *Mohicharan Shak* (Beng.), *Mono-eja-dar* (Chakma).

A perennial herb. *Representative specimen*: Moulvi Bazar: Madhabpunji, 4.5.2003, M. Khatun 277. *Area of major consumption:* Moulvi Bazar district.

**32. A. microcephalum** (D. Don) Hassan, Bangladesh J. Bot. 22(1): 4 (1993). Vernacular names: Hilly Smartweed (Eng.), *Madhusilum Shak* (Beng.), *Ambimikchip* (Garo).

A perennial herb. *Representative specimen*: Netrakona: Bijoypur, 8.10.2001, M. Khatun 06. *Area of major consumption:* Netrakona district.

**33. A. salarkhanii** Hassan, Bangladesh J. Bot. 20(2): 245 (1991). Vernacular names: Hilly Smartweed (Eng.), *Giri Shobhan Shak* (Beng.), *Lambak* (Khasia).

An undershrub. *Representative specimen*: Rangamati: Kaptai, Shilchari, 6.7.2003, M. Khatun 350. *Area of major consumption:* Rangamati district.

**34. Persicaria tomentosa** (Willd.) Sasaki, List Pl. Form.: 170 (1928). Vernacular names: Hairy Knotweed (Eng.), *Pani-bishkatali* (Beng.), *Hagra* (Mandi).

A perennial, aquatic herb. *Representative specimen*: Tangail: Madhupur, Pirgacha, 18.4.2002, M. Khatun 177. *Area of major consumption:* Tangail district.

**35. Polygonum effusum** Meissn. in DC., Prodr. 14: 93 (1857). Vernacular names: Knotweed (Eng.), *Raniphul, Chemtisag* (Beng.), *Kuttasunga Gas* (Koach).

An annual herb. *Representative specimen*: Gazipur: Gozaripara, 4.12.2009, M. Khatun 552. *Area of major consumption:* Gazipur district.

**36. P. plebeium** R. Br., Prodr.: 420 (1810). Vernacular names: Small Knotweed (Eng.), *Khudi-bishkatali* (Beng.), *Khumchak, Gang-sum* (Khasia).

A prostrate annual herb. *Representative specimen*: Sylhet: Madhabkundu, 4.5.2003; M. Khatun 279. *Area of major consumption:* Habiganj district.

**37. Rumex dentatus** L., Mant. Pl. 2: 226 (1771). Vernacular names: Toothed Dock (Eng.), *Bon Tak-palong* (Beng.).

An annual herb. *Representative specimen*: Rajshahi: Shaheb Bazar, 13.9.2001, M. Khatun 47. *Area of major consumption:* Rajshahi district.

**38. R. maritimus** L., Sp. Pl. 1: 335 (1753). Vernacular names: Golden Dock (Eng.), *Bon Palong* (Beng.).

An annual herb. *Representative specimen*: Natore: Singair, 14.9.2001, M. Khatun 61. *Area of major consumption:* Dhaka and Natore districts.

**39. R. vesicarius** L., Sp. Pl. 1: 336 (1753). Vernacular names: Rosy Dock (Eng.), *Tok Palong, Chuka Palong* (Beng.).

An annual, glabrous herb. *Representative specimen*: Rajshahi: Meharchandi, 13.9.2001, M. Khatun 53. *Area of major consumption:* Rajshahi and Sylhet districts.

## 12. Elaeocarpaceae A. P. de Candolle (1824)

**40. Elaeocarpus floribundus** Blume, Bijdr.: 120 (1825). Vernacular names: Indian Olive (Eng.), *Jalpai* (Beng.), *Uthethamo* (Marma).

A medium-sized to large tree. *Representative specimen*: Bandarban: Bandarban Sadar, 6.3.2012, M. Khatun 618. *Area of major consumption:* Bandarban district.

## 13. Tiliaceae A. L. de Jussieu (1789)

**41. Corchorus capsularis** L., Sp. Pl. : 529 (1753). Vernacular names: Jute (Eng.), *Deshi Pat, Bagi Pat* (Beng.).

An erect, annual herb. *Representative specimen*: Faridpur: Mokshedpur, 9.7.2008, M. Khatun 412. *Area of major consumption*: Faridpur district.

**42. C. olitorius** L., Sp. Pl. : 529 (1753). Vernacular names: Tossa Jute (Eng.), *Tosha Pat, Lalita Pat, Deo Pat* (Beng.).

An annual, erect herb. *Representative specimen*: Dhaka: Bosila, 15.10.2002, M. Khatun 238. *Area of major consumption:* Dhaka district.

## 14. Sterculiaceae Bartling (1830)

**43. Sterculia villosa** Roxb. *ex* Smith in Rees, Cyc. 34: No. 16 (1816). Vernacular names: *Udal* (Beng.), *Lambuk* (Tripura).

A large, deciduous tree. *Representative specimen*: Moulvi Bazar: Madhabpungi, Madhabkundu, 5.5.2003, M. Khatun 280. *Area of major consumption:* Bandarban and Moulvi Bazar districts.

## 15. Malvaceae A. L. de Jussieu (1789)

**44. Hibiscus cannabinus** L., Syst. Nat. ed. 10, 2: 1149 (1759). Vernacular names: Kenaf Hemp, Decan Hemp (Eng.), *Mesta Pat* (Beng.), *Dare Kudrum* (Garo).

A herb with prickly stem. *Representative specimen*: Patuakhali: Kuakata, 23.5.2008, M. Khatun 485. *Area of major consumption:* Patuakhali district.

**45. H. sabdariffa** L., Sp. Pl.: 695 (1753). Vernacular names: Kenaf Hemp (Eng.), *Mesta Pat* (Beng.), *Chukair* (Garo), *Arak* (Rakhain).

An erect, annual herb. *Representative specimen*: Tangail: Madhupur, 18.4.2002, M. Khatun 175. *Area of major consumption:* Tangail district.

**46. H. surattensis** L., Sp. Pl.: 696 (1753). Vernacular names: Wild Sour (Eng.), *Mikhri* (Garo), *Maik Shak* (Khasia).

An annual herb. *Representative specimen*: Tangail: Madhupur, Chonia, 18.4.2002, M. Khatun 175. *Area of major consumption:* Tangail district.

**47. Malva verticillata** L., Sp. Pl.: 689 (1753). var. **rafiqii** Abedin in Nasir & Ali (Eds), Fl. W. Pak. 130: 43 (1979). Vernacular names: Whorled Malva (Eng.), *Napha, Napa* (Beng.).

An annual herb. *Representative specimen*: Dinajpur: Dinajpur proper, 16.9.2001, M. Khatun 83. *Area of major consumption:* Dinajpur and Rangpur districts.

## 16. Cucurbitaceae A. L. de Jussieu (1789)

**48. Benincasa hispida** (Thunb.) Cogn. in DC., Monogr. Phan. 3: 513 (1881). Vernacular names: Wax Gourd (Eng.), *Chalkumra* (Beng.).

An annual, hispid, climbing herb. *Representative specimen*: Jessore: Birampur, 5.1.2004, M. Khatun 391. *Area of major consumption:* Dhaka and Jessore districts.

**49. Coccinia grandis** (L.) Voigt, Hort. Suburb. Calc.: 59 (1854). Vernacular names: Ivy Gourd (Eng.), *Telakucha* (Beng.).

A perennial, climbing herb. *Representative specimen*: Natore: Shingra, 14.4.2001, M. Khatun 57. *Area of major consumption:* Gazipur district.

**50. Cucurbita maxima** Duch. *ex* Lamk., Encycl. 2: 151 (1786). Vernacular names: Pumpkin (Eng.), *Mistikumra, Mithakumra* (Beng.).

An annual, climbing herb. *Representative specimen*: Rajshahi: Binodpur, 13.9.2001, M. Khatun 41. *Area of major consumption:* Rajshahi district.

**51. Lagenaria siceraria** (Molina) Standl., Publ. Field Mus. Nat. Hist. Chicago, B. Ser. 3: 435 (1930). Vernacular names: Bottle Gourd (Eng.), *Lau* (Beng.), *Boo-sthie* (Rakhain).

A large, annual, climbing herb. *Representative specimen*: Dhaka: Uttarbadda, 15.10.2002, M. Khatun 235. *Area of major consumption*: Dhaka district.

**52. Luffa cylindrica** (L.) M. Roem., Synops. 2: 63 (1846). Vernacular names: Sponge Gourd (Eng.), *Dhundul* (Beng.), *Mree-u-sthie* (Rakhain).

An extensive climbing herb. *Representative specimen*: Jessore: Khoertola, 5.1.2004, M. Khatun 392. *Area of major consumption:* Jessore district.

**53. Momordica charantia** L. var. **muricata** (Willd.) Chakravarty, Fasc. Fl. Ind. 2: 92 (1982). Vernacular names: Bitter Melon (Eng.), *Uchchhey* (Beng.), *Gang-khera-apang* (Rakhain).

An annual, climbing herb. *Representative specimen*: Gopalganj: Vatiapara, 3.3.2007, M. Khatun 426. *Area of major consumption:* Gopalganj district.

**54. Trichosanthes anguina** L., Sp. Pl. 1: 1008 (1753). Vernacular names: Snake Gourd (Eng.), *Chichinga* (Beng.), *Mring-bawn* (Rakhain).

An annual, climbing herb. *Representative specimen*: Noakhali: Begumganj, 7.5.2003, M. Khatun 241. *Area of major consumption:* Noakhali and Jessore districts.

**55. T. dioica** Roxb., Fl. Ind. 3: 701 (1832). Vernacular names: Pointed Gourd (Eng.), *Patal* (Beng.), *Pee-tho-sthie* (Rakhain).

An annual, climbing herb. *Representative specimen*: Rajshahi: Binodpur, 13.9.2001, M. Khatun 44. *Area of major consumption:* Rajshahi district.

**56. Zehneria japonica** (Thunb.) H. Y. Liu., Bull. Nat. Mus. Nat. (Taiwan) 1: 40 (1989). Vernacular name: White-fruited Creeping Cucumber (Eng.), *Herana Shak* (Chakma, Khasia).

An annual, climbing herb. *Representative specimen*: Rangamati: Kaptai, Digholchari, 7.7.2003, M. Khatun 346. *Area of major consumption:* Rangamati district.

**57. Z. scabra** (L. f.) Sond. in Harv. & Sond., Fl. Cap. 2: 486 (1862). Vernacular names: South African Zehneria (Eng.), *Rakhal Sasa* (Beng.), *Kolakachu* (Koch).

An annual, climbing herb. *Representative specimen*: Sherpur: Halchati, 1.11.2009, M. Khatun 527. *Area of major consumption:* Sherpur district.

## 17. Begoniaceae C. A. Agardh (1825)

**58. Begonia barbata** Wall. *ex*. A. DC., Prodr. 15(1) : 348 (1864). Vernacular name: *Tokpata* (Tripura).

A herb with creeping rootstock. *Representative specimen*: Moulvi Bazar: Sreemangal, 6.5.2003, M. Khatun 304. *Area of major consumption:* Moulvi Bazar district.

## 18. Brassicaceae Burnett (1835)

**59. Brassica juncea** (L.) Czerniak., Consp. Fl. Chark. : 8 (1859). Vernacular names: Indian Mustard, Brown Mustard (Eng.), *Raisarisha, Jhuni, Chanchi* (Beng.).

An annual herb. *Representative specimen*: Manikganj: Singair, 13.9.2002, M. Khatun 184. *Area of major consumption:* Dhaka and Sylhet districts.

**60. B. napus** L., Sp. Pl. 2: 666 (1753). Vernacular names: Colza, Rape (Eng.), *Magi, Togi, Sarisha* (Beng.).

An annual herb. *Representative specimen*: Rajshahi: Meharchandi, 13.9.2001, M. Khatun 41. *Area of major consumption:* Dhaka district.

**61. B. oleracea** var. **capitata** L., Sp. Pl. 2: 667 (1753). Vernacular names: Cabbage, Headed Cabbage (Eng.), *Bandhakapi* (Beng.).

An annual or biennial (in cold areas) herb. *Representative specimen*: Dhaka: Ashulia, 15.1.2002, M. Khatun 132. *Area of major consumption:* Dhaka, also all over the country.

**62. B. rapa** L., Sp. Pl.: 666 (1753). Vernacular names: Turnip (Eng.), *Shalgam* (Beng.).

An annual or biannual herb. *Representative specimen*: Rajshahi: Meherchandi, 13.9.2001, M. Khatun 55. *Area of major consumption:* Rajshahi district.

**63. Raphanus sativus** L., Sp. Pl. 2: 669 (1753). Vernacular names: Raddish (Eng.), *Mula* (Beng.), *Mou-laa* (Rakhain).

An annual herb. *Representative specimen*: Barisal: Barisal Sadar, 11.4.2010, M. Khatun 578. *Area of major consumption:* Barisal and Jessore districts.

## 19. Moringaceae Dumortier (1829)

**64. Moringa oleifera** Lamk., Encycl. 1(2): 398 (1785). Vernacular names: Drumstick Tree, Horseradish Tree (Eng.), *Sajna, Sojne* (Beng.), *Pepan-yuw-maa* (Marma).

A medium-sized tree. *Representative specimen*: Cox's Bazar: Kolatoli, 2.8.2009, M. Khatun 505. *Area of major consumption:* Cox's Bazar district.

## 20. Mimosaceae R. Brown (1814)

**65. Albizia procera** (Roxb.) Benth. in Hook., London J. Bot. 3: 89 (1844). Vernacular names: White Siris (Eng.), *Sada Koroi* (Beng.), *Fonagula* (Chakma).

A large, deciduous tree. *Representative specimen*: Chittagong: Alutila, 8.7.2003, M. Khatun 364. *Area of major consumption:* Khagrachari district.

## 21. Caesalpiniaceae R. Brown (1814)

**66. Bauhinia acuminata** L., Sp. Pl. : 375 (1753). Vernacular names: White Bauhinia, Mountain Ebony (Eng.), *Shada Kanchon* (Beng.), *Hingshiara* (Garo), *Jalong* (Khasia).

A large shrub or small tree. *Representative specimen*: Moulvi Bazar: Madhabkundu, 5.5.2003, M. Khatun 287. *Area of major consumption:* Moulvi Bazar and Netrakona districts.

**67. Caesalpinia digyna** Rottler, Neue Schriften Ges. Naturf. Freunde Berlin. 4: 200 (1803). Vernacular names: *Kamuno* (Marma)*, Loho* (Tanchangya).

A large, straggling, scandent shrub. *Representative specimen*: Bandarban: Balaghata, 5.3.2012, M. Khatun 621. *Area of major consumption:* Bandarban district.

**68. Cassia fistula** L., Sp. Pl.: 377 (1753). Vernacular names: Golden Shower Tree, Purging Cassia (Eng.), *Sonalu, Bandar Lati* (Beng.), *Shumrol* (Khasia).

A medium-sized deciduous tree. *Representative specimen*: Jessore: Hashimpur, 5.1.2004, M. Khatun 389. *Area of major consumption:* Jessore and Moulvi Bazar districts.

**69. Senna obtusifolia** (L.) Irwin & Barneby, Mem. N. Y. Bot. Gard. 35: 252 (1982). Vernacular names: Java Bean (Eng.), *Chakunda* (Beng.), *Dang Geya* (Marma).

An erect herb or undershrub. *Representative specimen*: Bandarban: Lama, 3.4.2011, M. Khatun 615. *Area of major consumption*: Bandarban district.

**70. S. sophera** (L.) Roxb., Fl. Ind. 2: 347 (1832). Vernacular names: Pepper-leaved Senna (Eng.), *Kalkashunda, Kasundi* (Beng.), *Eshi Shak* (Rakhain).

An undershrub or shrub. *Representative specimen*: Cox's Bazar: Kolatoli, 2.8.2009, M. Khatun 503. *Area of major consumption:* Manikganj and Dinajpur districts.

**71. S. tora** (L.) Roxb., Fl. Ind. 2: 340 (1832). Vernacular names: Sickle Senna (Eng.), *Chakunda, Kalkasham* (Beng.), *Sa Lai Pa* (Marma).

An erect foetid herb or undershrub. *Representative specimen*: Cox's Bazar: Kolatoli, 2.8.2010, M. Khatun 529. *Area of major consumption:* Rangamati district.

## 22. Fabaceae Lindley (1836)

**72. Cajanus cajan** (L.) Millsp., Publ. Field. Columb. Mus. Bot. Ser. 2: 53 (1900). Vernacular names: Pigeon Pea (Eng.), *Arhar* (Beng.), *Rahar* (Santal).

A shrub. *Representative specimen*: Khulna: Rupdia, 13.9.2007, M. Khatun 443. *Area of major consumption:* Moulvi Bazar district.

**73. Cicer arietinum** L., Sp. Pl. 2: 738 (1753). Vernacular names: Chickpea, Bengal Gram (Eng.), *Chola, But, Chana* (Beng.).

An annual herb. *Representative specimen*: Natore: Singair, 14.9.2001, M. Khatun 58. *Area of major consumption:* Natore district.

**74. Erythrina stricta** Roxb., Fl. Ind. 3: 251 (1832). Vernacular names: *Mandar, Teliamandar* (Beng.), *Kosano* (Marma, Tanchangya).

A large, deciduous tree. *Representative specimen*: Bandarban: Kibukpara, 5.3.2012, M. Khatun 619. *Area of major consumption:* Bandarban district.

**75. Lablab purpureus** (L.) Sweet., Hort. Brit. Ed. 1: 481 (1827). Vernacular names: Hyacinth Bean, Lablab (Eng.), *Sheem, Urshi, Ushi* (Beng.).

A perennial or annual climbing herb. *Representative specimen*: Cox's Bazar: St. Martin's Island, 18.2.2011, M. Khatun 595. *Area of major consumption:* Cox's Bazar district.

**76. Lathyrus sativus** L., Sp. Pl. : 730 (1753). Vernacular names: Grass Pea (Eng.), *Khesari* (Beng.).

A much branched annual herb. *Representative specimen*: Natore: Singair, 14.9.2001, M. Khatun 59. *Area of major consumption:* Rajshahi district.

**77. Phaseolus vulgaris** L., Sp. Pl. 1: 723 (1753). Vernacular names: Common Bean, Kidney Bean (Eng.), *Felong Dal* (Marma, Chakma).

A climbing herb. *Representative specimen*: Bandarban: Bandarban Sadar, 5.3.2012, M. Khatun 620. *Area of major consumption:* Bandarban district.

**78. Pisum sativum** L., Sp. Pl.: 727 (1753). Vernacular names: Garden Pea, Pea (Eng.), *Motor, Motorshuti* (Beng.).

A short-lived, climbing annual herb. *Representative specimen*: Pirojpur: Nazirpur, 11.4.2010, M. Khatun 583. *Area of major consumption:* Greater Faridpur district.

**79. Sesbania grandiflora** (L.) Poir. in Lamk., Encycl. Met. 7: 127 (1806). Vernacular names: *Bakful* (Beng.), *Agasta* (Rakhain).

A soft-wooded small tree. *Representative specimen*: Patuakhali: Kuakata, 23.5.2000, M. Khatun 482. *Area of major consumption:* Khagrachari and Patuakhali districts.

**80. Trigonella foenum-graecum** L., Sp. Pl. 3: 777 (1753). Vernacular names: Fenugreek (Eng.), *Methi* (Beng.).

An annual, robust, aromatic herb. *Representative specimen*: Rajshahi: Kazla, 13.9.2001, M. Khatun 45. *Area of major consumption:* Dhaka district.

**81. Vigna mungo** (L.) Happer. in Kew Bull. 11: 128 (1956). Vernacular names: Black Gram (Eng.), *Mashkalai* (Beng.), *Pee-shee-kanshi-deal* (Rakhain).

An annual herb. *Representative specimen*: Chapai Nawabganj: Moharajpur, 15.9.2001, M. Khatun 71. *Area of major consumption:* Chapai Nawabganj district.

## 23. Onagraceae A. L. de Jussieu (1789)

**82. Ludwigia adscendens** (L.) Hara, J. Jap. Bot. 28: 290 (1953). Vernacular names: *Kesardam* (Beng.), *Mulsishak* (Garo), *Gandu-pawn* (Rakhain).

An aquatic herb. *Representative specimen*: Natore: Natore Sadar, 14.9.2001, M. Khatun 60. *Area of major consumption*: Netrakona district.

## 24. Melastomataceae A. L. de Jussieu (1789)

**83. Osbeckia stellata** Buch.-Ham. *ex* Ker-Gawl., Bot. Reg. 8: 674 (1822). Vernacular names: Star Osbeckia (Eng.), *Gaichi* (Beng.), *Chakum* (Khasia).

A shrub. *Representative specimen*: Moulvi Bazar: Madhabkundu, 3.5.2003, M. Khatun 273. *Area of major consumption*: Moulvi Bazar district.

## 25. Euphorbiaceae A. L. de Jussieu (1789)

**84. Antidesma acidum** Retz., Obs. Bot. 5: 30 (1788). Vernacular names: Indian Laurel (Eng.), *Multa* (Beng.), *Mokhichikra* (Tripura).

A large shrub or small tree. *Representative specimen*: Habiganj: Chunarughat, 6.5.2003, M. Khatun 307. *Area of major consumption:* Habiganj district.

**85. Manihot esculenta** Crantz, Inst. 1: 167 (1766). Vernacular names: Cassava (Eng.), *Simul-alu* (Garo), *Kepalli Nolpai* (Marma).

A shrub. *Representative specimen*: Tangail: Madhupur, 18.4.2002, M. Khatun 172. *Area of major consumption:* Tangail and Sherpur districts.

**86. Ricinus communis** L., Sp. Pl.: 1007 (1753). Vernacular names: Castor (Eng.), *Venna, Rerhi, Bherenda* (Beng.), *Crusuba* (Marma, Tanchangya).

A shrubby or tree-like, somewhat herb. *Representative specimen*: Bandarban: Bandarban Sadar, 7.3.2012, M. Khatun 617. *Area of major consumption:* Bandarban district.

## 26. Vitaceae A. L. de Jussieu (1789)

**87. Cissus adnata** Roxb., Fl. Ind. ed. Carey 1: 405 (1820). Vernacular names: *Alianga-lata, Bhatia-lata* (Beng.) *Chuka-blei* (Tripura).

A slender, woody climber. *Representative specimen*: Cox's Bazar: Kolatoli, 2.8.2009, M. Khatun 508. *Area of major consumption:* Cox's Bazar district.

**88. C. assamica** (M. Lawson) Craib in Kew Bull.: 31 (1911). Vernacular names: *Amasha-pata* (Beng.), *Gelia Bleli* (Tripura).

A large, woody climber. *Representative specimen*: Moulvi Bazar: Sreemangal, 2.5.2003, M. Khatun 283. *Area of major consumption*: Moulvi Bazar district.

**89. C. elongata** Roxb., Fl. Ind. 1: 411 (1832). Vernacular names: *Dhemna, Chemna* (Beng.).

A large, climbing herb. *Representative specimen*: Bandarban: Lama, 3.4.2011, M. Khatun 615. *Area of major consumption:* Bandarban district.

**90. C. quadrangularis** L., Syst. Nat. ed. 12(2): 124 (1767). Vernacular names: *Harjora Lata, Harbhanga Lata* (Beng), *Marang Gach* (Santal), *Moi-bhanga Lota* (Garo).

A large climber. *Representative specimen*: Patuakhali: Kalachanpara, 23.5.2008, M. Khatun 495. *Area of major consumption:* Patuakhali district.

**91. C. repens** Lamk., Encycl. Math. Bot. 1: 31 (1783). Vernacular names: *Marnaria Pata* (Beng.), *Marmaria Lata* (Koch)

A large, herbaceous climber. *Representative specimen*: Chittagong: Jamtoli, 8.7.2003, M. Khatun 374. *Area of major consumption:* Sherpur district.

**92. Tetrastigma angustifolium** (Roxb.) Planch. in DC., Monogr. Phan. 5: 439 (1887). Vernacular name: *Nekung Rubi* (Beng.).

A large, glabrous, herbaceous climber. *Representative specimen*: Rangamati: Kaptai, Digholchari, 7.7.2003, M. Khatun 359. *Area of major consumption:* Rangamati district.

## 27. Sapindaceae A. L. de Jussieu (1789)

**93. Cardiospermum halicacabum** L., Sp. Pl.: 366 (1753). Vernacular names: Balloon Vine, Pigeon's Knee (Eng.), *Phutka, Lataphutki* (Beng.), *Kataboksa Shak* (Chakma).

An annual or perennial climbing herb. *Representative specimen*: Rangamati: Kaptai, Digholchari, 7.7.2003, M. Khatun 377. *Area of major consumption:* Rangamati district.

## 28. Anacardiaceae Lindley (1830)

**94. Mangifera indica** L., Sp. Pl.: 200 (1753). Vernacular names: Mango (Eng.), *Aam* (Beng.), *Kharai* (Khasia), *Tharaapang* (Rakhain).

A large tree. *Representative specimen*: Moulvi Bazar: Adampur, 3.5.2003, M. Khatun 267. *Area of major consumption:* Moulvi Bazar district.

**95. Spondias pinnata** (L. f.) Kurz, Pegu. Rep.: 44 (1875). Vernacular names: Hog Plum (Eng.), *Deshi-amra* (Beng.), *Thai-toui* (Tripura), *Soh-awla* (Khasia).

A medium-sized to large tree. *Representative specimen*: Habiganj: Chunarughat, 6.5.2003, M. Khatun 313. *Area of major consumption:* Moulvi Bazar district.

## 29. Meliaceae A. L. de Jussieu (1789).

**96. Azadirachta indica** A. Juss. in Mem. Mus. Nat. Hist. Paris 19: 221, t. 13 (1830). Vernacular names: Margosa Tree, Neem Tree (Eng.), *Neem* (Beng.), *Hoppa* (Rakhain).

A medium-sized to large tree. *Representative specimen*: Patuakhali: Kuakata, 24.5.2008, M. Khatun 486. *Area of major consumption*: Patuakhali district.

## 30. Rutaceae A. L. de Jussieu (1789)

**97. Aegle marmelos** (L.) Corr., Trans. Linn. Soc. Lond. 5: 223 (1800). Vernacular names: Bengal Quince (Eng.), *Bel* (Beng.), *War-e-si-apang* (Marma).

A deciduous tree. *Representative specimen*: Bandarban: Balaghata, 5.3.2012, M. Khatun 616. *Area of major consumption:* Bandarban and Khagrachari districts.

**98. Clausena excavata** Burm. f., Fl. Ind.: 87, t. 29, 2 (1768). Vernacular names: Clausena (Eng.), *Pan Karpur* (Beng.), *Pankauri* (Chakma).

An aromatic shrub. *Representative specimen*: Chittagong: Sitakunda, 7.7.2003, M. Khatun 285. *Area of major consumption:* Rangamati district.

**99. Zanthoxylum rhetsa** (Roxb.) DC., Prodr. 1: 728 (1824). Vernacular names: Indian Ivy-rue (Eng.), *Bajna, Kantahorina, Tambol* (Beng.), *Khazai* (Garo).

A medium-sized, deciduous tree. *Representative specimen*: Tangail: Madhupur, 18.4.2002, M. Khatun 178. *Area of major consumption:* Tangail and Netrakona districts.

## 31. Oxalidaceae R. Brown (1817)

**100. Oxalis corniculata** L., Sp. Pl.: 435 (1753). Vernacular names: Indian Sorrel (Eng.), *Amrul, Amrul Shak* (Beng.), *Amila Pata* (Chakma, Khasia).

A small perennial herb. *Representative specimen*: Rangamati: Kaptai, Digholchari, 7.7.2003, M. Khatun 348. *Area of major consumption:* Rangamati district.

## 32. Apiaceae Lindley (1836)

**101. Centella asiatica** (L.) Urban in Mart. & Eichler, Fl. Brasil. 11 (1): 287 (1879). Vernacular names: Indian Pennywort (Eng.), *Thankuni, Brahmabuti* (Beng.).

A perennial herb. *Representative specimen*: Dhaka: Bocila, 15.1.2002, M. Khatun 167. *Area of major consumption*: Dhaka district.

**102. Coriandrum sativum** L., Sp. Pl. 1: 256 (1753). Vernacular names: Coriander (Eng.), *Dhonay, Dhonia* (Beng.).

An annual herb. *Representative specimen*: Chittagong: Adampur, 6.7.2003, M. Khatun 372. *Area of major consumption:* Dhaka district.

**103. Eryngium foetidum** L., Sp. Pl. 1: 232 (1753). Vernacular names: Wild Coriander (Eng.), *Mysapagur* (Chakma).

An erect, biennial herb. *Representative specimen*: Rangamati: Digholchari, 7.7.2003, M. Khatun 351. *Area of major consumption:* Rangamati district.

**104. Foeniculum vulgare** (L.) Miller, Gard. Diet. ed. 8, no. 1 (1768). Vernacular names: Funnel (Eng.), *Pan-mohuri* (Beng.), *Moroi* (Marma)

A robust, aromatic herb. *Representative specimen*: Bandarban: Bandarban Sadar, 7.3.2012, M. Khatun 616. *Area of major consumption:* Bandarban district.

**105. Hydrocotyle sibthorpioides** Lamk., Enc. 3: 153 (1789). Vernacular names: Lawn Marsh Pennywort (Eng.), *Sakumubakla* (Marma).

A perennial, slender herb. *Representative specimen*: Khagrachari: Bottoli, 6.7.2003, M. Khatun 339. *Area of major consumption*: Rangamati district.

**106. Oenanthe benghalensis** (Roxb.) Kurz, J. Asiat. Soc. Beng. 2: 115 (1877). Vernacular names: Water Celery (Eng.), *Bandhunia* (Khasia).

A perennial, glabrous herb. *Representative specimen*: Rangamati: Kaptai, Bangchari, 6.7.2003, M. Khatun 343. *Area of major consumption*: Rangamati district.

**107. O. javanica** (Blume) DC., Prodr. 4: 138 (1830). Vernacular names: Water Dropwort, Java Waterdropwort (Eng.), *Pan-turasi* (Beng.), *Branju* (Rakhain).

A perennial herb. *Representative specimen*: Patuakhali: Kalachandpara, 23.5.2008, M. Khatun 491. *Area of major consumption*: Patuakhali district.

**108. Trachyspermum ammi** (L.) Sprague, Bull. Misc. Inform. Kew. 1929: 228 (1929). Vernacular names: Ajowan Caraway (Eng.), *Jawan* (Beng.), *Fuchi Shak* (Chakma, Marma).

An annual herb. *Representative specimen*: Khagrachari: Golabari, 6.7.2003, M. Khatun 326. *Area of major consumption*: Khagrachari district.

**109. T. roxburghianum** (DC.) H. Wolff in Engl., Pfl. Umbellif. Apioid-Ammin.: 129 (1927). Vernacular names: *Radhuni, Chanu* (Beng.), *Rajani* (Marma).

An annual, aromatic herb. *Representative specimen*: Dhaka: Bosila, 15.10.2002, M. Khatun 233. *Area of major consumption:* Bandarban district.

## 33. Solanaceae A. L. de Jussieu (1789)

**110. Capsicum frutescens** L., Sp. Pl.: 189 (1753). Vernacular names: Spur Pepper, Cayenne Pepper (Eng.), *Kacha Morich* (Beng.).

A herb. *Representative specimen*: Patuakhali: Patuakhali Sadar, 23.5.2008, M. Khatun 493. *Area of major consumption:* Patuakhali district.

**111. Physalis angulata** L., Sp. Pl. : 183 (1753). Vernacular names: Hogweed, Balloon Cherry (Eng.), *Potka* (Beng.), *Ambichok* (Garo).

An annual, much branched herb. *Representative specimen*: Pabna: Thanapara, 5.3.2007, M. Khatun 467. *Area of major consumption*: Pabna district.

**112. Solanum americanum** Mill., Gard. Dict. ed. 8, No. 5 (1768). Vernacular names: Glossy Nightshade (Eng.), *Titbegun* (Beng.).

An erect annual herb. *Representative specimen*: Habiganj: Chunarughat, 6.5.2003, M. Khatun 316. *Area of major consumption:* Habiganj district.

**113. S. tuberosum** L., Sp. Pl.: 185 (1753). Vernacular names: Potato (Eng.), *Alu, Gol Alu* (Beng.), *Mraa-u-shey* (Rakhain).

A viscoid herb. *Representative specimen*: Munshiganj: Gozaria, 18.1.2010, M. Khatun 566. *Area of major consumption:* Cox's Bazar district.

**114. S. villosum** Mill., Gard. Dict. ed. 8, no. 2 (1768). Vernacular names: Orange Nightshade (Eng.), *Titbegun, Kakmachi* (Beng.), *Kha-rey-je-key* (Rakhain).

An annual herb. *Representative specimen*: Patuakhali: Keranipara, 23.5.2008, M. Khatun 490. *Area of major consumption:* Chapai Nawabganj district.

## 34. Convolvulaceae A. L. de Jussieu (1789)

**115. Hewittia sublobata** (L. f.) O. Kuntze, Rev. Gen. Pl. 2 : 441 (1891). Vernacular name: *Dhudla Shak* (Beng).

A twining perennial herb. *Representative specimen*: Rajbari: Salki, 12.3.2006, M. Khatun 403. *Area of major consumption*: Rajbari and Feni districts.

**116. Ipomoea aquatica** Forssk., Fl. Aeg.-Arab. : 44 (1755). Vernacular names: Water Spinach (Eng.), *Kalmi Shak* (Beng.).

An aquatic herb. *Representative specimen*: Gopalganj: Digholia, 3.3.2007, M. Khatun 424. *Area of major consumption:* Gopalganj district.

**117. I. batatus** (L.) Lamk., Tabl. Encycl. 1: 465 (1791). Vernacular names: Sweet Potato (Eng.), *Misti Alu, Ranga Alu* (Beng.).

A perennial herb. *Representative specimen*: Dhaka: Diabari, 15.1.2002, M. Khatun 134. *Area of major consumption:* Dhaka district.

**118. Operculina turpethum** (L.) S. Manso., Enum. Subst. Bras.: 16 (1836). Vernacular names: Turpeth Root (Eng.), *Dudh Kalmi* (Beng.).

A glabrous twiner. *Representative specimen*: Patuakhali: Kolapara, 23.5.2008, M. Khatun 479. *Area of major consumption:* Gazipur and Patuakhali districts.

## 35. Verbenaceae Jaume St.- Hilaire (1805)

**119. Clerodendrum inerme** (L.) Gaertn., Fruct. Sem. Pl. 1: 271 (1788). Vernacular names: Glory Bower (Eng.), *Banjui, Batraj, Koklata* (Beng.), *Jarems* (Khasia).

An erect to scandent shrub. *Representative specimen*: Moulvi Bazar: Madhabpunji, 4.5.2003, M. Khatun 282. *Area of major consumption*: Moulvi Bazar district.

**120. Premna benghalensis** C. B. Clarke in Hook. f., Fl. Brit. Ind. 4: 577 (1885). Vernacular names: *Pakhirhar* (Beng.), *Koya Jarul* (Khasia).

A medium-sized evergreen tree. *Representative specimen*: Habiganj: Chunarughat, 6.5.2003, M. Khatun 317. *Area of major consumption:* Habiganj district.

**121. P. esculenta** Roxb., Fl. Ind. ed. 2, 3: 81 (1832). Vernacular names: *Lalong, Lalana* (Beng.), *Gun-duri, Darkakha* (Khasia).

A small shrub. *Representative specimen*: Habiganj: Chunarughat, Kalenga, 6.5.2003, M. Khatun 312. *Area of major consumption:* Habiganj district.

**122. P. mucronata** Roxb., Fl. Ind. ed. 2, 3: 635 (1832). Vernacular name: *Khatamuri* (Koch).

A shrub or small tree. *Representative specimen*: Sherpur: Runctia, Kochpara, 1.11.2009, M. Khatun 531. *Area of major consumption:* Sherpur district.

**123. P. obtusifolia** R. Br., Prod. Fl. Nov. Holl. 1: 512 (1810). Vernacular names: *Gambari, Bhuttsirabi* (Beng.), *Lalom Pata* (Khasia, Chakma).

A shrub to small evergreen tree. *Representative specimen*: Khagrachari: Golabari, 6.7.2003, M. Khatun 318. *Area of major consumption:* Khagrachari district.

## 36. Lamiaceae Lindley (1836).

**124. Ajuga macrosperma** Wall. *ex* Benth. in Wall., Pl. As. Rar. 1: 58 (1830). Vernacular names: *Sabarang* (Chakma, Marma).

A herb. *Representative specimen*: Khagrachari: Golabari, 6.7.2003, M. Khatun 337. *Area of major consumption:* Khagrachari and Rangamati districts.

**125. Leucas aspera** (Willd.) Link, Enum. Hort. Berol. 2: 113 (1822). Vernacular names: *Dondokalosh* (Beng.), *Dong-Ke-La* (Coach).

A stout, erect or diffuse, annual herb. *Representative specimen*: Bandarban: Lama, 3.4.2011, M. Khatun 609. *Area of major consumption:* Chittagong district.

**126. L. cephalotes** (Roth.) Spreng., Syst. 2: 743 (1825). Vernacular names: *Bara Kalkus* (Beng.), *Thaelsi* (Garo).

A stout, erect, annual herb. *Representative specimen*:: Netrakona: Bijoypur, 8.10.2000, M. Khatun 05. *Area of major consumption:* Netrakona district.

**127. Ocimum americanum** L., Cent. Pl. 1: 15 (1755). Vernacular names: Rosary Ocimum (Eng.), *Tulshi* (Beng.), *Bontulsi* (Coach).

An erect, annual, aromatic herb. *Representative specimen*: Rangamati: Kaptai, Koblachara, 7.7.2003, M. Khatun 347. *Area of major consumption:* Rangamati district.

**128. Pogostemon benghalensis** (Burm. f.) O. Kuntze, Rev. Gen. Pl. 2: 529 (1891). Vernacular name: *Lomboi Shak* (Chakma, Marma, Khasia).

An erect, stout, aromatic undershrub. *Representative specimen*: Rangamati: Kaptai, Digholchari, 6.7.2003, M. Khatun 338. *Area of major consumption:* Rangamati district.

### 37. Scrophulariaceae A. L. de Jussieu (1789)

**129. Bacopa monnieri** (L.) Pennell in Pflanzenfam. 4(3b): 77 (1891). Vernacular names: Water Hyssop (Eng.), *Brammi, Brammi Shak, Dupkalmini* (Beng.)

An annual herb. *Representative specimen*: Cox's Bazar: St. Martin's Island, 18.2.2011, M. Khatun 596. *Area of major consumption:* Patuakhali and Cox's Bazar districts.

**130. Scoparia dulcis** L., Sp. Pl.: 116 (1753). Vernacular names: Goat Weed (Eng.), *Bondhone* (Beng.), *Shamgaldak* (Garo).

An erect, much branched, perennial herb. *Representative specimen*: Dinajpur: Mukundupur, 16.9.2001, M. Khatun 87. *Area of major consumption:* Dinajpur district.

### 38. Acanthaceae A. L. de Jussieu (1789)

**131. Hygrophila polysperma** (Roxb.) T. Anders., Journ. Linn. Soc. Bot. 9: 456 (1867). Vernacular names: Dwarf Hygrophila (Eng.), *Puinnya Shak* (Beng.), *Puinna Shak* (Rakhain).

A small, much branched herb. *Representative specimen*: Patuakhali: Kalapara, 23.5.2008, M. Khatun 496. *Area of major consumption*: Patuakhali district.

**132. H. schulli** (Buch.-Ham.) M. R. & S. N. Almeida, Journ. Bomb. Nat. Hist. Soc. 83 (Suppl.): 221 (1986). Vernacular names: Star Thorn (Eng.), *Talmakhna, Kulekhara* (Beng.).

An annual, erect herb. *Representative specimen:* Jhalakathi: Jhalakathi proper, 11.4.2010, M. Khatun 577. *Area of major consumption:* Barisal and Rajbari districts.

**133. Nelsonia canescens** (Lamk.) Spreng., Syst. 1 : 42 (1824). Vernacular names: *Paramul* (Beng.), *Khaia Shak* (Khasia).

A trailing or diffuse herb. *Representative specimen*: Habiganj: Chunarughat, 6.5.2003, M. Khatun 302. *Area of major consumption*: Habiganj district.

**134. Ruellia tuberosa** L., Sp. Pl.: 635 (1753). Vernacular names: Blue Bell (Eng.), *Chatpotey* (Beng.), *Charasak* (Khasia).

A perennial erect herb. *Representative specimen*: Moulvi Bazar: Lawacherra, 2.5.2003, M. Khatun 249. *Area of major consumption:* Habiganj and Moulvi Bazar districts.

### 39. Bignoniaceae A. L. de Jussieu (1789)

**135. Oroxylum indicum** (L.) Kurz, For. Fl. Brit. Burm. 2: 237 (1877). Vernacular names: Midnight Horror (Eng.), *Kanaidingi* (Beng.), *Kharam-sha-bawn* (Rakhain), *Fona-gulogach* (Chakma).

A medium-sized tree. *Representative specimen*: Patuakhali: Kuakata, 24.5.2008, M. Khatun 478. *Area of major consumption:* Patuakhali and Rangamati districts.

### 40. Sphenocleaceae Lindly (1829)

**136. Sphenoclea zeylanica** Gaertn., Fruct. Sem. Pl. 1: 113 (1788). Vernacular names: *Jhill Mirich* (Beng.), *Vui Shak* (Chakma), *Radai* (Mog).

A robust herb. *Representative specimen*: Rangamati: Kaptai, Debachari, 7.7.2003, M. Khatun 356. *Area of major consumption:* Khagrachari district.

## 41. Rubiaceae A. L. de Jussieu (1789)

**137. Hedyotis corymbosa** (L.) Lamk., Tab. Encycl. 1: 272 (1791). Vernacular names: *Khetpapra, Panki* (Beng.), *Dimatita* (Marma).

An annual, diffuse or prostrate herb. *Representative specimen*: Dinajpur: Mukundopur, 16.9.2001, M. Khatun 79. *Area of major consumption:* Dinajpur district.

**138. Morinda citrifolia** L., Sp. Pl.: 176 (1753). *Vernacular names: Ach, Banach, Tufania* (Beng.), *Ken-thug-blag, Chirasak* (Khasia).

A shrub or small tree. *Representative specimen*: Cox's Bazar: Kolatoli, 2.8.2009, M. Khatun 506. *Area of major consumption:* Chittagong district.

**139. Paederia foetida** L., Mant. 1: 52 (1767). Vernacular names: *Gandha Bhaduli, Badali* (Beng.), *Padbaj Ludi* (Chakma).

A twining, glabrous shrub. *Representative specimen*: Netrakona: Polastola, 8.10.2001, M. Khatun 14. *Area of major consumption:* Netrakona district.

**140. Spermacoce articularis** L. f., Sppl. Pl: 119 (1782). Vernacular names: *Horinshing, Usni* (Santal).

A procumbent, perennial, mat forming herb. *Representative specimen*: Dinajpur: Noyabad, 16.9.2001, M. Khatun 86. *Area of major consumption:* Dinajpur district.

**141. S. latifolia** Aublet, Hist. Pl. Guiane Fr. 1: 55, t. 194 (1755). Vernacular name: *Ghuiojhil Shak* (Tanchangya).

A prostrate or decumbent herb. *Representative specimen*: Bandarban: Lama, 3.4.2011, M. Khatun 612. *Area of major consumption:* Bandarban district.

**142. S. stricta** L. f., Suppl. Pl.: 120 (1781). Vernacular names: *Narkel Jhuri Shak* (Rakhain), *Mijlick* (Chakma).

An erect or rarely prostrate, annual herb. *Representative specimen*: Patuakhali: Keranipara, 23.5.2008, M. Khatun 492. *Area of major consumption*: Patuakhali and Sylhet districts.

## 42. Caprifoliaceae A. L. de Jussieu (1789)

**143. Sambucus javanica** Reinw. *ex* Blume, Bijdr.: 657 (1826). Venacular names: Javanese Elder (Eng.), *Hoklati* (Beng.), *Maytraba* (Chakma).

A large shrub or small tree. *Representative specimen*: Rangamati: Rangamati sadar, 10.7.2003, M. Khatun 333. *Area of major consumption:* Bandarban and Rangamati districts.

## 43. Asteraceae Dumortier (1822)

**144. Ageratum conyzoides** L., Sp. Pl.: 839 (1753). Vernacular names: Tropical White Weed (Eng.), *Ozone Shak* (Chakma), *Mukri* (Tripura), *Hinor* (Khasia).

An annual herb. *Representative specimen*: Khagrachari: Bot-toli, 6.7.2003, M. Khatun 323. *Area of major consumption*: Khagrachari district.

**145. Blumea lacera** (Burm. f. ) DC. in Wight, Contr. Bot. Ind.: 14 (1834). Vernacular names: *Barakukshim, Kukurshunga, Kuksung* (Beng.), *Leikhamal* (Manipuri).

An annual aromatic herb. *Representative specimen*: Moulvi Bazar: Komolganj, 3.5.2003, M. Khatun 271. *Area of major consumption:* Feni and Moulvi Bazar districts.

**146. Elephantopus scaber** L., Sp. Pl.: 814 (1753). Vernacular names: *Banmula* (Beng.), *Gejia Shak* (Koch).

A perennial herb. *Representative specimen*: Noakhali: Begumganj, 7.5.2003, M. Khatun 321. *Area of major consumption:* Noakhali district.

**147. Emilia sonchifolia** (L.) DC. in Wight, Contrib. : 24 (1834). Vernacular names: Lilac Tassel Flower (Eng.), *Mechitra, Sadusi* (Beng.), *Miam Shak* (Chakma).

An annual branched herb. *Representative specimen*: Khagrachari: Dokkhin Golabari, 8.7.2003, M. Khatun 376. *Area of major consumption:* Khagrachari district.

**148. Enhydra fluctuans** Lour., Fl. Chchinch.: 511 (1790). Vernacular names: *Helencha, Hingcha, Harhach* (Beng.).

An annual, aquatic herb. *Representative specimen*: Gopalganj: Tungipara, 4.3.2007, M. Khatun 299. *Area of major consumption:* Gopalganj district.

**149. Lactuca scariola** L. var. **sativa** Hook. f., Fl. Brit. Ind. 3 : 404 (1881). Vernacular names: Garden Lettuce (Eng.), *Latus Pata* (Beng.).

An annual or biennial herb. *Representative specimen*: Rajshahi: Meharchandi, 13.9.2001, M. Khatun 36. *Area of major consumption:* Dhaka district.

**150. Spilanthes calva** DC. in Wight, Contrib. : 19 (1834). Vernacular names: Paracress (Eng.), *Surfa, Kannya* (Beng.), *Marhatitiga* (Marma)

An annual herb. *Representative specimen*: Khagrachari: Shomoboy market, 6.7.2003, M. Khatun 340. *Area of major consumption:* Rangamati district.

**151. Synedrella nodiflora** (L.) Gaertn., Fruct. 2: 456, t. 171 (1791). Vernacular name: *Hamfui* (Marma, Chakma).

An annual pubescent herb. *Representative specimen*: Khagrachari: Pouromarket, 6.7.2003, M. Khatun 344. *Area of major consumption:* Khagrachari district.

**152. Vernonia cinerea** (L.) Less., Linnaea 4 (1): 291 (1829). Vernacular names: Little Ironweed (Eng.), *Kuksim* (Beng.), *Kalojira* (Garo).

An erect, annual or perennial herb. *Representative specimen*: Dinajpur: Noiyabad, 16.9.2001, M. Khatun 98. *Area of major consumption:* Netrakona district.

**153. Xanthium indicum** Koen. *ex* Roxb., Fl. Ind. 3: 601 (1832). Vernacular names: Rough Cocklebur (Eng.), *Ghagra, Ban-okra* (Beng.), *Baksala* (Hajong).

An annual herb. *Representative specimen*: Munshiganj: Mahakali, 18.1.2010, M. Khatun 568. *Area of major consumption:* Barisal, Faridpur, Madaripur, Rajshahi and Chittagong districts.

*Note:* Seedlings of this species are highly poisonous, thus should never be plucked for vegetables in this stage.

## LILIOPSIDA
### 1. Arecaceae C.H. Schultz-Schultzen. (1832)

**154. Calamus tenuis** Roxb., Fl. Ind. 3: 780 (1832). Vernacular names: Rattan (Eng.), *Bet* (Beng.), *Khring* (Marma).

A thicket forming climber. *Representative specimen*: Khagrachari: Bot-toli, 6.7.2003, M. Khatun 330. *Area of major consumption:* Khagrachari district.

## 2. Araceae A.L. de Jussieu (1789)

**155. Amorphophallus bulbifer** (Roxb.) Blume, Rumph. 1: 148 (1847). Vernacular names: Voodoo Lily (Eng.), *Jongli Ol* (Beng.), *Chungmuru* (Garo).

A herb. *Representative specimen*: Sherpur: Runctia, 31.10.2009, M. Khatun 539. *Area of major consumption:* Sherpur and Moulvi Bazar districts.

**156. A. nepalensis** (Wall.) Bogner & Mayo, Aroideana 8(1): 19 (1985). Vernacular names: *Khar Kochu* (Tripuri), *Dadonga* (Koch).

A tuberous herb. *Representative specimen*: Mymensingh: Haluaghat, 6.10.2009, M. Khatun 115. *Area of major consumption:* Mymensingh district.

**157. Colocasia esculenta** (L.) Schott in Schott & Endl., Melet. Bot. 1: 18 (1832). Vernacular names: Taro, Coco-yam (Eng.), *Kachu* (Beng.).

A perennial herb. *Representative specimen*: Jessore: Monirampur, 5.1.2004, M. Khatun 382. *Area of major consumption:* Jessore and Dhaka districts.

**158. C. gigantea** (Blume) Hook. f., Fl. Brit. Ind. 6: 524 (1893). Vernacular names: Giant Elephant Ear (Eng.), *Salad Kachu* (Beng.), *Chinjapang* (Khasia).

A perennial herb. *Representative specimen*: Khagrachari: Silchari, 8.7.2003, M. Khatun 334. *Area of major consumption:* Khagrachari and Sylhet districts.

**159. Homalomena aromatica** (Roxb. *ex* Sims.) Schott in Schott & Endl., Melet. Bot. 1: 20 (1832). Vernacular names: *Bonkachu* (Beng.), *Kachu Gondhobi* (Khasia), *Chikon Shak* (Marma).

A perennial herb. *Representative specimen*: Moulvi Bazar: Adampur bit, 3.5.2003, M. Khatun 268. *Area of major consumption:* Moulvi Bazar district.

**160. Lasia spinosa** (L.) Thwait., Enum. Pl. Zeyl. 1: 336 (1864). Vernacular names: *Kanta Kachu* (Beng.), *Bonadia* (Garo).

A perennial, stout herb. *Representative specimen*: Netrakona: Durgapur, 8.10.2000, M. Khatun 04. *Area of major consumption:* Netrakona and Habiganj districts.

**161. Typhonium trilobatum** (L.) Schott, Wien. Zeitschr. 3: 72 (1829). Vernacular names: *Ghetkachu, Ghekul* (Beng.), *Kharkon* (Santal).

A small, tuberous, terrestrial herb. *Representative specimen*: Gazipur: Rajendrapur, 4.12.2009, M. Khatun 554. *Representative specimen*: Dhaka and Gazipur districts.

**162. Xanthosoma sagittifolium** (L.) Schott in Schott & Endl., Melet. Bot.: 19 (1832) ('*sagittaefolium*'). Vernacular names: Tannia, Tanier (Eng.), *Mukhikachu* (Beng.).

A perennial herb. *Representative specimen*: Rajbari: Salmara, 12.3.2006, M. Khatun 401. *Area of major consumption*: Faridpur district.

**163. X. violaceum** Schott, Oesterr. Bot. Wochenbl. 3: 370 (1853). Vernacular names: Blue Taro, Purple-stem Taro (Eng.), *Dudhkachu* (Beng.).

A perennial herb. *Representative specimen*: Jessore: Hashimpur, 5.1.2004, M. Khatun 385. *Area of major consumption*: Jessore district.

## 3. Commelinaceae R. Brown (1810)

**164. Commelina benghalensis** L., Sp. Pl.: 41 (1753). Vernacular names: Blue Commelina, Bengal Dayflower (Eng.), *Kanchira* (Beng.), *Piachara* (Chakma).

A small herb. *Representative specimen*: Chapai Nawabganj: Moharajpur, 12.9.2001, M. Khatun 66. *Area of major consumption:* Chapai Nawabganj district.

## 4. Poaceae Barnhart (1895)

**165. Melocanna baccifera** (Roxb.) Kurz, Prelim. Rep. For. Veg. Pegu, App. B.: 94 (1875). Vernacular names: Berry Bamboo (Eng.), *Bajali Muli, Tarai* (Beng.), *Wa-thui, Muiya* (Chakma, Marma, Khasia).

A diffusely clumped, sympodial bamboo. *Representative specimen*: Khagrachari: Bot-toli, 6.7.2003, M. Khatun 341. *Area of major consumption:* Chittagong Hill Tracts.

## 5. Zingiberaceae Lindley (1835)

**166. Curcuma longa** L., Sp. Pl. 1: 2 (1753). Vernacular names: Turmeric (Eng.), *Halud, Haldi* (Beng.).

A rhizomatous herb. *Representative specimen*: Cox's Bazar: Kolatoli, 2.8.2009, M. Khatun 513. *Area of major consumption:* Cox's Bazar district.

**167. Globba marantina** L., Mant. Alt. : 170 (1771). Vernacular names: Yellow Dancing Girl (Eng.), *Holi Shak* (Khasia).

A small, annual herb. *Representative specimen*: Rangamati: Kaptai, Bangchari, 7.7.2003, M. Khatun 357. *Area of major consumption*: Khagrachari and Rangamati districts.

**168. Zingiber officinale** Rosc., Trans. Linn. Soc. Lond. 8: 348 (1807). Vernacular names: Ginger (Eng.), *Ada* (Beng.), *Shaen-pang* (Rakhain).

A small, rhizomatous herb. *Representative specimen*: Cox's Bazar: Kolatoli pahar, 2.8.2009, M. Khatun 501. *Area of major consumption:* Cox's Bazar district.

## 6. Pontederiaceae Kunth (1816)

**169. Monochoria hastata** (L.) Solms in A. DC., Monogr. Phaner. 4: 523 (1883). Vernacular names: Arrowleaf False Pickereweed (Eng.), *Baranukha* (Beng.), *Chichir* (Garo), *Projukti Shak* (Tripura).

A perennial, robust herb. *Representative specimen*: Netrakona: Bijoypur, 8.10.2001, M. Khatun 13. *Area of major consumption:* Netrakona district.

**170. M. vaginalis** (Burm. f.) Presl, Rel. Haenk. 1: 128 (1827). Vernacular names: Heartshape False Pickereweed (Eng.), *Nukha* (Beng.), *Kusrisha* (Khasia).

A slender, perennial herb. *Representative specimen*: Habiganj: Chunarughat, 6.5.2003, M. Khatun 314. *Area of major consumption:* Khagrachari and Moulvi Bazar districts.

## 7. Liliaceae A.L. de Jussieu (1789)

**171. Allium cepa** L., Sp. Pl. ed.1: 300 (1753). Vernacular names: Onion (Eng.), *Piaz* (Beng.).

An annual herb. *Representative specimen*: Jessore: Hashimpur, 5.7.2004, M. Khatun 394. *Area of major consumption:* Dhaka district.

**172. A. sativum** L., Sp. Pl. 1: 297 (1753). Vernacular names: Garlic (Eng.), *Rashun* (Beng.).

An erect herb. *Representative specimen*: Rajbari: Salmara, 8.2.2008, M. Khatun 417. *Area of major consumption:* Rajbari district.

## 8. Dioscoriaceae R. Brown (1810)

**173. Dioscorea pentaphylla** L., Sp. Pl.: 1032 (1753). Vernacular names: Five-leaf Yam (Eng.), *Jhum Alu, Kanta Alu* (Beng.), *Khaiamor* (Mog), *Patil Alu* (Coach, Rajbonshi).

A twining herb. *Representative specimen*: Bandarban: Lama, 3.4.2011, M. Khatun 607. *Area of major consumption:* Gazipur and Bandarban districts.

# PTERIDOPHYTA
## 1. Angiopteridaceae Fee *ex* Bonner (1867)

**174. Angiopteris evecta** (Forst.) Hoffm., Comm. Soc. Reg. Gott. 12: 29, t. 5 (1796). Vernacular names: King Fern (Eng.), *Dhekia Shak* (Beng.), *Siblu* (Marma).

A large, semi-erect tree fern. *Representative specimen*: Bandarban: Lama, 3.4.2011, M. Khatun 616. *Area of major consumption:* Bandarban and Cox's Bazar districts.

## 2. Athyriaceae Pichi Sermolli (1970)

**175. Diplazium esculentum** (Retz.) Sw., Schrad. J. 1801 (2): 312 (1803). Vernacular names: Edible Fern (Eng.), *Dhekia Shak* (Beng.), *Teria Shak* (Tripuri).

A terrestrial fern. *Representative specimen*: Habiganj: Rema-Kalenga, 6.5.2003, M. Khatun 303. *Area of major consumption*: Chittagong and Sylhet districts.

**176. D. polypodioides** Bl., En. Pl. Jav.: 194 (1828). Vernacular names: Fern (Eng.), *Dhekia* (Beng.).

A fern. *Representative specimen*: Chittagong: Bangchari, 8.7.2003, M. Khatun 361. *Area of major consumption:* Chittagong and Mymensingh districts.

## 3. Blechnaceae (Presl) Copel. (1947)

**177. Blechnum orientale** L., Sp. Pl. 2: 1077 (1753). Vernacular name: *Boro Dhekia Shak* (Garo).

A large, terrestrial fern. *Representative specimen*: Cox's Bazar: Himchari, 2.8.2009, M. Khatun 517. *Area of major consumption:* Mymensingh and Cox's Bazar districts.

## 4. Dennstaedtiaceae Pichi Sermolli (1970, 1977)

**178. Microlepia strigosa** (Thunb.) Presl, Epim.: 95 (1849). Vernacular names: Lacy Fern (Eng.), *Fita Dhekia* (Beng.), *Dheki Shak* (Coach and Garo).

A tufted fern. *Representative specimen*: Sylhet: Satchari forest, 17.5.2005, Momtaz Mahal Mirza 526 (DACB). *Area of major consumption:* Habiganj and Sylhet districts.

## 5. Helminthostachyaceae Ching (1941)

**179. Helminthostachys zeylanica** (L.) Hook., Gen. Fil.: t. : 47 (1840). Vernacular names: Fern (Eng.), *Shada Dhekia* (Garo).

A terrestrial fern. *Representative specimen*: Tangail: Pirgacha, 18.4.2002, M. Khatun 173. *Area of major consumption:* Mymensingh district.

## 6. Marsileaceae Mirbel (1802)

**180. Marsilea minuta** (L.) Mant.: 308 (1771). Vernacular names: Marshy Fern (Eng.), *Susni Shak* (Beng.).

A small fern-allies. *Representative specimen*: Dinajpur: Noyabad, 16.9.2001, M. Khatun 84. *Area of major consumption:* Natore and Dinajpur districts.

**181. M. quadrifolia** L., Sp. Pl. 2: 1099 (1753). Vernacular names: Water Clover (Eng.), *Susni Shak* (Beng.).

A fern-allies. *Representative specimen*: Rajshahi: Godagari, 13.9.2001, M. Khatun 46. *Area of major consumption:* Rajshahi district.

## 7. Ophioglossaceae (R. Br.) Agardh (1882)

**182. Ophioglossum reticulatum** L., Sp. Pl. 2: 1063 (1753). Vernacular names: Adder's Tongue (Eng.), *Sharpa Jihba* (Beng.).

A terrestrial fern. *Representative specimen*: Mymensingh: Valukapara, 6.10.2001, M. Khatun 119. *Area of major consumption:* Mymensingh district.

## 8. Parkeriaceae Hook. (1825)

**183. Ceratopteris pteridoides** (Hook.) Hiern., Bot. Jahrb. 34: 561 (1905). Vernacular names: Aquatic Fern (Eng.), *Pani Fern* (Beng.), *Pani Dhekia Shak* (Garo).

An aquatic fern. *Representative specimen*: No specimen was collected by the author, but this species has been reported to be used as leafy vegetable by Sarker and Hossain (2009). *Area of major consumption:* Mymensingh district.

**184. C. thalictroides** (L.) Brongn., Bull. Soc. Phil. 1821: 186 (1822). Vernacular names: Water Fern (Eng.), *Pani Dhekia* (Beng.), *Keng Khah* (Garo).

An aquatic fern. *Representative specimen*: Bandarban: Lama, 3.4.2011, M. Khatun 610. *Area of major consumption*: Bandarban district.

## 9. Stenochlaenaceae Ching (1970)

**185. Stenochlaena palustris** (Burm. f.) Bedd., Ferns Brit. India (Suppl.) : 26 (1876). Vernacular names: Climbing Fern (Eng.), *Lata Dhekia* (Beng.).

A climbing terrestrial fern. *Representative specimen*: Dacca: Bander, 17.8.1941, S. K. Sen and Atul 638 (DUSH). *Area of major consumption*: Chittagong and Dhaka districts.

## 10. Thelypteridaceae Ching (1970)

**186. Ampelopteris prolifera** (Retz.) Copel., Gen. Fil : 144 (1947). Vernacular names: Walking Fern (Eng.), *Dhekia Shak* (Beng.).

A creeping pteridophyte. *Representative specimen*: Chittagong: Foy's lake, 12.7.2004, Momtaz Mahal Mirza 421 (DACB). *Area of major consumption:* Mymensingh and Chittagong districts.

Taxonomic study on the leafy vegetables is the first in its nature in Bangladesh. The present study shows that Amaranthaceae is the largest family in Magnoliopsida represented by 14 species followed by Asteraceae, Cucurbitaceae and Fabaceae comprising 10 species each. In Liliopsida, Araceae stands the highest position with 9 species followed by Zingiberaceae with 3 species. Out of 10 families of pteridophytes the families Athyriaceae, Marsileaceae and Parkeriaceae possess 2 species each. Twenty seven familes are represented by a single species used as leafy vegetables of which 20 families belong to angiosperms and 7 families to pteridophytes. Fifteen largest families of leafy vegetables reported to be found in Bangladesh are shown in Fig. 1.

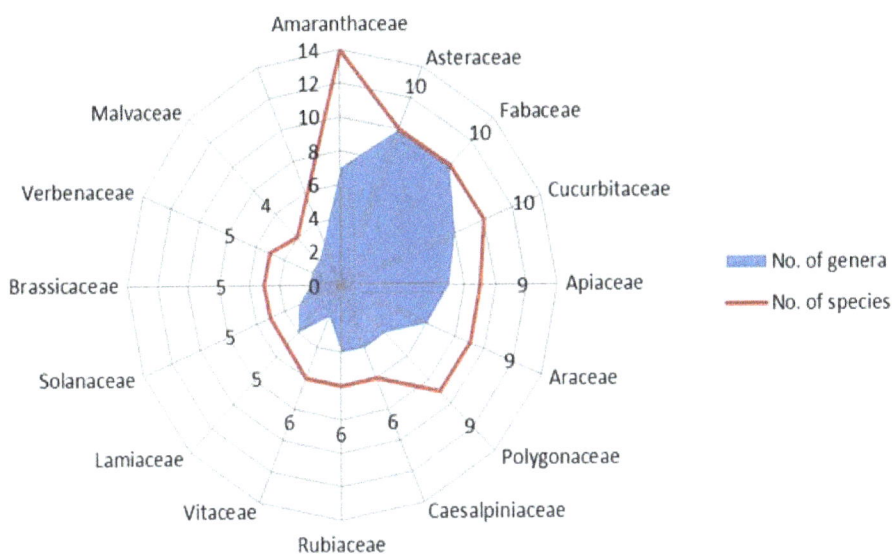

Fig. 1. Radder diagram showing 15 largest families of leafy vegetables in Bangladesh.

The study reveals the identification of 61 newly documented leafy vegetables for Bangladesh (Table 2). The majority of these newly documented leafy vegetables are consumed by the people of the hilly areas, especially in the Chittagong Hill Tracts and greater Sylhet district.

Out of 186 leafy vegetables recorded for Bangladesh, 140 taxa are wild and 46 are cultivated. Among the cultivated ones, 16 taxa are cultivated only for leafy vegetables, *viz., Amaranthus tricolor, A. viridis, Basella alba, Benincasa hispida, Brassica oleracea* var. *capitata, Celosia argentea, Coriandrum sativum, Corchorus capsularis, C. olitorius, Ipomoea aquatica, Lactuca scariola* var. *sativa, Malva verticillata, Raphanus sativus, Spinacea oleracea, Trachyspermum ammi* and *Trigonella foenum-graceum*. Most of these species are cultivated throughout the country except *Celosia argentia, Malva verticillata* and *Trachyspermum ammi. Celosia argentia* is cultivated in Rangamati (Kaptai), Sylhet and Patuakhali districts, *Malva verticillata* is grown in Dinajpur and Rangpur districts, and *Trachyspermum ammi* is cultivated only in Khagrachari (Golabari) district. Thirty taxa are cultivated for other purposes *viz.* spice, pulse but leaves of them are also used as vegetables, *viz., Allium cepa, A. sativum, Beta vulgaris, Brassica juncea, B. napus, B. rapa, Cajanus cajan, Capsicum frutescens, Cicer arietinum, Colocasia esculenta, Cucurbita maxima, Curcuma longa, Hibiscus cannabinus, Ipomoea batatus, Lablab purpureus, Lageneria siceraria, Lathyrus sativus, Luffa cylindrica, Manihot esculenta, Melocanna baccifiera, Momordica charantia* var. *muricata, Phaseolus vulgaris, Pisum sativum, Solanum tuberosum, Trachyspermum roxburghianum, Trichosanthes anguina, T. dioica, Vigna mungo, Xanthosoma sagittifolium* and *Zingiber officinale*. In case of tree species usually young leaves are used as vegetables.

Indigenous leafy vegetables can play an important role to alleviate hunger and malnutrition, but they are often neglected in research. They are important sources of micronutrients including Vitamin A and C, iron and other nutrients and are sometimes better nutritional sources than the modern vegetables. Wild leafy vegetables do not warrant any health hazard as they are free from any insecticide, herbicide and pesticides as well as free from the application of chemical fertilizers. Therefore, wild leafy vegetables are superior to the cultivated ones, if they are more or

**Table 2. Newly documented leafy vegetables of Bangladesh.**

| No. | Scientific name | Family | Local name/Ethnic name | Main consumption area |
|---|---|---|---|---|
| 1. | *Achyranthes aspera* L. | Amaranthaceae | Apang, Longra (Santal) | Mukundupur, Dinajpur; Comilla; Mymensingh |
| 2. | *Aegle marmelos* (L.) Corr. | Rutaceae | Bel, War-e-si-apang (Marma) | Bandarban; Khagrachari |
| 3. | *Aerva sanguinolenta* (L.) Blume | Amaranthaceae | Lal Apang, Nenga (Chakma) | Moulvi Bazar; Sunamganj |
| 4. | *Ajuga macrosperma* Wall. *ex* Benth. | Lamiaceae | Sabarang (Chakma, Marma) | Khagrachari; Rangamati |
| 5. | *Ageratum conyzoides* L. | Asteraceae | Ozone Shak (Chakma), Hinor (Khasia), Mukri (Tripura) | Khagrachari |
| 6. | *Albizia procera* Benth. | Mimosaceae | Sada Koroi, Fonagula (Chakma) | Alutila pahar, Khagrachari |
| 7. | *Antidesma acidum* Retz. | Euphorbiaceae | Multa, Mokhichikra (Tripura) | Chunarughat, Habiganj |
| 8. | *Bauhinia acuminata* L. | Caesalpiniaceae | Sada Kanchan, Jalong (Khasia), Hingshiara (Garo) | Moulvi Bazar; Netrakona |
| 9. | *Begonia barbata* Wall. *ex* A. DC. | Begoniaceae | Tokpata (Tripura) | Moulvi Bazar |
| 10. | *Blumea lacera* (Burm. f.) DC. | Asteraceae | Barakukshim, Leikhamal (Manipuri) | Feni; Moulvi Bazar |
| 11. | *Caesalpinia digyna* Rottler | Caesalpiniaceae | Kamuno (Marma) | Bandarban |
| 12. | *Cajanus cajan* L. Millsp. | Fabaceae | Arhar (Beng.), Rahar (Santal) | Sreemangal, Moulvi Bazar |
| 13 | *Cardiospermum helicacabum* L. | Sapindaceae | Kataboksa Shak (Chakma) | Rangamati |
| 14. | *Cassia fistula* L. | Caesalpiniaceae | Sonalu, Shumrol (Khasia) | Madhabkundu, Moulvi Bazar; Jessore |
| 15. | *Celosia cristata* L. | Amaranthaceae | Moragphul, Shibjota (Garo) | Netrakona; Patuakhali |
| 16. | *Cissus quadrangularis* L. | Vitaceae | Harjora Lata, Marang Gach (Santal), Moi-bhanga Lata (Garo) | Patuakhali |
| 17. | *C. repens* Lamk. | Vitaceae | Marmaria Lata (Koch) | Gajni, Sherpur |
| 18. | *Clausena excavata* Brum. f. | Rutaceae | Pan-karpur, Pankauri (Chakma) | Kamlachari, Rangamati |
| 19. | *Clerodendrum inerme* (L.) Gaertn. | Verbenaceae | Koklata, Jarems (Khasia) | Madhubpungi, Moulvi Bazar |
| 20. | *Colocasia gigantea* (Blume) Hook. f. | Araceae | Salad Kachu, Chinjapang (Khasia) | Khagrachari; Sylhet |
| 21. | *Elaeocarpus floribundus* Blume | Elaeocarpaceae | Jalpai, Uthethamo (Marma) | Bandarban |
| 22. | *Elephantopus scaber* L. | Asteraceae | Banmula, Gejiashak (Koch) | Noakhali |

**Table 2 contd.**

| No. | Scientific name | Family | Local name/Ethnic name | Main consumption area |
|---|---|---|---|---|
| 23. | *Emilia sonchifolia* (L.) DC. | Asteraceae | Mechitra, Sadusi, Miam Shak (Chakma) | Khagrachari |
| 24. | *Foeniculum vulgare* Miller | Apiaceae | Moroi (Marma) | Bandarban |
| 25. | *Ficus carica* L. | Moraceae | Dumur, Soluya (Khasia) | Madhabpunji, Moulvi Bazar |
| 26. | *Globba marantina* L. | Zingiberaceae | Globba, Holishak (Khasia) | Khagrachari; Rangamati |
| 27 | *Hewittia sublobata* (L. f.) O. Kuntze | Convolvulaceae | Dhudla Shak | Feni; Rajbari |
| 28. | *Hygrophila schulli* (Buch-Ham.) M. R. & S. N. Almeida | Acanthaceae | Kulekhara, Surmadhani Shak | Barisal; Rajbari |
| 29. | *Hygrophila polysperma* (Roxb.) T. Anders. | Acanthaceae | Puinna Shak (Rakhain) | Patuakhali |
| 30 | *Ludwigia adscendens* (L.) Hara | Onagraceae | Kesardam, Mulsishak (Garo) | Bijoypur, Netrakona |
| 31. | *Luffa cylindrica* (L.) M. Roem. | Cucurbitaceae | Dhundal, Mree-u-sthie (Rakhain) | Jessore |
| 32. | *Mangifera indica* L. | Anacardiaceae | Aam, Kharai (Khasia) | Madhabkundu, Moulvi Bazar |
| 33. | *Monochoria hastata* (L.) Solms. | Pontederiaceae | Baranukha, Chichir (Garo) | Netrakona |
| 34. | *M. vaginalis* (Burm. f.) Presl. | Pontederiaceae | Bhainsakachu, Kusrisha (Khasia) | Khagrachari; Moulvi Bazar |
| 35. | *Nelsonia canescens* (Lamk.) Spreng. | Acanthaceae | Paramul, Khaia Shak (Khasia) | Chunarughat, Habiganj |
| 36. | *Oenanthe javanica* (Blume) DC. | Apiaceae | Pan-turasi, Branju (Rakhain) | Patuakahli |
| 37. | *Operculina turpethum* (L.) Silva-Manso | Convolvulaceae | Dudhkalmi, Noapata | Gazipur; Patuakhali |
| 38. | *Oroxylum indicum* L. | Bignoniaceae | Kanaidingi, Fona-gulogach (Chakma) | Patuakhali; Rangamati |
| 39. | *Peperomia pellucida* (L.) H. B. & K. | Piperaceae | Luchi Pata, Samol-hapang (Garo) | Netrakona |
| 40. | *Phaseolus vulgaris* L. | Fabaceae | Felong Dal (Marma, Chakma) | Bandarban |
| 41. | *Piper longum* Roxb. | Piperaceae | Pipul | Rajbari |
| 42. | *Pogostemon benghalensis* (Burm. f.) O. Kuntze | Lamiaceae | Lomboi Shak (Chakma, Marma, Khasia) | Kaptai, Rangamati |
| 43. | *Premna mucronata* Roxb. | Verbenaceae | Katha Muri (Koch) | Sherpur |
| 44. | *Ruellia tuberosa* L. | Acanthaceae | Chatpoty, Charashak (Khasia) | Lawachara, Habiganj |

**Table 2 contd.**

| No. | Scientific name | Family | Local name/Ethnic name | Main consumption area |
|---|---|---|---|---|
| 45. | *Ricinus communis* L. | Euphorbiaceae | Crusuba (Marma, Tanchangya) | Bandarban |
| 46. | *Sambucus javanica* Reinw. | Caprifoliaceae | Maytraba (Chakma) | Bandarban; Rangamati |
| 47. | *Senna sophera* L. | Caesalpiniaceae | Kalkasunda, Eshi Shak (Rakhain) | Dinajpur; Manikgonj |
| 48. | *S. tora* (L.) Roxb. | Caesalpiniaceae | Chakunda, Kalkasham, Sa Lai Pa (Marma) | Kaptai, Rangamati |
| 49. | *Sesbania grandiflora* (L.) Poir. | Fabaceae | Bakphul, Agasta (Rakhain) | Khagrachari; Patuakhali |
| 50. | *Spermacoce stricta* L. f. | Rubiaceae | Narikeljuri Shak (Rakhain), Mijlick (Chakma) | Chittagong; Patuakhali; Sylhet |
| 51. | *Spondias pinnata* (L. f.) Kurz | Anacardiaceae | Amra, Thai-toui.(Tripura), Soh-awla (Khasia) | Madhabpunji, Moulvi Bazar |
| 52. | *Spilanthes calva* DC. | Asteraceae | Marhati-tiga (Marma) | Rangamati |
| 53. | *Sphenoclea zeylanica* Gaertn. | Sphenocleaceae | Jhilmarich, Radai (Mog), Vui-shak (Chakma) | Khagrachari |
| 54. | *Sterculia villosa* Roxb. | Sterculiaceae | Udal, Lambuk, (Tripura) | Bandarban, Moulvi Bazar |
| 55. | *Synedrella nodiflora* (L.) Gaertn. | Asteraceae | Hamfui (Marma, Chakma) | Khagrachari |
| 56. | *Tetrastigma angustifolium* (Roxb.) Planch. | Vitaceae | Nekung Rubi | Cox's Bazar |
| 57. | *Trachyspermum ammi* (L.) Sprague | Apiaceae | Jawan, Fuchi Shak (Chakma, Marma) | Golabari, Khagrachari |
| 58. | *T. roxburghianum* (DC.) H. Wolff | Apiaceae | Chanu, Rajani (Marma) | Bandarban |
| 59. | *Vernonia cinerea* (L.) Less | Asteraceae | Kuksim, Kalojira (Garo) | Netrakona |
| 60. | *Zehneria japonica* (Thunb.) H.Y. Liu | Cucurbitaceae | Herana Shak (Chakma, Khasia) | Rangamati |
| 61. | *Z. scabra* (L. f.) Sond. | Cucurbitaceae | Rakhal Sasa, Kola-kachu (Koch) | Sherpur |

less similar in nutritional values. Despite leafy vegetables have continuously been neglected for long time by the elite people, specially of urban societies, and has been referred to as "*Shak*, the poor men's food", now with the advancement of scientific research on their nutritive and medicinal values, have become an important item of our daily diet. Edible plants are thought not to contain any risk factors, therefore research for wild edible plants is necessary specially for famine situation and for the people living in rural and forest areas.

The present study on the leafy vegetables of Bangladesh is the most comprehensive study in the country. Extensive botanical exploration throughout the country over ten years resulted in documentation of 61 new reports as leafy vegetables for Bangladesh presenting 30.8% new addition to the species used as leafy vegetables. This provides a baseline study that could through more light on further research, particularly on nutra-medicinal aspects including determination of proximate nutrients, vitamins, micronutrients, macronutrients. Considring the present study as a baseline, if the nutrient compositions and other nutra-medicinal properties (i.e. antidiabetic, anti-cancerous, antibacterial and antioxidant) of the leafy vegetables, particularly the newly documented species could be determined, it would be possible to alleviate poverty and malnutrition in Bangladesh through the increased production and consumption of nutritious and health-promoting leafy vegetables.

## Acknowledgements

The authors thank the Director, Bangladesh National Herbarium (DACB) for allowing to work in the herbarium. Thanks are also due to the informants for their help during field visits. A fellowship offered to the first author by the Ministry of Science and Technology is gratefully acknowledged.

## References

Ahmed, Z.U., Hassan, M.A., Begum, Z.N.T., Khondker, M., Kabir, S.M.H., Ahmad, M., Ahmed, A.T.A., Rahman, A.K.A. and Haque, E.U. (Eds). 2008-2009. Encyclopedia of Flora and Fauna of Bangladesh, Vols. **6-10, 12**. Asiatic Society of Bangladesh, Dhaka.

Ali, S.M.K., Malek, M.A., Jahan, K. and Salamtullah, Q. (eds). 1977 (reprint 1992). *Deshio Khadyo-drobber Pustiman* (Nutritional value of local foods). Institute of Nutrition and Food Science, University of Dhaka, Dhaka.

Cronquist, A. 1981. An Integrated System of Classification of Flowering Plants. Columbia University Press, New York, 1262 pp.

Dassanayake, M.D. and Fosberg, F.R. (Eds). 1980-1985. A Revised Handbook to the Flora of Ceylon, Vols.**1-6**. Amerind Publishing Co. Pvt. Ltd., New Delhi.

Dini, I., Tenore, G.C. and Dini, A. 2005. Nutritional and antinutritional composition of Kancolla seeds: an interesting and underexploited andine food plant. Food Chemistry **92**(1): 125-132.

FAO. 2012. The State of Food Insecurity in the World 2012. Rome, Italy.

Fasuyi, A.O. 2006. Nutritional potentials of some tropical vegetable leaf meals: chemical characterization and functional properties. Afr. J. Biotechnol. **5**(1): 49-53.

Hassan, M.A. 2010. *Deshio Shak Shobjir Pusti Upadhan, Veshojgun o Patthaya Bichar*. The Royal Publishers, pp. 1-127.

Hooker, J.D. 1872-1897. The Flora of British India, Vols. **1-7**. L. Reeve & Co. Ltd. Kent, London.

Hyland, B.P.M. 1972. A technique for collecting botanical specimens in rain forests. Flora Malesiana Bulletin **26**: 2038-2040.

Kawatra, A., Singh, G. and Sehgal, S. 2001. Nutrition composition of selected green leafy vegetables, hervs and carrots. Plant Foods for Human Nutrition **56**: 359-365.

Khan, M.S. (Ed.) 1972-1987. Flora of Bangladesh. Nos. **1-39**. Bangladesh National Herbarium, BARC, Dhaka.

Khan, M.S. and Halim, M. 1987. Aquatic Angiosperms of Bangladesh. Bangladesh National Herbarium, BARC, Dhaka. pp. 1-120.

Khan, M.S. and Rahman, M.M. (Eds). 1989-2002. Flora of Bangladesh, Nos. **40-53**. Bangladesh National Herbarium, Dhaka.

Kimura, M. and Rodriguez-Amaya, D.B. 2003. Carotenoid composition of hydroponic leafy vegetables. J. Agric. & Food Chem. **51**: 2603-2607.

Kmiecik, W., Lisiewska, Z. and Jaworska, G. 2001. Effect of storage conditions on the technological value of dill (*Anethum graveolens* L.). Folia Horticulturea **13**: 33-43.

Orech, F.O., Christensen, D.L., Lasen, T., Friis, H., Aagaard-Hansen, J. and Estambale, B.A. 2007. Mineral content of traditional leafy vegetables from western Kenya. Inter. J. Food Sci. & Nutr. **58**(8): 595-602.

Prain, D. 1903 (rep. 1963). Bengal Plants, Vols. **1-2**. Botanical Survey of India, Calcutta. pp. 1-1013.

Rashid, M.E. and Rahman, M.A. 2011. Updated nomenclature and taxonomic status of the plants of Bangladesh included in Hook. f., The Flora of British India: Volume-I. Bangladesh J. Plant Taxon. **18**(2): 177-197.

Rashid, M.E. and Rahman, M.A. 2012. Updated nomenclature and taxonomic status of the plants of Bangladesh included in Hook. f., The Flora of British India: Volume-II. Bangladesh J. Plant Taxon. **19**(2): 173-190.

Reddy, C.V.K. 1999. Greens for Good Health. Nutrition **33**(3): 9.

Sarker, S.K. and Hossain, A.B.M.E. 2009. Pteridophytes of greater Mymensingh district of Bangladesh used as vegetables and medicines. Bangladesh J. Plant Taxon. **16**(1): 47-56.

Saxena, R. 1999. How green is your diet? Nutrition **33**(3): 9.

Siddiqui, K.U., Islam, M.A., Ahmed, Z.U., Begum, Z.N.T., Hassan, M.A., Khondker, M., Rahman, M.M., Kabir, S.M.H., Ahmed, A.T.A., Rahman, A.K.A. and Haque, E.U. (eds). 2007-2008. Encyclopedia of Flora and Fauna of Bangladesh, Vols. **5, 11**. Asiatic Society of Bangladesh, Dhaka.

Su, Q., Rowley, K.G., Itsiopoulos, C. and O'Dea, K. 2002. Identification and quantitation of major carotenoids in select components of the Mediterranean diet: Green leafy vegetables, figs and olive oil. European J. Clin. Nutr. **56**: 1149-1154.

Sundriyal, M. and Sundriyal, R.C. 2001. Wild edible plants of the Sikkim Himalaya: Nutritive values of selected species. Economic Botany **55**: 377-390.

# 4

# ANGIOSPERM FLORA OF SADAR UPAZILA OF MUNSHIGANJ DISTRICT, BANGLADESH

M. Oliur Rahman[1], Momtaz Begum and Md. Wajib Ullah

*Department of Botany, University of Dhaka, Dhaka 1000, Bangladesh*

*Keywords:* Floristics; Angiosperms; Munshiganj; Taxonomy.

## Abstract

Investigation to make inventory of the angiosperm species diversity in the local flora of Sadar Upazila of Munshiganj district has been made. A total of 240 taxa in 186 genera under 68 families are recognized, and enumerated citing each species with updated nomenclature, Bangla names, habit, habitat, phenology, potential value, status of occurrence in the area and voucher specimens. Of these 240 taxa, Magnoliopsida is represented by 195 taxa in 146 genera and 55 families, whereas Liliopsida by 45 taxa under 40 genera and 13 families. The local people of the area use over 50 medicinal plants as sources of medicine for their primary health care. Some species are assessed as rare to this local flora which need to be brought under conservation management for environmental sustainability of the area.

## Introduction

Munshiganj Sadar, an administrative Upazila of Munshiganj district comprises an area of 160.79 sq. km., and is consisted of 9 administrative unions, namely Adhara, Bajra Jogini, Char Kewar, Char Siloi, Mahakali, Mollakandi, Panchasar, Rampal and Rekabi Bazar. The Upazila witnesses the same climatic condition as other parts of the district. The hot summer, the long rainy season and the pleasant spring-cum-winter are the main noticeable seasons prevailing in the locality. The temperature of the area fluctuates between 12.7°C and 33.7°C throughout the year. Monthly average relative humidity varies from 62 to 83%, and monthly rainfall ranges from 7.7 to 373.1 mm throughout the year (BBS, 2011). There are three types of soil in the adjoining areas of Munshiganj Sadar Upazila, *viz.,* heavy clayey soil that prevails over the major part of the study area, light clayey soil which occupies the second position in the area, and heavy loamy soil prevails over small area. The Upazila presents diverse habitats including scrub jungles, homesteads, *char* lands, riparian, roadsides and wetlands.

Over the last few decades several attempts have been made on the floristic studies in Bangladesh, particularly in the forest and protected areas (Khan and Afza, 1968; Khan and Banu, 1972; Khan and Hassan, 1984; Rahman and Hassan, 1995; Rahman and Uddin, 1997; Uddin and Rahman, 1999; Khan and Huq, 2001; Uddin and Hassan, 2004; Tutul *et al.*, 2009, 2010; Arefin *et al.*, 2011; Uddin and Hassan, 2012). Studies on angiosperm flora in different Upazilas of Bangladesh are limited (Islam *et al.*, 2009; Rahman *et al.*, 2012; Moniruzzaman *et al.*, 2012; Rahman and Alam, 2013), however, there has been no floristic study on Munshigonj Sadar Upazila. The main objectives of the present study are to explore, identify and document the angiosperms of Munshiganj Sadar Upazila.

[1]Corresponding author. Email: dr_oliur@yahoo.com

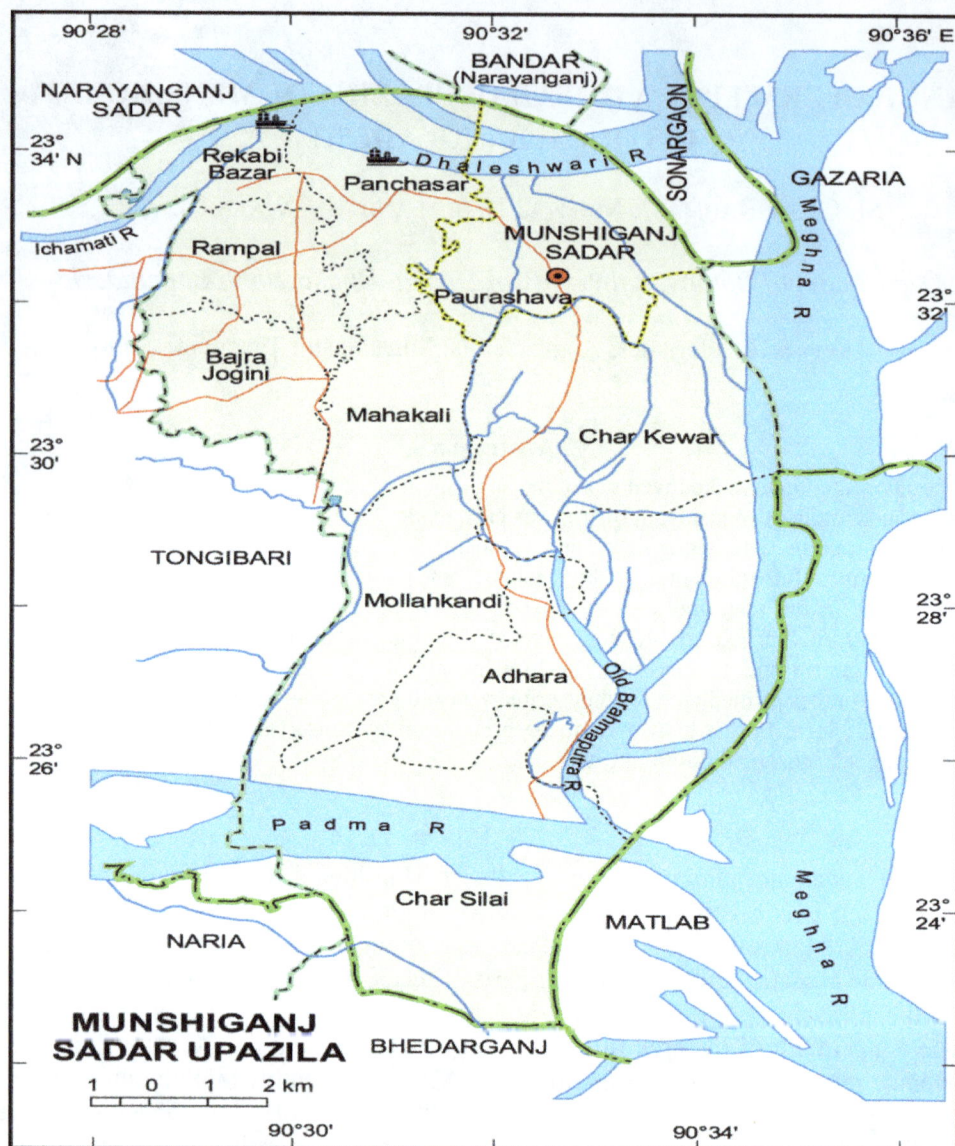

Fig. 1. Map of Munshiganj Sadar Upazila showing the sampling sites of different unions.

## Materials and Methods

The work is based on fresh materials collected from Sadar Upazila of Munshiganj district from May 2012 to April 2013 (Fig. 1). Plant specimens were collected from different areas within nine unions of the Upazila and processed using standard herbarium techniques (Hyland, 1972). Collected plant specimens were critically studied, examined and identified at the department of Botany, University of Dhaka. Identification was confirmed by experts, by comparing with herbarium specimens deposited both at Dhaka University Salar Khan Herbarium (DUSH) and Bangladesh National Herbarium (DACB), and by consulting standard floras and literature, *viz.*, Hooker (1872-1897), Prain (1903), Khan (1972-1987), Dassanayake and Fosberg (1980-1985), Khan and Rahman (1989-2002), Siddiqui *et al.* (2007), Ahmed *et al.* (2008a,b, 2009a,b,c,d) and Rashid and Rahman (2011, 2012). The identified families are arranged according to Cronquist's system of plant classification (Cronquist, 1981), and the genera and species under each family are arranged in an alphabetical order. Each species is furnished with updated nomenclature, Bangla names (wherever available), habit, habitat, phenology, potential value, status of occurrence in the area and voucher specimens. Voucher specimens are deposited in DUSH.

## Results and Discussion

Taxonomic study of angiosperm flora of Munshiganj Sadar Upazila has revealed a total of 240 angiosperm taxa under 186 genera and 68 families (Table 1). Among them Magnoliopsida (dicotyledons) is represented by 195 taxa belonging to 146 genera and 55 families, whereas Liliopsida (monocotyledons) having comparatively less representation, only 45 species under 40 genera and 13 families. The study has revealed that Magnoliopsida constitutes about 81% of the total, while Liliopsida constitutes 19% of the total angiosperm flora. In Magnoliopsida, Asteraceae appears as the largest family comprising 13 species under 13 genera followed by Euphorbiaceae, Moraceae, Fabaceae and Mimosaceae. In Liliopsida, Poaceae is the largest family with 23 species under 19 genera followed by Araceae, Arecaceae, Commelinaceae and Cyperaceae. In the study area the number of species in the families varies from 1 to 23. Out of 68 families recorded, each of 27 families is represented by a single species. Five largest genera of dicotyledons are *Ficus* (7 species), *Amaranthus* (4 species), *Persicaria* (4 species), *Syzygium* (4 species) and *Ipomoea* (3 species); while that of monocotyledons are *Commelina*, *Echinochloa*, *Leptochloa* and *Brachiaria*, each with 2 species.

Among the total flora herbs are represented by 141 taxa, shrubs by 26 and trees by 73. Out of 68 families recorded, 10 dominant families are Poaceae, Asteraceae, Euphorbiaceae, Moraceae, Fabaceae, Mimosaceae, Solanaceae, Myrtaceae, Malvaceae and Rubiaceae. The dominant families along with the number of species and genera are shown in Figure 2. These ten families comprise 105 species that represent about 44% of the total species identified. The remaining 58 families with a total 135 species represent 56% of the total.

One of the important phenomena of the study area is that the Sadar Upazila presents *char* land. The major angiosperms of the *char* are *Alternanthera paronychyoides* St. Hill., *A. sessilis* (L.) R. Br. *ex.* Roem & Schult., *Brassica napus* L. (cultivated), *Chenopodium album* L., *Citrullus lanatus* (Thub.) Matsumura & Nakai (cultivated), *Cyanotis cristata* (L.) D. Don, *Glinus oppositifolius* (L.) A. DC., *Gnaphalium luteo-album* L., *Grangea maderaspatana* (L.) Poir. and *Oxalis corniculata* L. The common riparian plants are *Brachiaria decumbens* Stapf, *Coix aquatica* Roxb., *Crateva magna* (Lour.) DC., *Ipomoea fistulosa* Mart. *ex* Choisy, *Operculina turpethum* (L.) S. Manso, *Persicaria barbata* (L.) Hara, *P. lapathifolia* (L.) S. F. Gray, *Phragmites karka* (Retz.) Trin. *ex* Steud. and *Saccharum spontaneum* L. The common roadside plants include *Albizia lebbeck* (L.) Benth. & Hook., *A. procera* (Roxb.) Benth., *Commelina benghalensis* L., *Euphorbia hirta* L., *Ficus benghalensis* L., *F. racemosa* L., *Sida acuta* Burm. f. and *Spilanthes calva* DC. The upazila provides several aquatic habitats including ponds, *beels*, *jheels*, rivers, etc. which offer luxuriant formation of angiosperm flora. Some common aquatic angiosperms are *Barringtonia acutangula* (L.) Gaertn., *Eichhornia crassipes* (Mart.) Solms, *Enhydra fluctuans* Lour., *Hygrorhyza aristata* (Retz.) Nees, *Ipomoea aquatica* Forssk., *Lemna perpusilla* Torrey, *Ludwigia adscendens* (L.) Hara, *L. hyssopifolia* (G. Don) Exell Apud A. & R. Fern., *Nymphaea nouchali* Burm. f., *N. pubescens* Willd., *N. rubra* Roxb. *et* Andr., *Nymphoides hydrophylla* (Lour.) O. Kuntze, *Ottelia alismoides* (L.) Pers., *Panicum paludosum* Roxb., *Persicaria barbata* (L.) Hara, *P. lanata* (Roxb.) Hassan, *P. orientalis* (L.) Spach, *Phragmites karka* (Retz.) Trin. *ex* Steud., *Pistia stratiotes* L. and *Vallisneria spiralis* L.

**Table 1. Enumeration of angiosperm taxa of Munshiganj Sadar Upazila.**

| Taxa | Bangla name | Habit | Habitat | Phenology | Potential value | Status of occurrence | Voucher specimens |
|---|---|---|---|---|---|---|---|
| **Magnoliopsida** | | | | | | | |
| **1. Annonaceae** | | | | | | | |
| 1. *Annona reticulata* L. | Sharifa | Tree | H | Oct - Jan | Fruits edible | Cultivated | W 303 |
| 2. *Annona squamosa* L. | Ata | Tree | H | Mar - Oct | Fruits edible | Cultivated | W 323 |
| 3. *Polyalthia longifolia* (Sonn.) Thw. | Debdaru | Tree | R | Mar - Oct | Timber, ornamental | Cultivated | W 266 |
| **2. Piperaceae** | | | | | | | |
| 4. *Peperomia pellucida* (L.) H. B. & K. | Luchipata | Herb | W | Jan - Dec | Medicinal | Common | W 69 |
| **3. Nymphaeaceae** | | | | | | | |
| 5. *Nymphaea nouchali* Burm. f. | Nil shapla | Herb | A | Jan - Dec | Vegetable | Common | W 29 |
| 6. *N. pubescens* Willd. | Shapla | Herb | A | Aug - Jan | Vegetable, medicinal | Common | W 34 |
| 7. *N. rubra* Roxb. et Andr. | Lal shapla | Herb | A | Jan - Dec | Vegetable | Common | W 111 |
| **4. Ranunculaceae** | | | | | | | |
| 8. *Ranunculus sceleratus* L. | Jal-dhonia | Herb | C | Feb - Mar | Medicinal | Common | W 287 |
| **5. Menispermaceae** | | | | | | | |
| 9. *Cissampelos pareira* L. | Akandi | Herb | S | Feb - Jul | Medicinal | Rare | W231 |
| 10. *Cyclea barbata* Miers | Patalpur | Herb | S | Mar - Nov | – | Common | W 148 |
| 11. *Stephania japonica* (Thunb.) Miers | Maknadi | Herb | S | Mar - Sep | Medicinal | Common | W232 |
| 12. *Tiliacora acuminata* (Lamk.) Hook. f. & Thoms. | Bhag lata | Shrub | S | Mar - Dec | Thatching | Rare | W 77 |
| **6. Moraceae** | | | | | | | |
| 13. *Artocarpus heterophyllus* Lamk. | Kanthal | Tree | H, R | Feb - Jun | Fruits edible | Cultivated | W 271 |
| 14. *A. lacucha* Buch.-Ham. | Dewa | Tree | H, R | Feb - Sep | Fruits edible | Cultivated | W 263 |
| 15. *Ficus benghalensis* L. | Bot | Tree | R, Ri | Mar - Aug | Medicinal | Common | W 325 |
| 16. *F. elastica* Roxb. ex Hornem | Rubber gach | Tree | R, Ri | Mar - Apr | Rubber | Cultivated | W 19 |
| 17. *F. heterophylla* L. f. | Bhui-dumur | Shrub | R, S | Nov - Feb | Medicinal | Common | W 87 |
| 18. *F. hispida* L. f. | Dumur | Tree | R | Apr - Sep | Medicinal | Common | W 34 |
| 19. *F. racemosa* L. | Gulang dumur | Tree | R, Ri | Mar - Nov | Medicinal | Common | W 47 |

**Table 1 contd.**

| Taxa | Bangla name | Habit | Habitat | Phenology | Potential value | Status of occurrence | Voucher specimens |
|---|---|---|---|---|---|---|---|
| 20. *F. religiosa* L. | Panbot | Tree | R, Ri | Mat - Sep | Medicinal | Common | W 194 |
| 21. *F. rumphii* Bl. | Sunamjor | Tree | R | Mar - Nov | Fruits edible | Common | W 30 |
| 22. *Morus alba* L. | Tut | Tree | S | Jan - Dec | Fruits edible, medicinal | Cultivated | W 225 |
| 23. *Streblus asper* Lour. | Sheora | Tree | S | Apr - Nov | Medicinal | Common | W 32 |
| **7. Urticaceae** | | | | | | | |
| 24. *Laportea interrupta* (L.) Chew | Lal Bichuti | Herb | S | May - Sep | Medicinal | Common | W 350 |
| **8. Chenopodiaceae** | | | | | | | |
| 25. *Chenopodium album* L. | Bothua shak | Herb | C | Dec - Mar | Leafy vegetable, medicinal | Common | W 213 |
| 26. *C. ambrosioides* L. | Chandanbetu | Herb | C | Nov - Mar | Medicinal | Common | W 53 |
| **9. Amaranthaceae** | | | | | | | |
| 27. *Alternanthera paronychioides* St. Hill. | Juli-khata | Herb | W | Jan - May | Vegetable | Rare | W 54 |
| 28. *A. sessilis* (L.) R. Br. ex. Roem & Schult. | Sachi -shak | Herb | W | Nov - Jul | Medicinal, vegetable | Common | W 11 |
| 29. *Amaranthus dubius* Mart. | Not known | Herb | R, W | Jan - Dec | Vegetable | Common | W 127 |
| 30. *A. spinosus* L. | Kanta notey | Herb | R, W | Jan - Dec | Leafy vegetable, medicinal | Common | W 39 |
| 31. *A. tricolor* L. | Lalshak | Herb | R, W | Jan - Dec | Vegetable | Cultivated | W 239 |
| 32. *A. viridis* L. | Notey shak | Herb | R | Dec - Apr | Leafy vegetable | Common | W 26 |
| **10. Portulacaceae** | | | | | | | |
| 33. *Portulaca oleracea* L. | Nuntashak | Herb | W | Jan - Dec | Leafy vegetable, medicinal | Common | W253 |
| **11. Molluginaceae** | | | | | | | |
| 34. *Glinus oppositifolius* (L.) A. DC. | Gima shak | Herb | C | Oct - Jan | Leafy vegetable | Common | W 25 |
| **12. Polygonaceae** | | | | | | | |
| 35. *Persicaria barbata* (L.) Hara | Bishkatali | Herb | A | Aug - Apr | Medicinal | Rare | W 321 |
| 36. *P. lanata* (Roxb.) Hassan | Shet-panimorich | Herb | A | Apr - Sep | Aquatic ornamental | Rare | W 195 |
| 37. *P. lapathifolia* (L.) S. F. Gray | Panimorich | Herb | A | Feb - May | Antibaterial | Rare | W 185 |
| 38. *P. orientalis* (L.) Spach | Bara panimorich | Herb | A | Mar - Aug | Antibaterial | Common | W 08 |
| 39. *Polygonum effusum* Meissn. | Raniphul | Herb | C | Feb - May | Leafy vegetable | Rare | W 216 |
| 40. *P. plebeium* R. Br. | Chemtisag | Herb | Ri | Jan - Apr | Leafy vegetable, antibaterial | Common | W 315 |

**Table 1 Contd.**

| Taxa | Bangla name | Habit | Habitat | Phenology | Potential value | Status of occurrence | Voucher specimens |
|---|---|---|---|---|---|---|---|
| **13. Dilleniaceae** | | | | | | | |
| 41. *Dillenia indica* L. | Chalta | Tree | R, H | May - Feb | Timber, fruits edible | Cultivated | W 150 |
| **14. Elaeocarpaceae** | | | | | | | |
| 42. *Elaeocarpus floribundus* Blume | Jalpai | Tree | H | May - Jun | Fruits edible | Cultivated | W 246 |
| **15. Tiliaceae** | | | | | | | |
| 43. *Corchorus capsularis* L. | Deshi pat | Herb | W | Aug - Feb | Fibre | Cultivated | W 158 |
| **16. Sterculiaceae** | | | | | | | |
| 44. *Abroma augusta* (L.) L. f. | Ulotkombol | Shrub | R | Jun - Dec | Medicinal | Rare | W 84 |
| 45. *Melochia corchorifolia* L. | Tiki-okra | Herb | W | Mar - Jun | – | Common | W 152 |
| **17. Bombacaceae** | | | | | | | |
| 46. *Bombax ceiba* L. | Shimul | Tree | R, Ri | Feb - May | Fibre | Planted | W 214 |
| **18. Malvaceae** | | | | | | | |
| 47. *Abutilon indicum* (L.) Sweet | Petari | Herb | R, S | Jul - Apr | Medicinal | Common | W 90 |
| 48. *Gossypium arboreum* L. | Tula | Shrub | W | Oct - Jan | Fibre | Cultivated | W 259 |
| 49. *Malvaviscus arboreus* Cav. | Not known | Tree | H | Not recorded | Ornamental | Cultivated | W 91 |
| 50. *Sida acuta* Burm. f. | Kureta | shrub | R, W | Sep - May | Medicinal | Common | W 100 |
| 51. *S. rhombifolia* L. | Kureta | Herb | R, W | Jul - Dec | Medicinal | Common | W 31 |
| 52. *Urena lobata* L. | Ban-okhra | shrub | R, W | Oct - Apr | Fibre | Common | W 07 |
| **19. Lecythidaceae** | | | | | | | |
| 53. *Barringtonia acutangula* (L.) Gaertn. | Hijal | Tree | M | May - Sep | Medicinal | Common | W 46 |
| **20. Cucurbitaceae** | | | | | | | |
| 54. *Citrulus lanatus* (Thunb.) Matsumura & Nakai | Tarmuj | Herb | C | Mar - Sep | Fruits edible | Cultivated | W 65 |
| 55. *Coccinia grandis* (L.) Voigt. | Telakucha | Herb | S | Mar - Nov | Medicinal | Common | W 14 |
| 56. *Luffa cylindrica* (L.) M. Roem. | Dhundul | Herb | H | Jun - Dec | Vegetable | Cultivated | W 147 |
| 57. *Momordica charantia* L. var. *charantia* C.B. Clarke | Korolla | Herb | H | May - Oct | Vegetable, medicinal | Cultivated | W 173 |

Table 1 Contd.

| Taxa | Bangla name | Habit | Habitat | Phenology | Potential value | Status of occurrence | Voucher specimens |
|---|---|---|---|---|---|---|---|
| **21. Capparaceae** | | | | | | | |
| 58. *Cleome rutidosperma* DC. | Not known | Herb | R, W | Jan - Dec | – | Common | W 27 |
| 59. *Crateva magna* (Lour.) DC. | Barun | Tree | R, Ri | Feb - Apr | Medicinal | Common | W 79 |
| **22. Brassicaceae** | | | | | | | |
| 60. *Brassica napus* L. | Sarisha | Herb | C | Mar - Jul | Oil, vegetable | Cultivated | W 15 |
| 61. *Rorippa indica* (L.) Hiern | Bonsharisha | Herb | C | Jan - Jun | Medicinal | Common | W 38 |
| 62. *R. palustris* (L.) Bess. | Panisarisha | Herb | C, W | Mar - Oct | Antiscorbutic | Rare | W 289 |
| **23. Moringaceae** | | | | | | | |
| 63. *Moringa oleifera* Lamk. | Sajina | Tree | H | Oct - Mar | Vegetable, medicinal | Cultivated | W 333 |
| **24. Sapotaceae** | | | | | | | |
| 64. *Manilkara zapota* (L.) P. van Royen | Sofeda | Tree | H | Jan - Dec | Fruits edible | Cultivated | W 188 |
| **25. Ebenaceae** | | | | | | | |
| 65. *Diospyros malabarica* (Desr.) Kostel. | Gab | Tree | H | May - Aug | Medicinal | Common | W 178 |
| 66. *D. montata* Roxb. | Bon gab | Tree | H, S | Mar - May | Timber, fish poisoning | Rare | W 326 |
| **26. Mimosaceae** | | | | | | | |
| 67. *Acacia auriculiformis* A. Cunn. ex Benth. & Hook. | Akashmoni | Tree | R | Jun - Feb | Timber, ornamental | Planted | W 205 |
| 68. *A. mangium* Willd. | Mangium | Tree | R | May - Dec | Timber | Planted | W 306 |
| 69. *Albizia lebbeck* (L.) Benth. & Hook. | Kala koroi | Tree | H, R, Ri | May - Dec | Timber | Common | W 244 |
| 70. *A. lucidior* (Steud.) Nielsen | Sil koroi | Tree | H, R, Ri | May - Oct | Timber | Common | W 327 |
| 71. *A. procera* (Roxb.) Benth. | Koroi | Tree | H, R | May - Dec | Timber | Common | W 229 |
| 72. *A. richardiana* (Voigt.) King & Prain | Gagan siris | Tree | H, R | Aug - Dec | Timber | Rare | W 251 |
| 73. *Leucaena leucocephala* (Lamk) de Wit. | Ipil-ipil | Tree | H, R | Mar - Nov | Timber, dye, forage | Planted | W 75 |
| 74. *Mimosa pudica* L. | Lajjabati | Herb | W | Jan - Dec | Soil binder, medicinal | Common | W 162 |
| 75. *Neptunia oleracea* Lour. | Panilajak | Herb | A | Sep - Jul | Food, drink, medicinal | Rare | W 296 |
| 76. *Pithecellobium dulce* (Roxb.) Benth. | Jilapi | Shrub | R, W | Jan - Jul | Hedge plant, dye, drink | Planted | W 108 |

**Table 1 Contd.**

| Taxa | Bangla name | Habit | Habitat | Phenology | Potential value | Status of occurrence | Voucher specimens |
|---|---|---|---|---|---|---|---|
| **27. Caesalpiniaceae** | | | | | | | |
| 77. *Bauhinia purpurea* L. | Lalkanchan | Shrub | R | Nov - May | Ornamental | Planted | W 270 |
| 78. *Saraca asoca* (Roxb.) de Wild. | Ashok | Tree | R | Feb - Jun | Medicinal | Planted | W 17 |
| 79. *Senna occidentalis* Roxb. | Barakalkesunda | Herb | R, W | May - Oct | Ornamental | Common | W 144 |
| 80. *S. tora* (L.) Roxb. | Kalkasham | Herb | R, W | Jan - Dec | Medicinal | Common | W 59 |
| 81. *Tamarindus indica* L. | Tentul | Tree | H, Ri | Apr - Dec | Fruits edible, medicinal | Cultivated | W 302 |
| **28. Fabaceae** | | | | | | | |
| 82. *Butea monosperma* (Lamk.) Taub. | Palash | Tree | R | Jan - Apr | Ornamental | Common | W 329 |
| 83. *Cajanus cajan* (L.) Millsp. | Arhar | Shrub | R | Dec - Apr | Pulse | Cultivated | W 247 |
| 84. *Dalbergia sissoo* Roxb. | Shishu | Tree | R | Mar - Aug | Timber | Cultivated | W 206 |
| 85. *Desmodium heterocarpon* (L.) DC. | Not known | shrub | R | Mar - Jul | – | Rare | W 97 |
| 86. *Erythrina fusca* Lour. | Kanta mandar | Tree | R, W | Feb - May | Hedge plant, firewood | Common | W 273 |
| 87. *Lathyrus aphaca* L. | Jangli motor | Herb | C | Nov - Mar | Fodder, nitrogen fixation | Common | W 283 |
| 88. *L. sativus* L. | Khesari | Herb | C | Feb - Sep | Pulse | Cultivated | W 286 |
| 89. *Pisum sativum* L. | Motor | Herb | C | Nov - Mar | Pulse | Cultivated | W 284 |
| 90. *Sesbania bispinosa* (Jacq.) Wight. | Dhaincha | Herb | W | May - Oct | Fibre, nitrogen fixation | Cultivated | W 116 |
| 91. *Vicia sativa* L. | Ankari | Herb | C | Jul - Nov | Fodder, nitrogen fixation | Rare | W 285 |
| **29. Lythraceae** | | | | | | | |
| 92. *Lagerstroemia speciosa* (L.) Pers. | Jarul | Tree | R | Apr - Aug | Ornamental, timber | Cultivated | W 224 |
| 93. *Lawsonia inermis* L. | Mehedi | Shrub | H | Jun - Dec | Dye, medicinal | Planted | W 307 |
| **30. Myrtaceae** | | | | | | | |
| 94. *Eucalyptus camaldulensis* Dehnhardt | Eucalyptus | Tree | R | Jan - Dec | Timber, charcoal, post | Planted | W 19 |
| 95. *E. grandis* Hill ex Maiden | Eucalyptus | Tree | R | Jan - Dec | Ornamental, timber | Planted | W 249 |
| 96. *Psidium guajava* L. | Peyara | Tree | H | Jan - Dec | Fruits edible | Cultivated | W 260 |
| 97. *Syzygium balsameum* (Wight) Walp. | Buti jam | Tree | H | Oct - Mar | Fruits edible | Rare | W 330 |

**Table 1 Contd.**

| Taxa | Bangla name | Habit | Habitat | Phenology | Potential value | Status of occurrence | Voucher specimens |
|---|---|---|---|---|---|---|---|
| 98. *S. cumini* (L.) Skeels | Kalo-jam | Tree | H | Mar - Jun | Fruits edible, medicinal | Cultivated | W 190 |
| 99. *S. jambos* (L.) Alston | Gulab-jam | Tree | H | Mar - Jun | Fruits edible | Cultivated | W 254 |
| 100. *S. samarangense* (Blume) Merr. & Perry | Jamrul | Tree | H | Mar - Jul | Fruits edible | Cultivated | W 238 |
| **31. Punicaceae** | | | | | | | |
| 101. *Punica granatum* L. | Dalim | Shrub | H | Jan - Dec | Fruits edible, medicinal | Cultivated | W 274 |
| **32. Onagraceae** | | | | | | | |
| 102. *Ludwigia adscendens* (L.) Hara | Kesardam | Herb | A | Mar - Dec | Medicinal | Common | W 291 |
| 103. *L. hyssopifolia* (G. Don) Exell | Not known | Herb | A | Jan - Dec | – | Common | W 42 |
| 104. *L. prostrata* Roxb. | Not known | Herb | R | Jan - Sep | Forage | Common | W 36 |
| **33. Combretaceae** | | | | | | | |
| 105. *Terminalia arjuna* (Roxb. *ex* DC.) Wight & Arn. | Arjun | Tree | H, R | Apr - Oct | Medicinal | Planted | W 272 |
| 106. *T. bellirica* (Gaertn.) Roxb. | Bohera | Tree | H | Mar - Nov | Medicinal | Planted | W 340 |
| 107. *T. catappa* L. | Katbadam | Tree | H | Mar - Dec | Medicinal, dye, oil | Planted | W 341 |
| **34. Euphorbiaceae** | | | | | | | |
| 108. *Acalypha indica* L. | Muktajhuri | Herb | W | Dec - Apr | Medicinal | Rare | W 57 |
| 109. *Chrozophora rottleri* (Geiseler) A. Juss. ex Spreng. | Khudi okra | Herb | R, W | Mar - Oct | Medicinal | Common | W 60 |
| 110. *Codiaeum variegatum* (L.) A. Juss. | Patabahar | Herb | H | Jan - Dec | Ornamental | Planted | W 179 |
| 111. *Croton bonplandianus* Baill. | Croton | Herb | R, W | Apr - Sep | Antiseptic | Common | W 19 |
| 112. *Euphorbia hirta* L. | Dhudia | Herb | R, W | Feb - Mar | Medicinal | Common | W 23 |
| 113. *E. thymifolia* L. | Dhudia | Herb | R | Jan - Dec | Medicinal | Common | W 28 |
| 114. *Pedilanthus tithymaloides* Poit. | Rangchita | Shrub | W | Mar - Dec | Medicinal, hedge plant | Common | W 92 |
| 115. *Phyllanthus acidus* (L.) Skeels | Orboroi | Tree | H | Mar - Dec | Fruits edible | Planted | W 339 |
| 116. *P. niruri* L. | Bhui-amla | Herb | R, W | Aug - Oct | Medicinal | Common | W 74 |
| 117. *P. reticulatus* Poir. | Chitki | Shrub | R, W | Jun - Oct | Medicinal | Common | W 67 |
| 118. *Ricinus communis* L. | Venna | Shrub | R, S | Sep - Feb | Medicinal, oil | Common | W 18 |
| 119. *Trewia nudiflora* L. | Pitali | Tree | R | May - Oct | Medicinal | Common | W 136 |

**Table 1 Contd.**

| Taxa | Bangla name | Habit | Habitat | Phenology | Potential value | Status of occurrence | Voucher specimens |
|---|---|---|---|---|---|---|---|
| **35. Rhamnaceae** | | | | | | | |
| 120. *Ziziphus mauritiana* Lamk. | Boroi | Tree | H | Sep - Mar | Fruits edible | Cultivated | W 309 |
| 121. *Z. oenoplia* (L.) Mill. | Ban boroi | Tree | S | Aug - Dec | Medicinal | Common | W 318 |
| **36. Vitaceae** | | | | | | | |
| 122. *Cayratia trifolia* (L.) Domin | Amal lata | Herb | S | Not recorded | – | Rare | W 85 |
| **37. Sapindaceae** | | | | | | | |
| 123. *Cardiospermum helicacabum* L. | Phutca | Herb | R, W | Apr - Jun | Medicinal | Common | W 223 |
| 124. *Litchi chinensis* Somn. | Litchu | Tree | H | Apr - Jun | Fruits edible | Planted | W 345 |
| **38. Anacardiaceae** | | | | | | | |
| 125. *Mangifera indica* L. | Aam | Tree | H | Jan - Jun | Fruits edible | Planted | W 225 |
| 126. *Spondius pinnata* (L. f.) Kurtz | Amra | Tree | H | Feb - Aug | Fruits edible | Planted | W 201 |
| **39. Meliaceae** | | | | | | | |
| 127. *Aphanamixis polystachya* (Wall.) R. N. Parker | Pitraj | Tree | R | Feb - May | Timber, medicinal | Common | W 191 |
| 128. *Azadirachta indica* A. Juss. | Neem | Tree | H, R | Mar - Jul | Medicinal | Common | W 294 |
| 129. *Melia azedarach* L. | Goranim | Tree | R | Mar - Feb | Medicinal | Common | W 258 |
| 130. *Swietenia mahagoni* Jacq. | Mahogoni | Tree | H, R | Apr - Oct | Timber | Planted | W 242 |
| **40. Rutaceae** | | | | | | | |
| 131. *Aegle marmelos* (L.) Correa | Bel | Tree | H | May - Jul | Fruits edible, medicinal | Planted | W 317 |
| 132. *Citrus limon* (L.) Burm. f. | Goralebu | Tree | H | Mar - Nov | Fruits edible | Planted | W 256 |
| 133. *C. maxima* (Burm.) Merr. | Jambura | Shrub | H | Feb - Nov | Fruits edible | Planted | W 180 |
| 134. *Limonia acidissima* L. | Kothbel | Tree | H | Feb - Dec | Fruits edible | Planted | W 337 |
| 135. *Murraya paniculata* (L.) Jack | Kamini | Shrub | R | Mar - Jan | Ornamental, medicinal | Planted | W 269 |

**Table 1 Contd.**

| Taxa | Bangla name | Habit | Habitat | Phenology | Potential value | Status of occurrence | Voucher specimens |
|---|---|---|---|---|---|---|---|
| **41. Oxalidaceae** | | | | | | | |
| 136. *Averrhoa carambola* L. | Kamranga | Tree | H | Oct - Jul | Fruits edible | Planted | W 176 |
| 137. *Oxalis corniculata* L. | Amrul | Herb | W | Sep - May | Medicinal | Common | W 29 |
| **42. Apiaceae** | | | | | | | |
| 138. *Centella asiatica* (L.) Urban | Thankuni | Herb | W | Apr - Oct | Medicinal | Common | W 62 |
| **43. Apocynaceae** | | | | | | | |
| 139. *Alstonia scholaris* (L.) R. Br. | Chatim | Tree | R | Nov - May | Medicinal, Oil | Common | W 300 |
| 140. *Carissa carandas* L. | Karamcha | Tree | H | Mar - Nov | Fruits edible | Planted | W 342 |
| 141. *Tabernaemontana divaricata* (L.) R. Br. ex Roem. & Schult. | Tagar | Shrub | H | May - Dec | Ornamental | Common | W 345 |
| **44. Asclepiadaceae** | | | | | | | |
| 142. *Calotropis procera* (Ait.) R. Br. | Akando | Shrub | R, Ri | Jan - Dec | Medicinal | Common | W 50 |
| **45. Solanaceae** | | | | | | | |
| 143. *Capsicum frutescens* L. | Kacha morich | Herb | H | Jan - Dec | Spice | Planted | W 22 |
| 144. *Datura metel* L. | Dhutura | Herb | S, W | Apr - Sep | Medicinal | Common | W 09 |
| 145. *Nicotiana plumbaginifolia* Viv. | Ban tamak | Herb | W | Mar - Oct | – | Common | W 02 |
| 146. *Physalis angulata* L. | Fotka | Herb | W | Feb - Aug | Medicinal | Rare | W 82 |
| 147. *P. minima* L. | Fotka | Herb | W | Jan - Dec | Medicinal | Common | W 304 |
| 148. *Solanum torvum* Swartz | Gota begun | Herb | R, W | Jan - Dec | Tender fruits as vegetable | Common | W 51 |
| 149. *S. villosum* Mill. | Tit begun | Herb | R | Feb - Aug | – | Rare | W 35 |
| **46. Convolvulaceae** | | | | | | | |
| 150. *Calystegia hederacea* Wall. | Not known | Herb | W | Dec - Feb | – | Rare | W 252 |
| 151. *Ipomoea aquatica* Forssk. | Kolmishak | Herb | A | Oct - Feb | Leafy vegetable | Common | W 113 |
| 152. *I. batatas* (L.) Lamk. | Misti alu | Herb | W | Dec - May | Tubers edible | Cutivated | W 164 |

**Table 1** Contd.

| Taxa | Bangla name | Habit | Habitat | Phenology | Potential value | Status of occurrence | Voucher specimens |
|---|---|---|---|---|---|---|---|
| 153. *I. fistulosa* Mart. *ex* Choisy | Dholkolmi | Shrub | R, Ri | Jan - Dec | Hedge plant | Common | W 123 |
| 154. *Operculina turpethum* (L.) S. Manso | Dudh kalmi | Herb | Ri | Oct - May | Medicinal | Common | W 298 |
| **47. Cuscutaceae** | | | | | | | |
| 155. *Cuscuta reflexa* Roxb. | Swarnalata | Herb | S | Jan - Apr | Medicinal | Common | W 109 |
| **48. Menyanthaceae** | | | | | | | |
| 156. *Nymphoides hydrophylla* (Lour.) Kuntze | Ponchuli | Herb | A | Oct - Feb | Medicinal | Common | W 166 |
| **49. Boraginaceae** | | | | | | | |
| 157. *Heliotropium indicum* L. | Hatishur | Herb | R, W | Jan - Dec | Medicinal | Common | W 05 |
| **50. Verbenaceae** | | | | | | | |
| 158. *Gmelina arborea* Roxb. | Gamari | Tree | R | Feb - Sep | Timber | Planted | W 335 |
| 159. *Lippia alba* (Mill.) Briton *ex* Wilson | Pichas-lakri | Shrub | R, S | Jan - Dec | Medicinal | Common | W 06 |
| 160. *Phyla nodiflora* (L.) Greene | Bhuiokra | Herb | C | Jan - Dec | Medicinal | Common | W 24 |
| 161. *Tectona grandis* L. f. | Shegun | Tree | R | Jul - Nov | Timber | Planted | W 336 |
| 162. *Vitex negundo* L. | Nishinda | Shrub | R, S | Apr - Feb | Medicinal | Common | W 337 |
| **51. Lamiaceae** | | | | | | | |
| 163. *Hyptis suaveolens* (L.) Poir. | Tokma | Herb | W | Nov - Apr | Medicinal | Common | W 138 |
| 164. *Leonurus sibiricus* L. | Roktodron | Herb | R, W | Jan - Dec | Medicinal | Rare | W 01 |
| 165. *Leucas aspera* (Willd.) Link | Shetodron | Herb | W | Jan - Dec | Medicinal, leafy vegetable | Common | W 03 |
| **52. Scrophulariaceae** | | | | | | | |
| 166. *Lindernia antipoda* (L.) Alston | Not known | Herb | C | Jan - Dec | – | Common | W 83 |
| 167. *L. crustacea* (L.) F. Muell | Not known | Herb | C | Apr - Dec | Medicinal | Common | W 66 |
| 168. *L. hyssopioides* (L.) Haines | Not known | Herb | C | Aug - May | – | Rare | W 310 |
| 169. *L. multiflora* (Roxb.) Mukerjee | Not known | Herb | C | Jun - Nov | – | Common | W 314 |
| 170. *Mecardonia procumbens* (Mill.) Small | Not known | Herb | R | Feb - Jun | – | Common | W 313 |
| 171. *Scoparia dulcis* L. | Bondhone | Herb | R, W | Dec - Jan | Medicinal | Common | W 04 |

**Tabe 1 Contd.**

| Taxa | Bangla name | Habit | Habitat | Phenology | Potential value | Status of occurrence | Voucher specimens |
|---|---|---|---|---|---|---|---|
| **53. Acanthaceae** | | | | | | | |
| 172. *Dipteracanthus prostratus* (Poir.) Nees | Not known | Herb | W | May - Aug | Medicinal | Rare | W 139 |
| 173. *Hygrophila phlomoides* Nees | Lotpiple | Herb | C | Oct - Dec | Head-ache | Rare | W 267 |
| 174. *Justicia gendarussa* Burm. f. | Jagatmardan | Shrub | R | Dec - May | Medicinal, hedge plant | Common | W 80 |
| 175. *Ruellia tuberosa* L. | Chatpoty | Herb | W | Jan - Feb | Ornamental | Common | W 16 |
| 176. *Rungia pectinata* (L.) Nees | Pindi | Herb | W | Nov - May | Medicinal | Common | W 228 |
| **54. Rubiaceae** | | | | | | | |
| 177. *Dentella serpyllifolia* Wall. *ex* Craib | Bhuipat | Herb | W | Nov - Jul | – | Rare | W 12 |
| 178. *Hedyotes corymbosa* (L.) Lamk. | Panki | Herb | C, W | Jan - Dec | Medicinal | Common | W 44 |
| 179. *Hyptianthera stricta* (Roxb.) Wight & Arn. | Tahi seing | Tree | R | Feb - May | Firewood | Rare | W 102 |
| 180. *Neolamarckia cadamba* (Roxb.) Bosser | Kadam | Tree | R | May - Aug | Timber, ornamental | Planted | W 230 |
| 181. *Spermacoce latifolia* Aublet | Ghuiojhil shak | Herb | R | Dec - Apr | – | Common | W 98 |
| 182. *S. stricta* L. f. | Bishmijal | Herb | R | Dec - May | Antimicrobial | Rare | W 93 |
| **55. Asteraceae** | | | | | | | |
| 183. *Ageratum conyzoides* L. | Fulkuri | Herb | R, W | Sep - Jun | Medicinal | Common | W 61 |
| 184. *Blumea membranacea* Wall *ex* DC. | Not known | Herb | W | Jan - Mar | – | Common | W 125 |
| 185. *Eclipta alba* (L.) Hassk | Kesaraj | Herb | R, W | Jan - Dec | Medicinal | Common | W 41 |
| 186. *Enhydra fluctuans* Lour. | Helencha | Herb | A | Jan - Apr | Leafy vegetable, medicinal | Common | W 145 |
| 187. *Gnaphalium luteo-album* L. | Bara karma | Herb | C | Mar - Aug | Medicinal | Common | W 211 |
| 188. *Grangea maderaspatana* (L.) Poir. | Nemuti | Herb | C | Dec - May | Medicinal | Common | W 212 |
| 189. *Mikania cordata* (Burm. f.) Robinson | Tarulata | Herb | S | Oct - Feb | Medicinal | Common | W 13 |
| 190. *Sonchus oleraceus* L. | Totlea | Herb | W | Feb - May | – | Common | W 149 |
| 191. *Spilanthes calva* DC. | Marhatitga | Herb | R | Jan - Dec | Medicinal | Common | W 10 |
| 192. *Synedrella nodiflora* (L.) Gaertn. | Shialmoti | Herb | R, W | Jan - Dec | Medicinal | Common | W 78 |

**Table 1 Contd.**

| Taxa | Bangla name | Habit | Habitat | Phenology | Potential value | Status of occurrence | Voucher specimens |
|---|---|---|---|---|---|---|---|
| 193. *Vernonia cinerea* (L.) Less. | Shial lata | Herb | R, W | May - Sep | Medicinal | Common | W 20 |
| 194. *Xanthium indicum* Koen. ex Roxb. | Gaghra | Herb | W | Jan - Dec | Leafy vegetable, medicinal | Common | W 198 |
| 195. *Youngia japonica* (L.) DC. | Not known | Herb | R | Aug - Jan | – | Common | W 187 |
| **Liliopsida** | | | | | | | |
| **56. Alismataceae** | | | | | | | |
| 196. *Sagittaria sagittifolia* L. | Muyamuya | Herb | A | Oct - Dec | Fodder | Common | W 344 |
| **57. Hydrocharitaceae** | | | | | | | |
| 197. *Ottelia alismoides* (L.) Pers. | Panikola | Herb | A | Jan - Dec | Vegetable | Common | W 168 |
| 198. *Vallisneria spiralis* L. | Pat-seola | Herb | A | Oct - Mar | – | Common | W 135 |
| **58. Najadaceae** | | | | | | | |
| 199. *Najas indica* (Willd.) Cham | Not known | Herb | A | Jun - Sep | – | Common | W 132 |
| **59. Arecaceae** | | | | | | | |
| 200. *Areca catechu* L. | Supari | Tree | H | Jan - Dec | Masticatory, medicinal | Planted | W 334 |
| 201. *Borassus flabellifer* L. | Tal | Tree | H | Jan - Oct | Fruits edible, trunk for pillars and posts | Planted | W 349 |
| 202. *Phoenix sylvestris* Roxb. | Khejur | Tree | H, R | Dec - Jun | Fruits edible, leaves for mats and baskets | Planted | W 241 |
| **60. Araceae** | | | | | | | |
| 203. *Colocasia esculenta* (L.) Schott | Kochu | Herb | R, W | May - Oct | Vegetable, medicinal | Common | W 323 |
| 204. *Pistia stratiotes* L. | Topapana | Herb | A | Oct - Mar | Medicinal | Common | W 133 |
| 205. *Syngonium podophyllum* Schott | Bahari | Herb | W | Feb - Nov | Ornamental | Rare | W 40 |
| 206. *Typhonium flagelliforme* (Lodd.) Bl. | Ghetkachu | Herb | W | Apr - Oct | Medicinal | Common | W 167 |
| 207. *Xanthosoma violaceum* Schott | Kalokachu | Herb | W | Apr - Oct | Vegetable | Cultivated | W 248 |
| **61. Lemnaceae** | | | | | | | |
| 208. *Lemna perpusilla* Torrey | Khudipana | Herb | A | May - Sep | Water purifier | Common | W 316 |

**Table 1 Contd.**

| Taxa | Bangla name | Habit | Habitat | Phenology | Potential value | Status of occurrence | Voucher specimens |
|---|---|---|---|---|---|---|---|
| **62. Commelinaceae** | | | | | | | |
| 209. *Commelina benghalensis* L. | Dholpata | Herb | R, W | Feb - Dec | Green vegetable | Common | W 33 |
| 210. *C. longifolia* Lamk. | Pani-kanchira | Herb | W | Dec - Jan | – | Common | W 126 |
| 211. *Cyanotis cristata* (L.) D. Don | Kanua | Herb | C | Sep - Feb | Forage | Common | W 118 |
| **63. Cyperaceae** | | | | | | | |
| 212. *Cyperus rotundus* L. | Mutha | Herb | R, W | May - Sep | Medicinal | Common | W 217 |
| 213. *Kyllinga brevifolia* Rottb. | Not known | Herb | R, W | Mar - Dec | – | Common | W 343 |
| **64. Poaceae** | | | | | | | |
| 214. *Apluda mutica* L. | Not known | Herb | S | Jul - Sep | Medicinal | Common | W 210 |
| 215. *Bambusa balcooa* Roxb. | Barak bans | Herb | H | Mar - Sep | Household construction | Common | W 209 |
| 216. *Bothriochloa pertusa* (L.) A. Camus | Barboda ghas | Herb | W | Aug - Feb | Fodder | Rare | W 153 |
| 217. *Brachiaria decumbens* Stapf | Not known | Herb | R | Oct - Jan | Forage | Common | W 146 |
| 218. *B. kurzii* (Hook. f.) A. Camus | Not known | Herb | R, W | Jan - Dec | – | Rare | W 275 |
| 219. *Coix aquatica* Roxb. | Dhanga gurgas | Herb | R, W | Sep - Dec | – | Common | W 58 |
| 220. *Cynodon dactylon* L. Pers. | Durba ghas | Herb | R, W | Jul - Dec | Medicinal | Common | W 347 |
| 221. *Dactyloctenium aegyptium* (L.) P. Beauv. | Makra | Herb | W | Jan - Dec | – | Common | W 226 |
| 222. *Echinochloa colonum* (L.) Link | Shama ghas | Herb | C, W | Jan - Dec | Fodder | Common | W 277 |
| 223. *E. stagnina* (Retz.) P. Beauv. | Dul | Herb | C, W | Mar - Sep | Fodder | Common | W 264 |
| 224. *Eleusine indica* (L.) Gaertn. | Malankuri | Herb | R, W | Jun - Aug | Soil binder | Common | W 192 |
| 225. *Hygroryza aristata* (Retz.) Nees | Jangli dhan | Herb | A | Oct - Feb | Fodder, medicinal | Common | W 130 |
| 226. *Leptochloa chinensis* (L.) Nees | Not known | Herb | W | May - Dec | Fodder | Rare | W 200 |
| 227. *L. panicea* (Retz.) Ohwi | Not known | Herb | W | May - Oct | – | Rare | W 281 |
| 228. *Myriostachya wightiana* (Nees ex Steud.) Hook. f. | Dhansi | Herb | W | May - Nov | – | Common | W 117 |

**Table 1 Contd.**

| Taxa | Bangla name | Habit | Habitat | Phenology | Potential value | Status of occurrence | Voucher specimens |
|---|---|---|---|---|---|---|---|
| 229. *Oplismenus burmanii* (Retz.) P. Beauv. | Not known | Herb | R | Sep - Jan | Fodder | Rare | W 131 |
| 230. *Panicum paludosum* Roxb. | Barti | Herb | A | Jan - Dec | Fodder | Common | W 197 |
| 231. *P. repens* L. | Dhani ghas | Herb | C, R | Jan - Dec | Pasture | Common | W 280 |
| 232. *Pennisetum purpureum* Schum. | Not known | Herb | R | Sep - Feb | Fodder, pasture | Rare | W 106 |
| 233. *Phragmites karka* (Retz.) Trin. ex Steud. | Nal | Shrub | A | Jan - Dec | Soil binder, medicinal | Common | W 297 |
| 234. *Saccharum spontaneum* L. | Kash | Herb | Ri | Jan - Dec | Soil binder, fodder | Common | W 104 |
| 235. *Sclerostachya fusca* (Roxb.) A. Camus | Not known | Herb | W | Aug - Feb | – | Common | W 183 |
| 236. *Setaria glauca* (L.) P. Beauv. | Bajra | Herb | R | Sep - Dec | Forage, thatching, fencing, fuel | Common | W 282 |
| **65. Marantaceae** | | | | | | | |
| 237. *Schumannianthus dichotomus* (Roxb.) Gagnep. | Mukta pati | Shrub | W | Dec - Mar | Rough weaving fibre | Cultivated | W 88 |
| **66. Pontederiaceae** | | | | | | | |
| 238. *Eichhornia crassipes* (Mart.) Solms | Kochuripana | Herb | A | Jan - Dec | Manure, fodder | Common | W 220 |
| **67. Aloaceae** | | | | | | | |
| 239. *Aloe vera* (L.) Burm. f. | Gritakumari | Herb | H | Jan - Dec | Medicinal | Cultivated | W 208 |
| **68. Dioscoriaceae** | | | | | | | |
| 240. *Dioscorea alata* L. | Chupri alu | Herb | H | Oct - Dec | Tubers edible | Common | W 96 |

A = Aquatic, C = *Char* land, H = Homestead, M = Marsh land, R = Roadside, Ri = Riparian, S = Scrub jungle, W = Wasteland

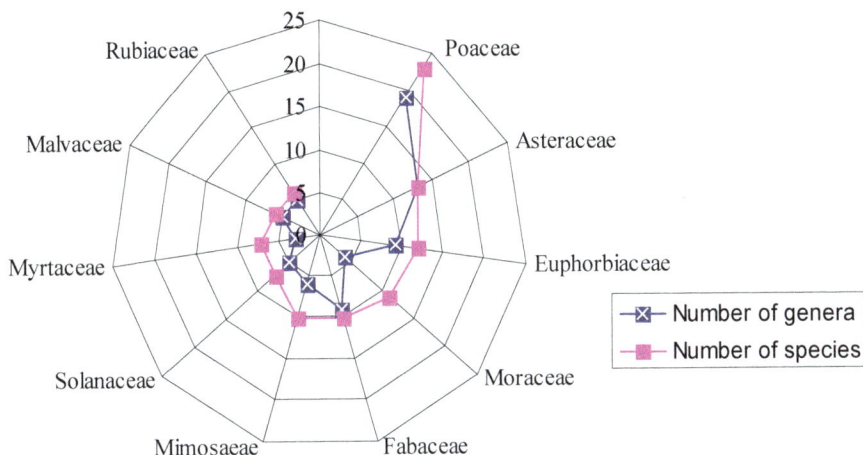

Fig. 2. Raddar diagram showing the 10 largest families in Munshiganj Sadar Upazila.

The present study identifies over 50 medicinal plants used by the local people of Munshiganj Sadar Upazila for their primary health care. They use the medicinal plants for treatment of several common diseases including dysentery, diarrhoea, diabetes, fever, cold and cough, asthma, ulcer, constipation, abdominal pain, indigestion, gonorrhoea, jaundice, stop bleeding, piles, scabies and rheumatic pain. Some of the important medicinal plants used by the local people are *Abroma augusta* (L.) L. f., *Acalypha indica* L., *Aloe vera* (L.) Burm. f., *Alstonia scholaris* (L.) R. Br., *Azadirachta indica* A. Juss., *Calotropis procera* (Ait.) R. Br., *Centella asiatica* (L.) Urban, *Coccinia grandis* (L.) Voigt., *Mikania cordata* (Burm. f.) Robinson, *Phyllanthus niruri* L., *Saraca asoca* (Roxb.) Willd., *Terminalia arjuna* (Roxb. *ex* DC.) Wight & Arn. and *Vitex negundo* L. Apart from medicinal uses some species are used by local people in their religious festivals, *viz.*, *Aegle marmelos* (L.) Correa, *Areca catechu* L., *Bauhinia purpurea* L., *Butea monosperma* (Lamk.) Taub. and *Cynodon dactylon* (L.) Pers. The study has also identified some rare plants in Munshiganj sadar upazila, i.e. *Alternanthera paronychyoides* St. Hill., *Diospyros montana* Roxb., *Dipteracanthus prostratus* (Poir.) Nees, *Operculina turpethum* (L.) S. Manso, *Persicaria lanata* (Roxb.) Hassan and *Tiliacora acuminata* (Lamk.) Hook. f. & Thoms.

Though the study area has a moderately rich resource of angiospermic flora, it witnesses some threats which might cause this resource to extinct. Observations and group discussion with local people during field works resulted in identifying some major threats which include urbanization, modern agriculture, lack of awareness, exotic plantation and river erosion. Therefore, efforts should be undertaken to safeguard the plants through *ex situ* and *in situ* approaches, public awareness should be built up, and protection of habitats of the species should be ensured.

### Acknowledgement

We thank Prof. Md. Abul Hassan of the Department of Botany, University of Dhaka for his help and cooperation during the course of this study.

### References

Ahmed, Z.U., Begum, Z.N.T., Hassan, M.A., Khondker, M., Kabir, S.M.H., Ahmad, M., Ahmed, A.T.A., Rahman, A.K.A. and Haque, E.U. (Eds) 2008a. Encyclopedia of Flora and Fauna of Bangladesh, Vol. **6**. Angiosperms: Dicotyledons (Acanthaceae – Asteraceae). Asiatic Society of Bangladesh, Dhaka, pp. 1-408.

Ahmed, Z.U., Hassan, M.A., Begum, Z.N.T., Khondker, M., Kabir, S.M.H., Ahmad, M., Ahmed, A.T.A., Rahman, A.K.A. and Haque, E.U. (Eds) 2008b. Encyclopedia of Flora and Fauna of Bangladesh, Vol. **12**. Angiosperms: Monocotyledons (Orchidaceae – Zingiberaceae). Asiatic Society of Bangladesh, Dhaka, pp. 1-552.

Ahmed, Z.U., Hassan, M.A., Begum, Z.N.T., Khondker, M., Kabir, S.M.H., Ahmad, M., Ahmed, A.T.A., Rahman, A.K.A. and Haque, E.U. (Eds) 2009a. Encyclopedia of Flora and Fauna of Bangladesh, Vol. **7**. Angiosperms: Dicotyledons (Balsaminaceae – Euphorbiaceae). Asiatic Society of Bangladesh, Dhaka, pp. 1-546.

Ahmed, Z.U., Hassan, M.A., Begum, Z.N.T., Khondker, M., Kabir, S.M.H., Ahmad, M., Ahmed, A.T.A., Rahman, A.K.A. and Haque, E.U. (Eds) 2009b. Encyclopedia of Flora and Fauna of Bangladesh, Vol. **8**. Angiosperms: Dicotyledons (Fabaceae – Lythraceae). Asiatic Society of Bangladesh, Dhaka, pp. 1-478.

Ahmed, Z.U., Hassan, M.A., Begum, Z.N.T., Khondker, M., Kabir, S.M.H., Ahmad, M. and Ahmed, A.T.A. (Eds) 2009c. Encyclopedia of Flora and Fauna of Bangladesh, Vol. **9**. Angiosperms: Dicotyledons (Magnoliaceae – Punicaceae). Asiatic Society of Bangladesh, Dhaka, pp. 1-488.

Ahmed, Z.U., Hassan, M.A., Begum, Z.N.T., Khondker, M., Kabir, S.M.H., Ahmad, M., and Ahmed, A.T.A. (Eds) 2009d. Encyclopedia of Flora and Fauna of Bangladesh, Vol. **10**. Angiosperms: Dicotyledons (Ranunculaceae – Zygophyllaceae). Asiatic Society of Bangladesh, Dhaka, pp. 1-580.

Arefin, M.K., Rahman, M.M., Uddin, M.Z. and Hassan, M.A. 2011. Angiosperm flora of Satchari National Park, Habiganj, Bangladesh. Bangladesh J. Plant Taxon. **18**(2): 117-140.

BBS (Bangladesh Bureau of Statistics) 2011. Monthly Statistical Bulletin. Statistics Division, Ministry of Planning, Government of the People's Republic of Bangladesh.

Cronquist, A. 1981. An Integrated System of Classification of Flowering Plants. Columbia Univ. Press, New York.

Dassanayake, M.D. and Fosberg, F.R. (Eds) 1980-1985. A Revised Handbook to the Flora of Ceylon, Vols. **1-5**. Amerind Publishing Co. Pvt. Ltd., New Delhi.

Hooker, J.D. 1872-1897. The Flora of British India, Vols. **1-7**. L. Reeve & Co. Ltd., Kent, England.

Hyland, B.P.M. 1972. A technique for collecting botanical specimens in rain forest. Flora Malesiana Bulletin **26**: 2038-2040.

Islam, M.R., Uddin, M.Z. and Hassan, M.A. 2009. An assessment of the angiospermic flora of Ramgarh upazila of Khagrachari district, Bangladesh. Bangladesh J. Plant Taxon. **16**(2): 115-140.

Khan, M.S. (Ed.) 1972-1987 Flora of Bangladesh. Nos. **1-39**. Bangladesh National Herbarium and Bangladesh Agricultural Research Council, Dhaka.

Khan, M.S. and Afza, S.K. 1968. A taxonomic report on the angiospermic flora of Teknaf and St. Martin's Island. Dhaka Univ. Studies, Part B. **16**: 35-37.

Khan, M.S. and Banu, F. 1972. A taxonomic report on the angispermic flora of Chittagong Hill Tracts-2. J. Asiatic Soc. Bangladesh **17**(2): 63-68.

Khan, M.S. and Hassan, M.A. 1984. A taxonomic report on the angiospermic flora of St. Martin's Island. Dhaka Univ. Studies, Part B. **32**(1): 76-78.

Khan, M.S. and Huq, A.M. 2001. The vascular flora of Chunati Wildlife Sanctuary in south Chittagong, Bangladesh. Bangladesh J. Plant Taxon. **8**(1): 47-64.

Khan, M.S. and Rahman, M.M. (Eds) 1989-2002. Flora of Bangladesh. Nos. **40-53.** Bangladesh National Herbarium and Bangladesh Agricultural Research Council, Dhaka.

Moniruzzaman, M., Hassan, M.A., Rahman, M.M., Layla, S. and Islam, M.R. 2012. A preliminary checklist of the angiospermic flora of Daulatpur Upazila in Kushtia district, Bangladesh. J. Asiat. Soc. Bangladesh, Sci. **38**(1): 53-65.

Prain, D. 1903. Bengal Plants, Vols. **1-2**. Botanical Survey of India, Calcutta.

Rahman, M.A. and Uddin, S.B. 1997. Assessment of plant diversity of Sitakunda in Chittagong. Bangladesh J. Plant Taxon. **4**(1): 17-36.

Rahman, M.O. and Hassan, M.A. 1995. Angiospermic flora of Bhawal National Park, Gazipur (Bangladesh). Bangladesh J. Plant Taxon. **2**(1&2): 47-79.

Rahman, M.O. and Alam, M.T. 2013. A taxonomic study on the angiosperm flora of Trishal upazila, Mymensingh. Dhaka Univ. J. Biol. Sci. **22**(1): 63-74.

Rahman, M.O., Antara, R.T., Begum, M. and Hassan, M.A. 2012. Floristic diversity of Dhamrai upazila of Dhaka, Bangladesh with emphasis on medicinal plants. Bangladesh J. Bot. **41**(1): 71-85.

Rashid, M.E. and Rahman, M.A. 2011. Updated nomenclature and taxonomic status of the plants of Bangladesh included in Hook. f., The Flora of British India: Volume-I. Bangladesh J. Plant Taxon. **18**(2): 177-197.

Rashid, M.E. and Rahman, M.A. 2012. Updated nomenclature and taxonomic status of the plants of Bangladesh included in Hook. f., The Flora of British India: Volume-II. Bangladesh J. Plant Taxon. **19**(2): 173-190.

Siddiqui, K.U., Islam, M.A., Ahmed, Z.U., Begum, Z.N.T., Hassan, M.A., Khondker, M., Rahman, M.M., Kabir, S.M.H., Ahmad, M., Ahmed, A.T.A., Rahman, A.K.A. and Haque, E.U. (Eds) 2007. Encyclopedia of Flora and Fauna of Bangladesh, Vol. **11**. Angiosperms: Monocotyledons (Agavaceae - Najadaceae). Asiatic Society of Bangladesh, Dhaka, pp. 1-399.

Tutul, E., Uddin, M.Z., Rahman, M.O. and Hassan, M.A. 2009. Angiospermic flora of Runctia Sal forest, Bangladesh. I. Liliopsida (Monocots). Bangladesh J. Plant Taxon. **16**(1): 83-90.

Tutul, E., Uddin, M.Z., Rahman, M.O. and Hassan, M.A. 2010. Angiospermic flora of Runctia Sal forest, Bangladesh. II. Magnoliopsida (Dicots). Bangladesh J. Plant Taxon. **17**(1): 33-53.

Uddin, M.Z. and Hassan, M.A. 2004. Flora of Rema-Kalenga Wildlife Sanctuary, IUCN Bangladesh Country Office, Dhaka, Bangladesh.

Uddin, S.B. and Rahman, M.A. 1999. Angiospermic flora of Himchari National Park, Cox's Bazar, Bangladesh. Bangladesh J. Plant Taxon. **6**(1): 31-68.

Uddin, S.N. and Hassan, M.A. 2012. Angiosperm flora of Rampahar Reserve Forest under Rangamati district in Bangladesh. I. Liliopsida (Monocots). Bangladesh J. Plant Taxon. **19**(1): 37-44.

# 5

# TAXONOMIC IDENTITY OF *THERIOPHONUM DANIELII* AND *T. MANICKAMII* (ARACEAE)

M. Sivadasan[1], V. Abdul Jaleel[2], Ahmed H. Alfarhan and
P. Lakshminarasimhan[3]

*Department of Botany and Microbiology, College of Science, King Saud University,
P. B. No. 2455, Riyadh 11451, Kingdom of Saudi Arabia*

*Keywords:* Aroideae; New synonymy; Endemic; India; Sri Lanka.

## Abstract

The genus *Theriophonum* (Araceae), represented by seasonally dormant tuberous perennials is endemic to India and Sri Lanka. Critical taxonomic appraisal of the constituent species supports existence of only five species, viz. *T. dalzellii, T. fischeri, T. infaustum, T. minutum* and *T. sivaganganum,* and all are with restricted distribution in India. *Theriophonum minutum* is the only species with extended distribution in Sri Lanka. The recently described *T. danielii* and *T. manickamii* are considered here as conspecific with *T. infaustum* and *T. fischeri*, respectively.

## Introduction

The genus *Theriophonum* Blume (1837) belonging to the subfamily Aroideae of Araceae comprises seasonally dormant tuberous perennials endemic to India and Sri Lanka. In India, the genus is represented by five species confined to the south and central parts, while there is only one species in Sri Lanka (Sivadasan and Nicolson, 1982). In the revision of the genus, Sivadasan and Nicolson (1982) provided a detailed account of the taxonomic history and stated that misidentifications have been frequent and mainly centered around Rheede's (1692) illustration of *Nelenschena minor* and the type of *Arum minutum* Willd. (1805) [=*Theriophonum minutum* (Willd.) Baill. (1895)]. Rheede's *Nir-tsjembu* (1692: 11: 33, t. 16) and *Nelenschena minor* (1692: 11: 33, t. 17) represent the first Pre-Linnean printed records of *Theriophonum*; both are identified as *T. infaustum* N. E. Br. (1880) (Sivadasan and Nicolson, 1982; Suresh *et al.*, 1983). The works of Schott (1860) and Engler (1879, 1920) are significant in recognition and delimitation of the species described until then. Engler (1920) recognized five species, which on scrutiny, were found to represent only three, viz. *T. dalzellii* Schott (1855), *T. infaustum* and *T. minutum*. After forty nine years, a fourth species, *T. sivaganganum* (Ramam. & Sebastine) Bogner (1969), was added to the genus by Bogner by transfer of *Pauella sivagangana* Ramamurthy & Sebastine (1967). Then *T. fischeri* Sivad. (Sivadasan and Nicolson, 1981) was added about 61 years after Engler's revision. Sivadasan and Nicolson (1982) recognized five species, viz. *T. dalzellii, T. fischeri, T. infaustum, T. minutum* and *T. sivaganganum* in their revision. Since then two more species, viz. *T. manickamii* Murugan & K. Natarajan (2008) and *T. danielii* Rajakumar *et al.* (2010) have been described. While reviewing the checklist of species of *Theriophonum*, the protologues of the above two species were studied which aroused suspicion as to their identity prompting reappraisal of the pertinent specimens, including types and protologues. The study revealed misidentifications.

---

[1]Corresponding author. E-mail: drmsivadasan@rediffmail.com
[2]Department of Post-graduate Studies and Research in Botany, Sir Syed College, Taliparamba, Kannur 670 142, Kerala, India
[3]Central National Herbarium, Botanical Survey of India, Botanic Garden P. O., Howrah 711 103, India

## Identity of *Theriophonum danielii* Rajakumar, Selvakumari., S. Murugesan & Chellaperumal (2010)

Rajakumar *et al.* (2010) described *Theriophonum danielii* based on specimens collected near Tisayanvillai, Tirunelveli district, Tamil Nadu, India. The publication of the new species was based on improper comparison of their specimens with *T. infaustum*. Critical examination of the morphological characters (Table 1 of the protologue reproduced here as Table 1, below) revealed that the tuber size, petiole length, lamina shape, spathe length and spadix length of the two species overlap, indicating clear range of variation in size and shape; and characters of the new species fell within the range of variation of *T. infaustum*. The size of neuter flowers of *T. danielii* as recorded by Rajakumar *et al.* (2010) represented that of a large specimen. The drawing of habit in Figure 1 of the protologue has an erroneous presentation of the leaves. Three petioles were clearly shown attached to the tuber, but the four leaf-laminae were shown above as if one petiole bifurcated producing an additional lamina. The shape of lamina was recorded as 'ovate' in contradiction to the shape of the majority of the leaves. The 'black dots' reported on staminate flowers and specified as a distinguishing character possibly might have been overlooked in *T. infaustum* by Brown (1880), being a trivial character and not clearly discernible in dried specimens.

**Table 1. Characters of *Theriophonum danielii* and *T. infaustum* (Reproduced from the protologue of *T. danielii*).**

| Characters | *T. infaustum* | *T. danielii* |
|---|---|---|
| Tuber ('Corm') size | 0.5-2.0 x 1-2 cm | 1.5-2.0 x 1.0 cm |
| Petiole length | 5.0-12.5 cm | 4-17 cm |
| Leaf shape | hastate-sagittate | Ovate |
| Spathe length | 2.0-5.5 cm | 3.5-4.5 cm |
| Spadix length | 4.0-4.5 cm | 3.0-3.5 cm |
| Neuter | 3.0-3.5 mm | 6 mm |
| Staminate flowers | black dots absent | black dotted |

A photograph (Cibachrome) of the type of *T. infaustum* (Fig. 1A) at Kew is with three specimens mounted on a single sheet clearly revealed variation in shape and size of leaves. Rheede (1692) provided illustrations of *Nir-tsjembu* and *Nelenschena minor* (Fig. 1B) which actually represented *T. infaustum* and the extreme variation in size of the two might have been the reason for describing them as distinct elements under separate names.

One of the authors (VAJ) visited MH and the herbarium of St. John's College [JCH, not in Index Herbariorum (http://sciweb.nybg.org/science2/IndexHerbariorum.asp), Palayamkottai, Tamil Nadu] in order to study the types of *T. danielii* which were reported to have been deposited in these institutions. But the types were not available in both the herbaria. Intensive search at JCH helped to locate a few un-mounted specimens bearing the same collection number (*1110*) as that of the holotype, but without designation as type. These specimens were studied in detail and identified as *T. infaustum*. Photograph of one of the specimens, presumably an isotype, is presented in Fig. 1C, and the similarity in nature and stature of the specimen tempt to assume it to be the specimen based on which the Fig. 1A of the protologue was prepared.

Based on all the above observations and study, *T. infaustum* and *T. danielii* are considered as conspecific. As per Articles 11.1 and 11.4 of ICN (McNeill *et al.*, 2012), the correct name for the taxon is *T. infaustum* and *T. danielii* is reduced to synonymy. Accordingly, we have:

***Theriophonum infaustum*** N. E. Br., J. Linn. Soc., Bot. 18: 260 (1880) ['1881', publ. 1880].

> *Type:* India, Kerala ('Malabar'), Paulghautcherry [Palghat?], *Wight 2775* (*Holotype*: K!).

*T. danielii* Rajakumar, Selvak., S. Murug. & Chellap., Indian J. Forest. 33(3): 447 (2010), pro syn.

> *Type:* 'India, Southern India, Tamil Nadu, Tirunelveli district, *Rajakumar, Selvakumari, Murugesan & Chellaperumal 1110* (*Holotype*: JCH, *Isotypes*: MH, JCH)' (extracted from protologue).

## Identity of *Theriophonum manickamii* Murugan & K. Natarajan (2008)

Murugan and Natarajan (2008) described *Theriophonum manickamii* based on specimens collected from Playamkottai taluk in Tirunelveli district, Tamil Nadu. While describing the species, they compared its characters with those of *T. sivaganganum*, a distant species. Sivadasan and Nicolson (1982) provided illustrations of spadices of *T. sivaganganum* and *T. fischeri* (Fig. 2A & 2B, respectively of their article), and illustration of spadix of *T. manickamii* is reminiscent of that of the latter species. A comparison of characters of spadices depicted by Sivadasan and Nicolson in Fig. 2B of their article with that provided by Murugan and Natarajan in Fig. 1B of the protologue revealed similarities between the two and brought out the erroneous conclusion on identity of the Tirunelveli specimens as belonging to a new species.

Sivadasan and Nicolson (1981) described *Theriophonum fischeri* as a new species solely based on herbarium specimens available at CAL, FRC and K. Owing to the non-availability of live specimens, details on variations and extent of variation in shape and size of juvenile and adult leaves were not recorded. Shape of juvenile leaves varied from linear-lanceolate to ovate-lanceolate, and that of mature leaves from narrowly hastate-sagittate to hastate-sagittate. The mature leaves of *T. manickamii* were described as narrowly-sagittate whereas in Tables 1 and 2 of the protologue, their shape was mentioned as 'narrowly hastate' which is same as that of *T. fischeri*. The characters of *T. fischeri* and *T. manickamii* given in Table 2 of the protologue of the latter are reproduced in Table 2 below, to show their general resemblance. Relatively bigger size of spathe and spadix of *T. fischeri* was due to the bigger size of the specimens studied, and in its protologue the range of size was not given; instead maximum sizes were given within which fall the sizes of spathe and spadix of *T. manickamii*. In both the species, the pistillate flowers were in 1-2 series. The number of ovules was almost the same. The shape and texture of stigma, and shape of filaments described in the protologue of *T. fischeri* were based on dried specimens and hence slight difference from that of the live specimens are possible. The relative positions of neuters and appendix were similar in both the species.

### *Discrepancies in Figure 1 of the protologue*

It is also to be pointed out that some of the illustrations in Fig. 1 of the protologue of *T. manickamii* were erroneous. The picture D of Fig. 1 representing longitudinal section of basal portion of spadix contained longitudinal sections of pistillate flowers on either side of spadix-axis, and ovules were shown as attached to roof of locule of ovary, thereby showing only apical placentation. The pictures F and G of Fig. 1 showed longitudinal section of pistillate flower and cross section of ovary, respectively. In Fig. 1F, three ovules were shown as pendent with apical placentation. In Fig. 1G, cross sections of three ovules were shown thereby representing the same pistillate flower with only three pendent ovules. One of the diagnostic characters of *Theriophonum* distinguishing from its closely related genus *Typhonium* Schott (1829) is having basal and apical placentation. But the drawings provided in the protologue by Murugan and Natarajan (2008) depicted only apical placentation.

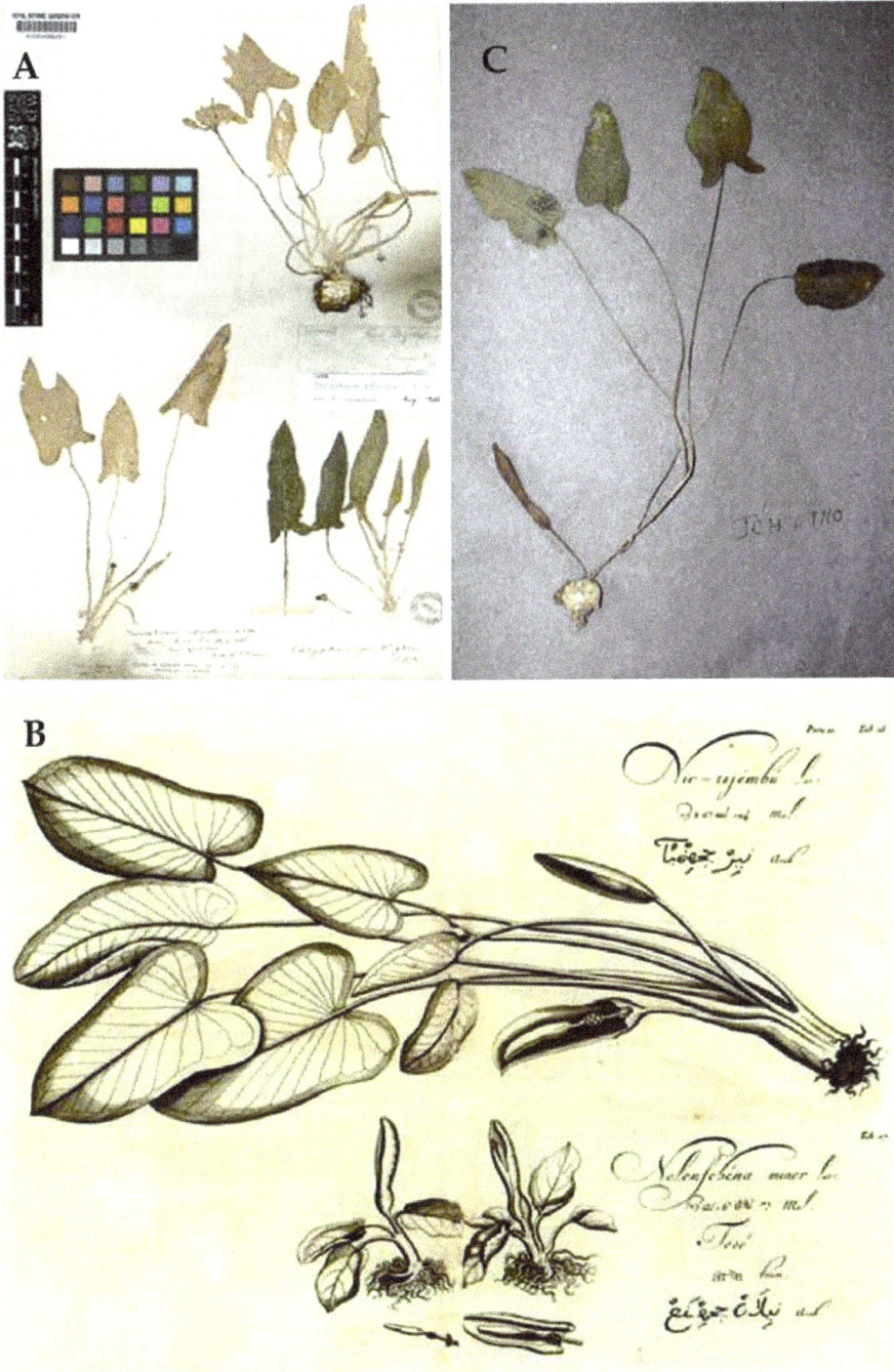

Fig. 1. A. Photograph (Cibachrome) of type of *Theriophonum infaustum* N. E. Br. at K. (© The Board of Trustees of the Royal Botanic Gardens, Kew; reproduced with consent); B. Photo of the double-page plate from Rheede's *Hortus Indicus Malabaricus* with illustrations of *Nir-tsjembu* and *Nelenschena minor* under Tab. 16 and Tab. 17 respectively; C. A specimen available at the herbarium ('JCH') of St. John's College, Palayamkottai, Tirunelveli with same number of holotype of '*Theriopnonnum danielii*'.

Fig. 2. *Theriophonum fischeri* Sivad. A. Photograph (Cibachrome) of holotype at K. (© The Board of Trustees of the Royal Botanic Gardens, Kew; reproduced with consent); B. Photograph (Cibachrome) of isotype (but labeled as holotype) at K. (© The Board of Trustees of the Royal Botanic Gardens, Kew; reproduced with consent); C. Plants under cultivation displaying variation in shape and size of leaves (Photo: C. N. Sunil).

One of the authors (VAJ) visited MH and XCH to study the types reported to have deposited there; but types were not available in either MH or in XCH and he was informed by the authorities that the types have not yet been deposited there, and assured to contact the authors in this regard. Photographs (Cibachrome) of holotype and isotype of *T. fischeri* (Holotype - *Fischer 2359*, Kew Negative No. 19850; Isotype - *Fischer 2359*, Kew Negative No. 19849 – but labeled as 'Holotype') obtained from K are presented below as Fig. 2A and Fig. 2B respectively to show variation in shape and size of leaves of the two specimens and to have an idea about the range of variation of mature leaves. Recently, Dr. C. N. Sunil, Department of Botany, S.N.M. College, Maliankara, Ernakulam, Kerala State collected specimens of *T. fischeri* from Ottappalam in Palakkad district, Kerala and is growing them in pots in the Botanic Garden of the College. A photograph of the plants (Fig. 2C) showed variation in shape and size of leaves.

**Table 2. Characters of *T. fischeri* and *T. manickamii* (Rreproduced from the protologue of *T. manickamii*).**

| Characters | *T. fischeri* | *T. manickamii* |
|---|---|---|
| Leaves | | |
| Juvenile | ovate-lanceolate | linear-lanceolate |
| Mature | hastate-sagittate | narrowly hastate-sagittate |
| Spathe | up to 12 cm long | 5-6 cm long |
| Spadix | c. 9.5 cm long | 4.5 cm long |
| Pistillate flowers | 1-2-seriate | 1-2-seriate |
| Ovules | 4-5 | 3-6 |
| Stigma | Discoid, smooth | Hemispherical, obscurely spinulose |
| Neuters | Adjacent to pistillate flowers and separated from staminate flowers | Adjacent to pistillate flowers and separated from staminate flowers |
| Filaments | Not beaked at apex | Obscurely beaked at apex |
| Appendix | Adjacent to staminate flowers | Adjacent to staminate flowers |

Based on all the above evidences, it is concluded that the recognition of *T. manickamii* as a new species by comparing characters of the specimens with that of *T. sivaganganum*, a very distant and dissimilar species rather than with *T. fischeri* has lead to the misidentification. *Theriophonum fischeri* has already been reported earlier from various localities in Tamil Nadu (Sivadasan and Nicolson, 1983; Daniel *et al.*, 1988; Kottaimuthu and Kumuthakavalli, 2011) including Tirunelveli district which is the type locality of *T. manickamii*. Therefore, *Theriophonum manickamii* is considered as conspecific to *T. fischeri*:

**Theriophonum fischeri** Sivad. in Sivadasan & Nicolson, Aroideana 4(2): 64 (1981).

*Type*: India, Kerala, Palghat district, Attappadi valley above Agali, 2000 ft.[600 m], *Fischer 2359* (*Holotype*: K!, *Isotypes*: CAL!, FRC!).

*T. manickamii* Murugan & K. Natarajan, J. Econ. Taxon. Bot. 32(3): 618 (2008), pro syn.

*Type:* 'India, Tamil Nadu, Tirunelveli district, Palayamkottai Taluk, on the way to Sasthakoil from Sivanthipatti village, 25.12. 2001. *Murugan 21277* (*Holotype*: MH, *Isotype*: XCH)' (extracted from protologue).

## Conclusion

Sivadasan and Nicolson (1982) recognized five species, *viz. Theriophonum dalzellii, T. fischeri, T. infaustum, T. minutum* and *T. sivaganganum* in their revision of the genus, and the number of species hold good even today since the two recently described species, viz. *T. danielii* and *T. manickamii* are unequivocally recognized as conspecific with *T. infaustum* and *T. fischeri*, respectively in the present taxonomic appraisal.

## Acknowledgements

The authors express their gratitude towards the directors of CAL, FRC, K, MH, XCH and Principals of St. John's College and St. Xavier's College, Palayamkottai, Tamil Nadu for providing permission and facilities to study the specimens at their herbaria. The Board of Trustees, Royal Botanic Gardens, Kew is thanked for granting permission for reproduction of images of type specimens available at K. The authors are grateful towards Dr. G. V. S. Murthy (MH), Dr. D. Narasimhan, Madras Christian College, Chennai, Messrs. Gopal Krishna (CAL), Jana Venkata Sudhakar (MH) and Gnanasekaran Gunadayalan (MH) for various help. The photograph provided by Dr. C. N. Sunil, S. N. M. College, Maliankara, Ernakulam is gratefully acknowledged. The first and third authors wish to thank the Deanship of Scientific Research, King Saud University for support through the research group project (No. RGP-VPP-135).

## References

Baillon, H.E. 1895. Monographie des Pandanacées, Cyclanthacées et Aracées. Histoire de Plantes, Vol. **13**. L. Hachette, Paris, pp. 1-523.

Blume, C.L. 1837. *Rumphia*, **1**. C. G. Sulpke, Leiden, Amsterdam, pp. 1-204.

Bogner, J. 1969. A new combination in *Theriophonum* Bl. (Araceae). Bull. Bot. Surv. India **10**: 244.

Brown, N.E. 1880. On some new Aroideae: with observations on other known forms. – Part I. J. Linn. Soc., Bot. **18**: 242-263.

Daniel, P., Rajendran, A. and Thiagaraj, J.G. 1988. On *Theriophonum fischeri* Sivadas. (Araceae) from the Tirunelveli plains, Tamil Nadu. Indian J. Forest. **11**: 163-165.

Engler, A. 1879. Araceae. *In*: Candolle, A. and Candolle C. de (Eds), Monographiae Phanerogamarum, Vol. **2**. G. Masson, Paris, pp. 1-681.

Engler, A. 1920. Araceae-Aroideae und Araceae-Pistioideae. *In*: Engler, A. (Ed.), Das Pflanzenreich, **IV-23F** (Heft 73). Wilhelm Engelmann, Berlin, pp.1-274.

Kottaimuthu, R. and Kumuthakavalli, R. 2011. Ethnobotany and taxonomy of *Theriophonum fischeri* Sivad. (Araceae). Life Sciences Leaflets **20**: 956-960.

McNeill, J., Barrie, F.R., Buck, W.R., Demoulin, V., Greuter, W., Hawksworth, D.L., Herendeen, P.S., Knapp, S., Marhold, K., Prado, J., Prud'homme van Reine, W.F., Smith, G.F., Wiersema, J.H. and Turland, N.J. 2012. International Code of Nomenclature for algae, fungi and plants (Melbourne Code) adopted by the Eighteenth International Botanical Congress Melbourne, Australia, July 2011. Regnum Vegetabile, **154**. Koeltz Scientific Books, Germany, pp. 1-240.

Murugan, C. and Natarajan, K. 2008. *Theriophonum manickamii* (Araceae) – A new plant species from the Tirunelveli district, Tamil Nadu, India. J. Econ. Taxon. Bot. **32**: 618-623.

Rajakumar, T.J.S., Selvakumari, R., Murugesan, S. and Chellaperumal, N. 2010. *Theriophonum danielii*, a new species of Araceae from Tirunelveli district, Tamil Nadu, India. Indian J. Forest. **33**: 447-448.

Ramamurthy, K. and Sebastine, K.M. 1967. A new genus of Araceae from Madras State, India. Bull. Bot. Surv. India **8**: 348-351.

Rheede tot Draakestein, H.A. van. 1692. Hortus Indicus Malabaricus, Vol. **11**. Johannis van Someren, *et* Joannis van Dyck, Amsterdam, pp. 1-134 + Tabs. 65.

Schott, H.W. 1829. *Typhonium*. Wiener Z. Kunst **1829**(3): 732.

Schott, H.W. 1855. Aroideae, Fasc. **3**. Caroli Gerald et filii, Vindobonae, pp. 15-20 + plates 21-30.

Schott, H.W. 1860. Prodromus Systematis Aroidearum. Mechitharists's Press, Vienna, pp. 1-602.

Sivadasan, M. and Nicolson, D.H. 1981. A new species of *Theriophonum* Bl. (Araceae) from India. Aroideana **4**: 64-67.

Sivadasan, M. and Nicolson, D.H. 1982. A revision of *Theriophonum* (Araceae). Kew Bull. **37**: 277-290.

Sivadasan, M. and Nicolson, D.H. 1983. Araceae. *In*: Matthew, K.M. (Ed.), The Flora of the Tamilnadu Carnatic, Vol. **3**. Rapinat Herbarium, Tiruchirapalli, India, pp. 1685-1704.

Suresh, C.R., Sivadasan, M. and Manilal, K.S. 1983. A commentary on Rheede's aroids. Taxon **32**: 126-132.

Willdenow, C.L. 1805. Caroli a Linné Species Plantarum, Vol. **4**(1). G. C. Nauk, Berlin, pp. 1-629.

# 6

# TAXONOMIC SIGNIFICANCE OF SPERMODERM PATTERN IN CUCURBITACEAE

M. Ajmal Ali[1], Fahad M.A. Al-Hemaid, Arun K. Pandey[2] and Joongku Lee[3]

*Department of Botany and Microbiology, College of Science, King Saud University, Riyadh-11451, Saudi Arabia.*

*Keywords:* Cucurbitaceae; SEM; Spermoderm; Testa.

## Abstract

Studies on spermoderm using scanning electron microscope (SEM) were undertaken in 12 taxa under 11 genera of the family Cucurbitaceae sampled from India, China and Korea. The spermoderm pattern in the studied taxa varies from rugulate, reticulate to colliculate type. The spermoderm shows rugulate type in *Benincasa hispida* and *Sicyos angulatus*; reticulate type in *Citrullus colocynthis, Cucumis melo* var. *agrestis, Diplocyclos palmatus, Hemsleya longivillosa, Luffa echinata, Momordica charantia, M. cymbalaria, Schizopepon bryoniifolius,* and *Trichosanthes cucumerina*; and colliculate type in *Gynostemma laxiflorum*. The present study clearly reveals that the testa features greatly varies across the genera which can be used as micromorphological markers for identification as well as character states for deducing relationship of the taxa within the family.

## Introduction

Spermoderm refers to the pattern present on the seed coat of mature seeds. Seed characteristic, particularly exomorphic features as revealed by scanning electron microscopy, have been used by many workers in resolving taxonomic problems (Koul *et al.*, 2000; Pandey and Ali, 2006) and evolutionary relationships (Kumar *et al.,* 1999; Segarra and Mateu, 2001).

Cucurbitaceae, with c. 800 species under 130 genera (Schaefer and Renner, 2011) are among the economically most important plant families (Kirtikar and Basu, 1975; Chakravarty, 1982; Ali and Pandey, 2006). Of the 130 genera, c. 50 contains single species, which illustrates the difficulties in deducing relationships within the family. Jeffrey (2005) divided the family Cucurbitaceae into 11 tribes under two subfamilies *viz.*, the Nhandiroboideae (Zanonioideae, with 60 species under 19 genera) and Cucurbitoideae (with c. 740 species under 111 genera). Nhandiroboideae are characterized by a gynoecium with three or rarely two, free styles, while Cucurbitoideae have the styles united into a single column. The most important diagnostic characters for the genera and tribes of Cucurbitaceae come from androecium and gynoecium morphology, type of tendril branching, pollen structure and seed coat (Jeffrey, 2005). Recently Schaefer and Renner (2011) have divided the family Cucurbitaceae into 95 genera in 15 tribes. The testa of Cucurbitaceae is formed by the outer integument and consists of a lignified epidermis, a hypodermis is of one or many layers of sclerotic cells, and an inner one-layered protective cover that in mature seeds is heavily lignified (Singh and Dathan, 2001).

---

[1]Corresponding author. Email: majmalali@rediffmail.com
[2]Department of Botany, University of Delhi, Delhi-110007, India.
[3]International Biological Material Research Center, Korea Research Institute of Bioscience and Biotechnology, Daejeon- 305806, South Korea.

Seed coat exhibits complex and highly diverse morphology and anatomy, providing valuable taxonomic characters. Despite seed coat morphology were studied in different groups of plants, no detailed work on spermoderm morphology in the systematics of Cucurbitaceae was conducted so far. The main objective of the present study is to evaluate taxonomic significance of spermoderm pattern in some members of the family Cucurbitaceae.

## Materials and Methods

The seeds for the present investigation were collected from nature during field trips or were procured from herbarium specimens (Table 1). Dry mature seeds were directly mounted on double-sided carbon tape which was affixed on aluminum stub. Seeds were then coated with very thin layer of gold in a sputter coater unit (Hitachi E-1010), and observed with a Hitachi S3400-N scanning electron microscope at 20 KV. Scanning Electron Microscopy was performed at Korea Research Institute of Bioscience and Biotechnology, Daejeon, South Korea. Renner and Pandey (2012), Lu *et al.* (2011) and Park (2007) were followed for taxon nomenclature. For terminology of spermoderm, Radford *et al.* (1974), Barthlot (1990) and Barthlot *et al.* (1998) were followed.

## Results and Discussion

In the present study, three different patterns of spermoderm (i.e. rugulate, reticulate and colliculate) were observed in the studied taxa (Table 1). The spermoderm pattern in *Benincasa hispida* was found rugulate, rugae were unevenly distributed and compactly arranged. Some rugae were raised at certain places and were larger or smaller in appearance (Fig. 1).

In *Citrullus colocynthis* the spermoderm pattern was reticulate with thin walled polygonal reticulae. The reticulae were compactly arranged and the testa cell surface showed transverse striation (Fig. 2). In *Cucumis melo* var. *agrestis* the spermoderm pattern was reticulate type. The testa cells were rectangular and compactly arranged. The flakes of waxy deposition were observed sporadically over the surface (Fig. 3).

*Diplocyclos palmatus* showed the reticulate pattern of spermoderm with distinct anticlinal and periclinal walls. The testa cells were polygonal with small protuberances covered with crystal like structures. These structures were unique to this species and were not found in any other cucurbits studied presently. Each testa cell presented small protuberances over which crystal like structures were formed (Fig. 4). In *Gynostemma laxiflorum* the spermoderm pattern was strongly colliculate type and the seed surface showed raised projections due to overgrowth of the testa cells and distributed throughout the seeds surface. The testa cells were smooth in appearance due to thin film of wax (Fig. 5). *Hemsleya longivillosa* exhibited reticulate type of spermoderm where the testa cells were thick walled, hexagonal, lack up of wax deposition, compactly arranged and the cell surface were granulated (Fig. 6). *Luffa echinata* presented reticulate type of spermoderm pattern. The reticulae were compactly arranged and at certain places they were superimposed in such a way that the spermoderm appeared to be rugulate (Fig. 7). In *Momordica charantia* the spermoderm was reticulate with thick walled testa cells. Testa cell wall was more or less filling the cells which gave a reticulate-punctate appearance (Fig. 8). *M. cymbalaria* shows reticulate type of spermoderm pattern. The testa cells were hexagonal, compactly arranged and were covered with thin layer of wax or sometimes globular granules spread over the testa cell surface or on the testa cell wall which masking the nature of spermoderm (Fig. 9). In *Schizopepon bryoniifolius* the spermoderm pattern was of reticulate type and the reticulae were thin walled. The testa cells were elongated and polygonal. Some of the testa cells were transversely septate (Fig. 10). *Sicyos angulatus* showed the rugulate type of spermoderm with prominent rugae which anastomose each other giving an interwoven appearance. The rugae were separated from one another by deep grooves. The spermoderm shows a few sporadically distributed waxy flakes (Fig. 11). In *Trichosanthes cucumerina* the spermoderm was reticulate and the reticulae were large, polygonal and prominent with thick layer of wax giving smooth appearance to the surface (Fig. 12).

**Table 1. Origin of taxa included in the present study and spermoderm pattern.**

| No. | Species | Locality | Voucher specimen | Spermoderm |
|---|---|---|---|---|
| 1 | *Benincasa hispida* (Thunb.) Cong | Bhagalpur, Bihar, India | Ali and Pandey 1001 (BHAG) | Rugulate |
| 2 | *Citrullus colocynthis* (L.) Schard. | Kishanganj, Bihar, India | Ali and Pandey 1050 (BHAG) | Reticulate |
| 3 | *Cucumis melo* var. *agrestis* Naud. | Purnia, Bihar, India | Ali and Pandey 1009 (BHAG) | Reticulate |
| 4 | *Diplocyclos palmatus* (L.) Jeffrey | Bhagalpur, Bihar, India | Ali and Pandey 1083 (BHAG) | Reticulate |
| 5 | *Gynostemma laxiflorum* C.Y. Wu & S.K. Chen | Anhui, China | X.F. Gao 390 (KUN) | Colliculate |
| 6 | *Hemsleya longivillosa* C.Y. Wu & C.L. Chen | Yunnan, China | *s.n.*, Acc. No. 0362096 (KUN) | Reticulate |
| 7 | *Luffa echinata* Roxb. | Katihar, Bihar, India | Ali and Pandey 1093 (BHAG) | Reticulate |
| 8 | *Momordica charantia* L. | Bhagalpur, Bihar, India | Ali and Pandey 1111 (BHAG) | Reticulate |
| 9 | *M. cymbalaria* Fenzl *ex* Naudin | Andhra Pradesh, India | S. Karuppusamy 28631 (SKU) | Reticulate |
| 10 | *Schizopepon bryoniifolius* Maxim. | Gangwon-do, Korea | Hyeong-Kyu Lee 00859 (KRIB) | Reticulate |
| 11 | *Sicyos angulatus* L. | Gyeonbsanguk-do, Korea | G.Y. Chung *s.n.* (KRIB) | Rugulate |
| 12 | *Trichosanthes cucumerina* L. | Bhagalpur, Bihar, India | Ali and Pandey 1113 (BHAG) | Reticulate |

Herbaria: BHAG (Department of Botany, Tilka Manjhi Bhagalpur University, Bhagalpur, Bihar, India); KRIB (Korea Research Institute of Bioscience and Biotechnology, Daejeon, South Korea); KUN (Kunming Institute of Botany, Chinese Academy of Sciences, China); SKU (Sri Krishnadevaraya University, Anantapur, Andhra Pradesh, India).

The systematic application of seed surface features, as observed under scanning electron microscope is tremendous considering that seed characters are only slightly influenced by environmental conditions. High structural diversity of seed provides most valuable criteria for classification at species and family level (Barthlott, 1984); therefore, spermoderm surface patterns have been extensively utilized as a secondary taxonomic characters (see Pandey and Ali, 2006). In the present study, the spermoderm pattern in the studied taxa varies from rugulate, reticulate to colliculate. The spermoderm shows rugulate pattern in *Benincasa hispida*, *Sicyos angulatus*; reticulate in *Citrullus colocynthis*, *Cucumis melo* var. *agrestis*, *Diplocyclos palmatus*, *Hemsleya longivillosa*, *Luffa echinata*, *Momordica charantia*, *M. cymbalaria*, *Schizopepon bryoniifolius*, *Trichosanthes cucumerina* to strongly colliculate in *Gynostemma laxiflorum*. The results of the present study reveals that even within the similar pattern of spermoderm, the testa features greatly varies from genera to genera which can be used as micromorphological markers for identification as well as character states for deducing generic and specific relationship within the family.

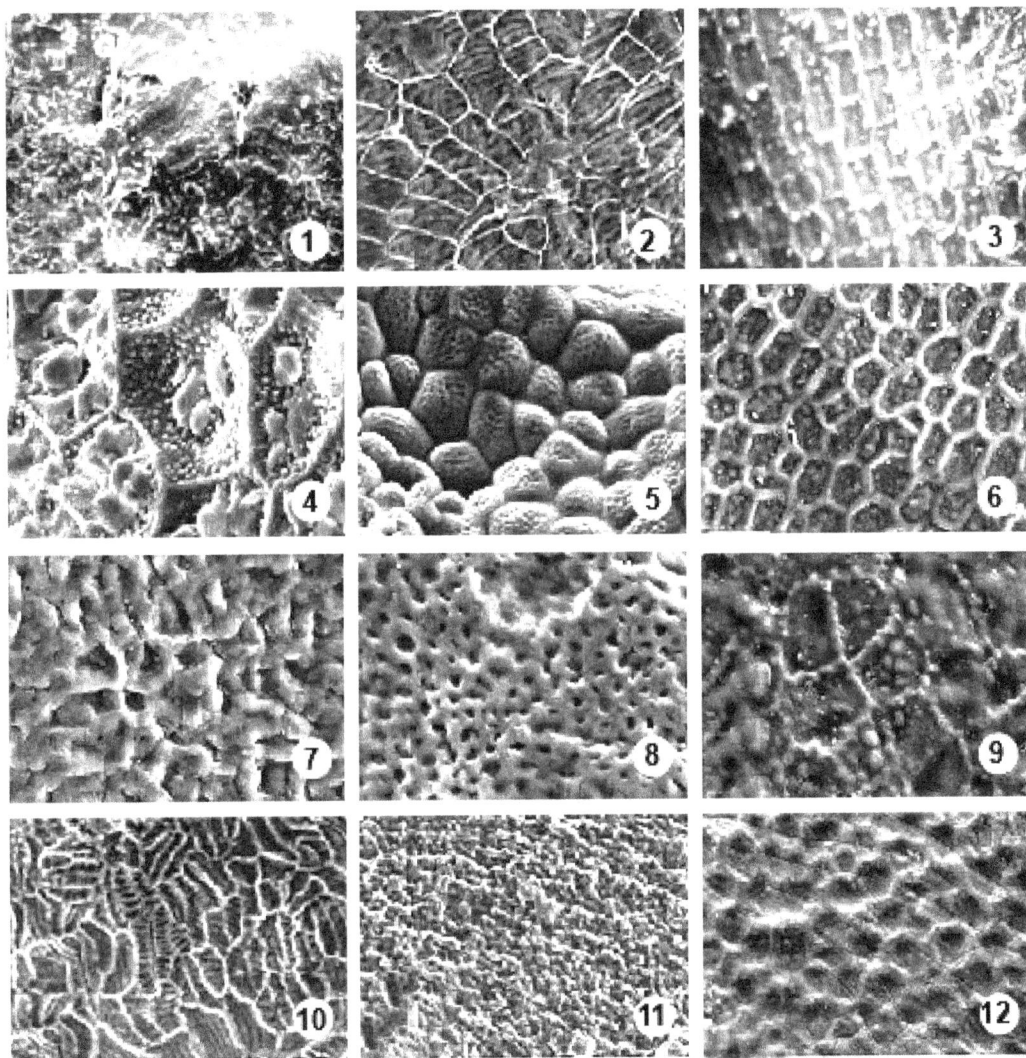

Figs 1-12. Scanning electron micrograph of the seed surface in Cucurbitaceae: 1. *Benincasa hispida* ×400 (rugulate); 2. *Citrullus colocynthis* ×400 (reticulate); 3. *Cucumis melo* var. *agrestis* ×400 (reticulate); 4. *Diplocyclos palmatus* ×1000 (reticulate); 5. *Gynostemma laxiflorum* ×600 (colliculate); 6. *Hemsleya longivillosa* ×400 (reticulate); 7. *Luffa echinata* ×1000 (reticulate); 8. *Momordica charantia* ×700 (reticulate); 9. *Momordica cymbalaria* ×1000 (reticulate); 10. *Schizopepon bryoniifolius* ×400 (reticulate); 11. *Sicyos angulatus* ×300 (rugulate); 12. *Trichosanthes cucumerina* ×320 (reticulate).

## Acknowledgements

The authors would like to extend their sincere appreciation to the Deanship of Scientific Research at King Saud University for its funding of this research through the Research Group Project No. RGP-VPP-195.

# References

Ali, M.A. and Pandey, A.K. 2006. Cucurbitaceae of Bihar: diversity and conservation. *In:* Trivedi, P.C. (Ed.), Global biodiversity status and conservation. Pointer Publisher Jaipur, India, pp. 250-260.

Barthlott, W. 1984. Microstructural features of seed surface. *In:* Heywood, V.H. and Moore, D.M. (Eds), Current Concept in Plant Taxonomy. Academic Press, London, pp. 95-105.

Barthlott, W. 1990. Scanning electron microscopy of the epidermal surface in plants. *In:* Claugher, D. (Ed.), Scanning Electron Microscopy in Taxonomy and Functional Morphology. Clarendon Press, Oxford, pp. 69-94.

Barthlott, W., Neinhuis C., Cutler, D., Ditsch, F., Meussel, I., Theisen, I., and Wilhelm, H. 1998. Classification and terminology of plant epicuticular waxes. Bot. J. Lin. Soc. **126**: 237-260.

Chakravarty, H.L. 1982. Cucurbitaceae *In:* Jain, S.K. (Ed.) Fascicles of Flora of India, No. **11**, Botanical Survey of India, Calcutta.

Jeffrey, C. 2005. A new system of Cucurbitaceae. Bot. Zhurn. **90**: 332-335.

Kirtikar, K. and Basu, B.D. 1975. Indian Medicinal Plants (Reprint Edition), Vol. **II**. Bishen Singh Mahendra Pal Singh, Dehra Dun, India, pp. 1106-1115.

Koul, K.K., Ranjan, N. and Raina, S.N. 2000. Seed coat microsculpturing in *Brassica* and allied genera (subtribe Brassicinae, Raphaninae, Moricandiinae). Ann. Bot. **86**: 385-395.

Kumar, P.P., Rao C.D., Rajasegar, G. and Rao, A.N. 1999. Seed surface architecture and random amplified polymorphic DNA profiles of *Paulownia fortunei, P. tomentosa* and their hybrid. Ann. Bot. **83**: 103-107.

Lu, A., Huang, L., Chen, S.K. and Jeffrey, C. 2011. Cucurbitaceae. *In*: Wu, Z.Y., Raven, P.H., Hong, D.Y. (Eds), Flora of China, Vol. **19**. Missouri Botanical Garden Press, St. Louis.

Pandey, A.K. and Ali, M.A. 2006. Testa topography in Papillionoidae and its taxonomic significance. *In:* Pandey, A.K., Wen, J., Dogra, J.V.V. (Eds), Plant Taxonomy: Advances and Relevance. CBS Publisher and Distributor, New Delhi, India, pp. 529-541.

Park, C. 2007. The Genera of Vascular Plants of Korea. Academy Publishing Co., Seoul, South Korea.

Radford, A.E., Dickison, W.C., Massey, J.R. and Bell, C.R. 1974. Vascular Plant Systematics. Harper and Row publishers, New York.

Renner, S.S. and Pandey, A.K. 2012. The Cucurbitaceae of India: Accepted names, synonyms, geographic distribution, and information on images and DNA sequences. PhytoKeys **20**: 53-118.

Schaefer, H. and Renner, S.S. 2011. Phylogenetic relationships in the order Cucurbitales and a new classification of the gourd family (Cucurbitaceae). Taxon **60**(1): 122-138.

Segarra, J.G. and Mateu, I. 2001. Seed morphology of *Linaria* species from eastern Spain: identification of species and taxonomic implications. Bot. J. Linn. Soc. **135**: 375-389.

Singh, D. and Dathan, A.S.R. 2001. Development and structure of seed coat in the Cucurbitaceae and its implications in systematics. *In:* Chauhan, S.V.S. and Chaturvedi, S.N. (Eds), Botanical Essays: Tribute to Professor Bahadur Singh. Printwell Publishers Distributors, Jaipur, India, pp. 87-114.

# 7

# A REVISED INFRAGENERIC CLASSIFICATION OF *DIMERIA* R. BR. (POACEAE: ANDROPOGONEAE)

M.S. Kiran Raj[1], M. Sivadasan[2,5], J.F. Veldkamp[3], A.H. Alfarhan[2]
and A.S.M. Amal Tamimi[4]

*Department of Botany, Sree Narayana College, Cherthala 688 582, Alappuzha, Kerala, India*

*Keywords*: Dimeriinae; New subspecies; New synonyms; Panicoideae; Sectional classification; Typification.

## Abstract

The four sections of the little known genus *Dimeria* R. Br. of the rather anomalous paleotropical subtribe Dimeriinae Hack. (Poaceae–Panicoideae–Andropogoneae) are revised. A key is provided. Three Peninsular Indian species, *viz. Dimeria sivarajanii, D. idukkiensis* and *D. borii* are treated here as subspecies of *D. bialata, D. kurumthotticalana* and *D. mooneyi* respectively; and five, *viz. D. chelariensis, D. copei, D. eradii, D. jayachandranii* and *D. kollimalayana* are reduced to synonymy.

## Introduction

*Dimeria* R. Br. (Poaceae) is a little known genus with about 65 species (Teerawatananon *et al.*, 2014). They are adapted to arid habitats and range from India to China, Korea, Indonesia, Micronesia, and northern Australia and to Sri Lanka and Madagascar (Bor, 1953; Clayton *et al.*, 2006; Kiran Raj and Sivadasan, 2008; Kiran Raj *et al.*, 2013a, b; Kiran Raj *et al.*, 2015) (Fig. 1).

It used to be the only genus of the rather anomalous paleotropical subtribe Dimeriinae Hackel (1889) until *Nanooravia* Kiran Raj & Sivad. (Kiran Raj *et al.*, 2013a, b) was described from India. The subtribe is distinguished by espatheate inflorescences consisting of 1–11 digitate racemes with tough raches and strongly laterally flattened solitary spikelets by which it differs from all other Andropogoneae Dumort. (Clayton, 1972). Brown (1810) placed it between *Imperata* Cirillo and *Ischaemum* L. Endlicher (1836) included it in the Andropogoneae between *Zoysia* Willd. and *Pleuroplitis* Trin. (= *Arthraxon* P. Beauv.). Steudel (1854) treated six species in Andropogoneae between *Euklaston* Steud. (= *Andropogon* L.) and *Pterygostachyum* Nees *ex* Steud., *Psilostachys* Steud. (synonyms of *Dimeria*) and *Amblyachyrum* Hochst. *ex* Steud. (= *Apocopis* Nees). Bentham (1881) included it in the subtribe Arthraxonae J. Presl together with *Apocopis* Nees and *Arthraxon* P. Beauv., but in 1883 in the Andropogoneae *s.l.* (Bentham, 1883).

Hackel (1889) had 12 species, 2 subspecies, and 10 varieties and regarded it as close to the Sacchareae Dumort., and very much to the Tristegineae Nees. For a discussion on the latter tribe see Veldkamp (2015).

[1]Formerly at: Department of Botany, University of Calicut, P.O. Box 673 635, Kerala, India
[2]Department of Botany & Microbiology, College of Science, King Saud University, P.O. Box 2455, Riyadh 11451, Kingdom of Saudi Arabia
[3]Naturalis Biodiversity Center, P.O. Box 9517, 2300 RA Leiden, The Netherlands
[4]Department of Biology, College of Science, Princess Nora bint Abdulrahman University, P.O. Box 87991, Riyadh 11652, Kingdom of Saudi Arabia
[5]Corresponding author. Email: drmsivadasan@gmail.com

Roberty (1960) found the genus *Dimeria* so homogeneous, that in his "cohors" *Dimeriastreae* he accepted only a single species, *D. avenacea* (Retz.) C.E.C. Fisch., with not less than 22 subvarieties all with invalid names, because he did not follow the Linnean infraspecific classification.

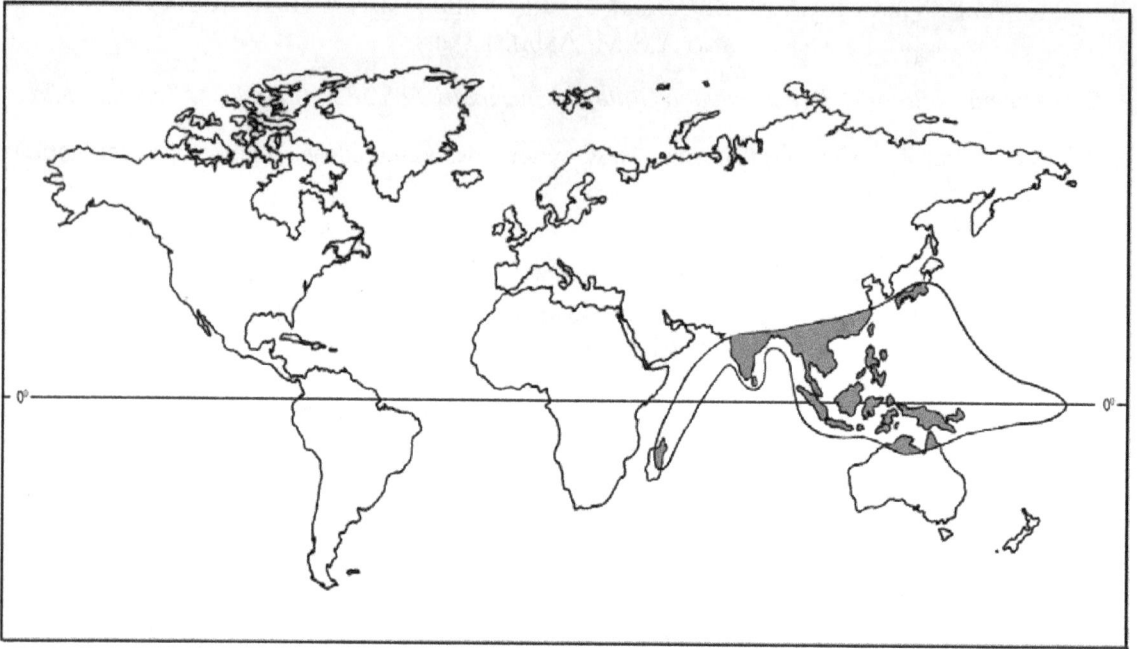

Fig. 1. Distribution of *Dimeria* R. Br.

Clayton and Renvoize (1986) regarded it as derived from the Ischaeminae J. Presl by suppression of the sessile spikelet. This might as well be loss of the pedicelled ones, as was observed by Miquel (1851: "rudimentary pedicels"), but by no one else. Kellogg and Watson (1993) in their phylogenetic analysis based on morphological data treated *Dimeria* as a sister group of *Cleistachne* Benth. of the subtribe Sorghinae Bluff *et al.* with both genera nesting in a clade. Estep *et al.* (2014) in a nuclear molecular study found *Dimeria* nested in a clade within *Ischaemum*, but with little basal support, so a reduction of *Dimeria* to *Ischaemum* seems premature.

The majority of the species (34 out of 65: Table 1) is confined to Peninsular India (Hackel, 1889; Hooker, 1896; Bor, 1953; Kiran Raj, 2008; Kiran Raj *et al.*, 2013a, b) indicating it to be at least a centre of speciation of the subtribe. In Southeast Asia, approximately 14 species have been reported for Indo-China, Malesia and China (Camus and Camus, 1922; Ridley, 1925; Jansen, 1953; Schmid, 1958; Henty, 1969; Gilliland, 1971; Lazarides, 1980; Chen and Phillips, 2006; Teerawatananon *et al.*, 2014).

### Sectional classification of Dimeria by Bor (1953)

The first infrageneric classification of the genus was by Bor (1953), who treated the species for India, Sri Lanka (Ceylon), and Myanmar (Burma) as belonging to three sections *viz. Dimeria* sect. *Annulares* Bor, sect. *Capillares* Bor, and sect. *Loriformes* Bor and the sections were recognized based on rachis and pedicel characters. As the type species, *D. acinaciformis* R. Br., is from Australia, he did not mention a section *Dimeria* in his treatise.

**Table 1. The sections of *Dimeria* R. Br. and their Peninsular Indian taxa.**

| I. *Dimeria* sect. *Dimeria* | *Dimeria acutipes* Bor |
|---|---|
| | *D. agasthyamalayana* Kiran Raj & Ravi |
| | *D. aristata* (Hack.) Senaratna |
| | *D. avenacea* (Retz.) C.E.C.Fisch. |
| | *D. connivens* Hack. |
| | *D. copeana* Sreek.,V.J.Nair & N.C. Nair. |
| | (= *D. chelariensis* Ravi, **syn. nov.**) |
| | *D. fuscescens* Trin. |
| | *D. kanjirapallilana* Jacob |
| | *D. lehmannii* Hack. (= *D. alata* Hook. f.) |
| | *D. orissae* Bor |
| | *D. ornithopoda* Trin. |
| | *D. trimenii* Hook. f. |
| II. *Dimeria* sect. *Annulares* | *D. veldkampii* Kiran Raj & Sivad. |
| | *D. woodrowii* Stapf |
| III. *Dimeria* sect. *Capillares* | *D. gracilis* Nees *ex* Steud. (= *D. laxiuscula* Thw. & Trimen) |
| | *D. hohenackeri* Hochst. *ex* Miq. |
| | *D. stapfiana* C.E. Hubb. *ex* Pilg. |
| | *D. stapfiana* var. *blatteri* (Bor) M.R. Almeida |
| IV. *Dimeria* sect. *Loriformes* | *D. balakrishnaniana* K. Ravik., Sreek. & V. Lakshm. |
| | *D. bialata* C.E.C. Fisch. |
| | *D. bialata* subsp. *sivarajanii* (N. Mohanan & Ravi) Kiran Raj & Sivad., **comb. & stat. nov.** (=*Dimeria sivarajanii* N. Mohanan & Ravi, Rheedea 6(2): 47.1996) |
| | *D. kalavoorensis* Ravi (=*D. copei* Ravi, **syn. nov.**) |
| | *D. deccanensis* Bor (=*D. kollimalayana* M. Mohanan & A.V.N. Rao, **syn. nov.**; =*D. jayachandranii* Arisdason & P. Daniel, **syn. nov.**) |
| | *D. fischeri* Bor |
| | *D. jainii* Sreek., V.J. Nair & N.C. Nair |
| | *D. josephii* Ravi & N. Mohanan |
| | *D. kurumthotticalana* Jacob (=*D. ceylanica* Bor; =*D. sreenarayanae* Ravi & Anil Kumar) |
| | *D. kurumthotticalana* subsp. *idukkiensis* (Ravi & Anil Kumar) Kiran Raj & Sivad., **comb. & stat. nov.** (*Dimeria idukkiensis* Ravi & Anil Kumar, Rheedea 2(2): 104. 1992) |
| | *D. kurzii* Hook. f. |
| | *D. lawsonii* (Hook. f.) C.E.C. Fisch. |
| | *D. mahendragiriensis* Ravi, H.O. Saxena & Brahmam |
| | *D. mooneyi* Raizada |
| | *D. mooneyi* subsp. *borii* (Sreek. et al.) Kiran Raj & Sivad., **comb. & stat. nov.** (*Dimeria borii* Sreek., V.J. Nair & N.C. Nair, J. Econ. Taxon. Bot. 3(2): 657.1982) |
| | *D. namboodiriana* Ravi & N. Mohanan |
| | *D. pubescens* Hack. |
| | *D. raizadae* V.J. Nair, Sreek. & N.C. Nair (=*Dimeria eradii* Ravi, **syn. nov.**) |
| | *D. raviana* Kiran Raj & Sivad. |
| | *D. thwaitesii* Hack. |

Bor (1953) pointed out the necessity of a further detailed study of more specimens of all species for a better understanding of diversity and extent of intraspecific variation. After 1953, a fairly large number of new species have been described, and it was found after field work and morphological examinations in the herbarium that some could not be properly assigned to a section. Also, the *Capillares* and *Loriformes* contained species with strictly triquetrous raceme-rachises, overlapping spikelets, and pedicels closely appressed to the rachis. *Dimeria acinaciformis* is characterized by the presence of triquetrous raceme-rachises and compressed pedicels.

Considering all the above aspects, a revised infrageneric classification of *Dimeria* is proposed here.

The raceme structure of the representative taxa of the sections is illustrated in Fig. 2 as an aid for easy understanding of the diagnostic characters.

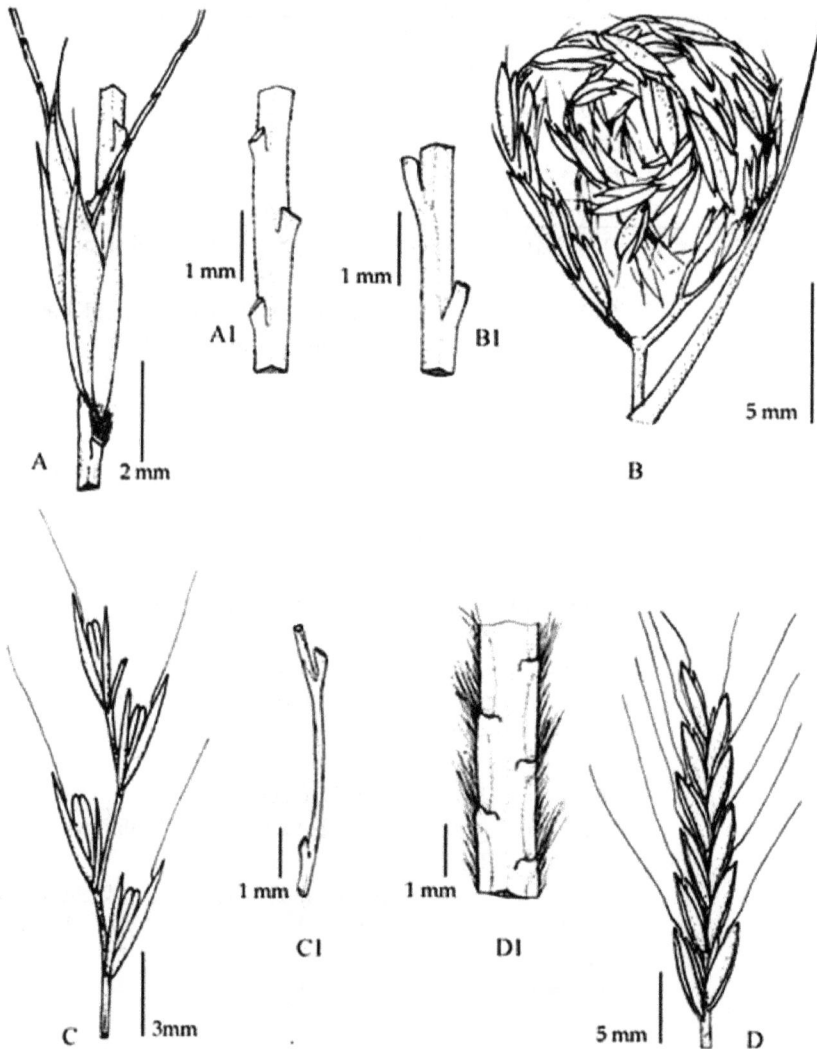

Fig. 2. Portions of racemes and rachises of representative taxa of the sections of *Dimeria*. A & A1. *Dimeria avenacea* (sect. *Dimeria*); B & B1. *D. woodrowii* (sect. *Annulares*); C & C1. *D. hohenackeri* (sect. *Capillares*); D & D1. *D. balakrishnaniana* (sect. *Loriformes*). (Drawings by M.S. Kiran Raj)

**I.   Dimeria** R. Br. sect. **Dimeria**

*Type*: *Dimeria acinaciformis* R. Br.

Annuals or perennials. Racemes 2 or 3, rarely 1; rachis of raceme triquetrous, trigonous in cross section, occasionally zigzag, usually wingless, if winged, only at the internodes; spikelets closely packed on the rachis, usually overlapping; raceme internodes c. 0.5 mm long; glumes slightly diverging at anthesis; pedicels 0.3–0.5 mm long, trigonous to flat, closely appressed to the rachis.

*Distribution*: Widely distributed in Tropical Asia to North Australia.

*Notes*: There are 12 species in Peninsular India of which seven, *viz. D. acutipes* Bor, *D. avenacea* (Retz.) C.E.C. Fisch., *D. connivens* Hack., *D. lehmannii* (Nees & Steud.) Hack., *D. ornithopoda* Trin., *D. orissae* Bor and *D. trimenii* Hook. f. were included by Bor (1953) in sect. *Loriformes. Dimeria chelariensis* Ravi (1995) is a synonym of *D. copeana* Sreek. *et al.* (Table 1).

According to Art. 22.1. of the ICN (McNeill *et al.*, 2012), an autonym is required and the correct name is *Dimeria* sect. *Dimeria*.

**II.   Dimeria** R. Br. sect. **Annulares** Bor

*Type*: *Dimeria woodrowii* Stapf

Annuals. Racemes 2 or 3, peduncle bent downwards or erect at maturity; rachis of raceme compressed, trigonous on one side and convex on the other, straight when young and curved at maturity to form a 'globule', or a single or double 'ringlet' carrying the spikelets along the inner side; raceme internodes up to 1 mm long; spikelets distantly arranged along the rachis; upper glume distinctly winged, or minutely winged, or wingless; pedicels terete, not compressed.

*Distribution*: Two species in Peninsular India (Table 1), and so far known only from lateritic plains of the Northern Western Ghats.

**III. Dimeria** R. Br. sect. **Capillares** Bor

*Lectotype*: *Dimeria hohenackeri* Hochst. *ex* Miq. (here designated)

Annuals or perennials. Racemes 3 to 5, rarely up to 11; rachis of raceme capillary and very thin, nearly triangular or circular in cross section, not winged; spikelets very distantly arranged along the rachis, late disarticulation from the pedicels; raceme internodes 2.5–3.5 mm long; glumes widely diverging at anthesis; pedicels 0.5–1.5 mm long, terete, not compressed.

*Distribution*: Restricted to Indian subcontinent (Western Ghats region of Peninsular India, Sri Lanka and Myanmar); three species in Peninsular India (Table 1).

*Notes*: Bor (1953) included seven species in *Dimeria.* sect. *Capillares,* and the type was not designated and hence the present lectotypification. In the present classification, three species are transferred to sect. *Dimeria* (Table 1).

**IV. Dimeria** R. Br. sect. **Loriformes** Bor

*Lectotype*: *Dimeria pubescens* Hack. (here designated)

Mostly annuals. Racemes 1 or 2, rarely 3; rachis of raceme compressed; dorsally flattened in cross section, winged; spikelets compactly arranged along the rachis, early disarticulation from the pedicels; raceme internodes 0.5–1.0 mm long; pedicels 0.3–0.5 mm long, distinctly compressed, flat, appressed to the wing and axis of rachis.

*Distribution*: Peninsular India, Myanmar and Sri Lanka; mostly occurring in Peninsular India with 17 species.

*Notes*: Bor (1953) did not designate a type for the section and hence it is lectotypified here. Seven species are excluded from Bor's *Loriformes*, and here placed in *Dimeria* sect. *Dimeria*.

Three Peninsular Indian species, *viz. Dimeria sivarajanii* N. Mohanan and Ravi (1996), *D. idukkiensis* Ravi and Anil Kumar (1992) and *D. borii* Sreek. *et al.* (1982) are reduced to subspecies of *D. bialata* C.E.C. Fisch. (1933), *D. kurumthotticalana* Jacob (1947) and *D. mooneyi* Raizada (1950), respectively (Table 1); two species, *viz. D. copei* Ravi (1996) and *D. eradii* Ravi (1995) are reduced to *D. kalavoorensis* Ravi (1996) and *D. raizadae* V.J. Nair *et al.* (1983), respectively; two species, *viz. D. kollimalayana* M. Mohanan and A.V.N. Rao (1984) and *D. jayachandranii* Arisdason and P. Daniel (2009), are regarded as conspecific with *D. deccanensis* Bor (1953), and they are treated as new synonyms (Table 1).

## Key to the Sections of Dimeria in Peninsular India

1. Racemes divergent; rachis of raceme always straight, never coiled; spikelets arranged along the outside and exposed.                                                 **2**

- Racemes non-divergent; rachis of raceme coiled to form a 'globule' or 'ringlet'; spikelets arranged along the inner side of rachis.                                                 **Dimeria** sect. **Annulares**

2. Rachis of raceme capillary and filiform, thin, wingless, angled to terete in cross section; spikelets distantly arranged on rachis, not readily disarticulating with pedicels; pedicles 1.0–1.5 mm long, terete, glabrous.                                                 **Dimeria** sect. **Capillares**

- Rachis of raceme not capillary, stout, winged or not, trigonous or compressed in cross section; spikelets compactly arranged on rachis, easily disarticulating with pedicels; pedicels 0.5–1.0 mm long, flat, often ciliate at the outer margin.                                                 **3**

3. Spikelet usually overlapping; rachis of raceme triquetrous, 0.5–0.7 mm wide, often minutely winged at the internode, scaberulous to sparsely ciliate along margin; pedicels compressed but not flat, completely appressed to the raceme-rachis.                                                 **Dimeria** sect. **Dimeria**

- Spikelets never overlapping; rachis of raceme abaxially flat, 0.8–1.5 mm wide, distinctly winged, glabrous to ciliate along margin; pedicels flat, basal half appressed to raceme-rachis and upper half attached to wing of rachis.                                                 **Dimeria** sect. **Loriformes**

## Acknowledgements

The first author is indebted to the Council of Scientific and Industrial Research (CSIR), New Delhi for the award of Senior Research Fellowship in 2001, the International Association for Plant Taxonomy (IAPT), Vienna for the Plant Systematics Research Grant Award in 2007 while working earlier at the University of Calicut, Kerala, India, and the University Grants Commission (UGC), New Delhi for granting a Minor Research Project in 2013. Sincere gratitude is expressed towards Prof. N. Ravi, an eminent agrostologist of Kerala and former Head of the Department of Botany, Sree Narayana College, Kollam, India for constant encouragements and comment on the manuscript. The authors are grateful towards the curators of AHMA, BM, BLAT, BSI, CAL, MH and K for permitting to study the specimens available at their respective herbaria, and providing necessary literature and cibachrome photographs of specimens. The second and fourth authors thankfully extend their appreciation to the Deanship of Scientific Research at the King Saud

University for encouragements and support extended through the research group project No. RGP-VPP-135.

## References

Arisdason, W. and Daniel, P. 2009. *Dimeria jayachandranii* (Poaceae), a new species from the Western Ghats, India. Kew Bull. **64**: 345–347.

Bentham, G. 1881. Notes on Gramineae. J. Linn. Soc., Bot. **19**: 67.

Bentham, G. 1883. Gramineae. *In*: Bentham, G. and Hooker, J.D., Genera Plantarum **3**: 1128. Reeve & Co., Williams & Norgate, London.

Bor, N.L. 1953. Notes on Asiatic grasses XI. The genus *Dimeria* R. Br. in India and Burma. Kew. Bull. **1952**(7): 553–592.

Brown, R. 1810. Prodromus florae Novae Hollandiae et insulae Van Diemen, **1**. J. Johnson, London, 204 pp.

Camus, E.G.and Camus, A. 1922.Graminées. *In*: Lecomte, H. and Humbert, H. (Eds), Flore Générale de L`Indo-Chine, **7**. Masson, Paris, pp. 202–650.

Chen, S.L. and Phillips, S.M. 2006. *Dimeria*. *In*: Zhengyi, W., Raven, P.H. and Hong, D.Y. (Eds), Flora of China, **22**. Science Press, Beijing and Peoples Republic of China and Missouri Botanical Garden Press, St. Louis, Missouri, pp. 614–616.

Clayton, W.D. 1972. Gramineae. *In*: Hutchinson, J. and Dalziel, J.M. (Eds), Flora of West Tropical Africa, **3**, 2nd edition, Crown Agents, London, pp. 413–414.

Clayton, W.D. and Renvoize, S.A. 1986. Genera Graminum. Grasses of the World. Kew Bull. Add. Ser. **XIII**. pp. 1–389.

Clayton, W.D., Vorontsova, M.S., Harman, K.T. and Williamson, H. 2006 (onwards). GrassBase - The Online World Grass Flora. <http://www.kew.org/data/grasses-db.html>. Retrieved on 2 March 2015.

Endlicher, S.L. 1836. Andropogoneae. *In*: Genera plantarum secundum ordines naturales disposita. F. Beck, Wien, pp. 106.

Estep, M.C., McKain, M.R., Diaz, D.V., Zhong, J., Hodge, J.G., Hodkinson, T.R., Layton, D.J., Malcomber, S.T., Pasqueth, R. and Kellogg, E.A. 2014. Allopolyploidy, diversification, and the Miocene grassland expansion. Proc. Natl. Acad. Sci. USA **111**(42): 15149–15154.

Fischer, C.E.C. 1933. *Dimeria bialata*. XLI - New or little-known plants from South India. Bull. Misc. Inform. Kew **1933**(7): 339–357.

Gilliland, H.B. 1971. A revised flora of Malaya. An illustrated systematic account of the Malayan flora, including commonly cultivated plants **3**. Government Printing Office, Singapore, pp. 1–319.

Hackel, E. 1889. Dimerieae. *In*: A. and C. de Candolle, Monographiae phanerogamarum **6**. G. Masson, Paris, pp. 76–90.

Henty, E.E. 1969. A manual of the grasses of New Guinea. Bot. Bull. **1**. Lae, New Guinea, pp. 1–215.

Hooker, J.D. 1896. *Dimeria*. *In*: Hooker, J.D. (Ed.), Flora of British India, **7**. Reeve & Co. Ltd., London, pp. 103–106.

Jacob, K.C. 1947. Some new species of South Indian Plants. J. Bombay Nat. Hist. Soc. **47**(1): 47–51.

Jansen, P. 1953. Notes on Malaysia grasses – I. Reinwardtia **2**: 265–267.

Kellogg, E.A. and Watson, L. 1993. Phylogenetic studies on a large data set. I. Bambusoideae, Andropogonodae and Pooideae (Gramineae). Bot. Rev. **59**: 273–343.

Kiran Raj, M.S. 2008. Taxonomic revision of the subtribe Dimeriinae Hack. of Andropogoneae (Poaceae - Panicoideae) in Peninsular India. Ph.D. thesis (Unpublished). University of Calicut, India, pp. 1–409.

Kiran Raj, M.S. and Sivadasan, M. 2008. A new species of *Dimeria* R. Br. (Poaceae) from Goa, India. Novon **18**: 183–186.

Kiran Raj, M.S., Sivadasan, M., Alfarhan, A.H. and Veldkamp, J.F. 2015. *Dimeria raviana* (Poaceae: Panicoideae), a new species from South Western Ghats, India. Phytotaxa **195**: 193–196.

Kiran Raj, M.S., Sivadasan, M., Veldkamp, J.F., Alfarhan, A.H. and Thomas, J. 2013a. *Nanooravia gen. nov.*, subtribe Dimeriinae (Poaceae-Panicoideae-Andropogoneae) from India. Nordic J. Bot. **31**: 161–165.

Kiran Raj, M.S., Sivadasan, M., Veldkamp, J.F., Alfarhan, A.H. and Thomas, J. 2013b. Validation of *Nanooravia santapaui* (Poaceae-Panicoideae-Andropogoneae-Dimeriinae). Nordic J. Bot. **31**: 638.

Lazarides, M. 1980. Phanerogamarum monographiae **XII**: the tropical grasses of Southeast Asia. pp. 1–225.

McNeill, J., Barrie, F.R., Buck, W.R., Demoulin, V., Greuter, W., Hawksworth, D.L., Herendeen, P.S., Knapp, S., Marhold, K., Prado, J., Prud'homme van Reine, W.F., Smith, G.F., Wiersema, J.H. and Turland, N.J. 2012. International Code of Nomenclature for algae, fungi and plants (Melbourne Code) adopted by the Eighteenth International Botanical Congress Melbourne, Australia, July 2011. Regnum Vegetabile **154**: 1–240.

Miquel, F.A.W. 1851. Analecta botanica indica. 2. Gramineae quaedam, praesertim Canaranae. Verhandelingen der EersteKlasse van het Koninklijk Nederlandsch Instituut van Wetenschappen III, **4**: 30–38, reprinted as Analecta Botanica Indica **2** (1851): 34–35.

Mohanan, M. and Rao, A.V.N. 1984 (1983, published in 1984). A new species of *Dimeria* R. Br. (Poaceae) from Kollimalai, South India. J. Bombay Nat. Hist. Soc. **80**(3): 615–617.

Mohanan, N. and Ravi, N. 1996. *Dimeria sivarajanii* (Poacaeae), a new species From Kerala, India. Rheedea **6**(2): 47–50.

Nair, V.J., Sreekumar, P.V. and Nair, N.C. 1983. *Dimeria raizadae* – a new species of Poaceae from Kerala, India. Indian J. For. **6**(2): 163–165.

Raizada, M.B. 1950. *Dimeria mooneyi. In*: Mooney, H., Supplement to the botany of Bihar and Orissa. Catholic Press, Ranchi, p. 263.

Ravi, N. 1995. Two new species of *Dimeria* R. Br. (Poaceae) from Kerala, India. Rheedea **5**(1): 37–42.

Ravi, N. 1996. Another two new species of *Dimeria* R. Br. (Poaceae) from Kerala, India. Blumea **41**(1): 251–256.

Ravi, N. and Anil Kumar, N. 1992. New and interesting species of *Dimeria* R. Br. (Poaceae) from Kerala, India. Rheedea **2**(2): 101–107.

Ridley, H.N. 1925. The Flora of the Malay Peninsula, **5**. Reeve & Co. Ltd., London, pp. 1–470.

Roberty, G. 1960. Monographie systématique des Andropogonées du globe. Boissiera **9**: 396–402.

Schmid, M. 1958. Flore agrostologique de l'Indochine. L'Agronomie Tropicale. Office de la Recherche Scientifique et Technique Outre-mer (ORSTOM), Paris, pp. 1–703.

Sreekumar, P.V., Nair, V.J. and Nair, N.C. 1982. *Dimeria borii* (Poaceae): a new species from Kerala, India. J. Econ. Taxon. Bot. **3**(2): 657–658.

Steudel, E.G. 1854. Synopsis plantarum glumacearum, **1**. J.B. Metzler, Stuttgart, pp. 1–474.

Teerawatananon, A., Boontia, V., Chantarasuwan, B., Hodkinson, T.R. and Sungkaew, S. 2014. A taxonomic revision of the genus *Dimeria* (Poaceae: Panicoideae) in Thailand. Phytotaxa **186**: 137–147.

Veldkamp, J.F. 2015. *Arundinella* (Gramineae) in Malesia with notes on other taxa and on aluminium accumulation. Blumea **59**: 167–179.

# 8

# CYPSELAR DIVERSITY IN FOUR SPECIES OF *SENECIO* L. (ASTERACEAE)

Tulika Talukdar[1] and Sobhan Kumar Mukherjee

*Department of Botany, University of Kalyani, Kalyani, Nadia 741235,
West Bengal, India*

*Keywords:* Senecio L.; Cypsela; Morphology; Pappus.

## Abstract

The genus *Senecio* L. is one of the largest genera of flowering plants and is an important member of the tribe Senecioneae (Asteraceae). Phenotypic information, including a broad range of morphological characters is very crucial for phylogenetic reconsideration of any family, tribe or genus. In the family Asteraceae, very little attention has been paid to cypselar diversity, though it is regarded as taxonomically valuable. A sincere attempt has been made to study detailed cypselas macro and micro-morphological features of four species of *Senecio* L. These diacritical features could be used to strengthen current inter-specific concept of *Senecio* L.

## Introduction

The genus *Senecio* L. belongs to the tribe Senecioneae of the family Asteraceae and considered as one of the largest genera of flowering plants carrying c.1000 species in strict sense (Nordenstam, 2007). The taxonomic foundation of the tribe Senecioneae was initiated by Cassini, who distinguished 17 'natural tribes' including Senecioneae ("Les Senecionees") in his third Memoire (Cassini, 1816) of the tribal classification of "Synantherees" i.e. family Asteraceae. The tribe is traditionally characterized by an epaleate receptacle and a pappus of capillary bristles in their cypselas. Phylogenetically, the tribe is proposed by Small (1919) as the most primitive tribe of the family Asteraceae. On the other hand, due to its moderately large size manifested by tremendous number of species and genera [c. more than 3000 species in 151 genera by Nordenstam (2007)], almost cosmopolitan distribution and incredible morphological diversity, it is reconsidered as the evolutionary successful one (Bremer, 1994). Although DNA data provide the most reliable information for estimating evolutionary relationships and distances between taxa, these data alone cannot explain how or why a particular plant evolved without phenotypic information, including a broad range of morphological and chemical characters (Calabria *et al.*, 2009). Therefore, a valid need of morphometric analysis cannot be ruled out.

It is really a fact that cypselar morphology in the family has not been received as much attention as it should be. According to Heywood *et al.* (1977), cypsela structure and anatomical features have been studied in details in only a few groups such as Anthemideae and Cardueae and found to be taxonomically valuable. They opined that "It is difficult to believe that carpological features will prove to be of lesser value in all the remaining tribes."

In this context, our present investigation deals with detailed studies of cypselas macro as well as micro-morphological features of four species of *Senecio* L., namely *S. aegyptius* L., *S. alpinus* (L.) Scop., *S. aquaticus* Hill. and *S. viscosus* L. of the tribe Senecioneae primarily using Light Microscope. Special emphasises have been given to traditional characters such as size and shape of cypselas, nature and distribution of ribs and furrows, nature of surface pubescence, structure of stylopodium, carpopodium, pappus etc. These diverse cypselar features could be utilized to construct an artificial key and to evaluate infra-generic phylogeny of *Senecio*.

[1]Department of Botany, A.P.C. Roy Govt. College, Siliguri, Darjeeling, West Bengal, India.
E-mail: talukdartulip12@gmail.com

**Materials and Methods**

Plant materials (cypselas) for the present investigation were obtained from Hortus Botanicus Hauniensis, Denmark (DK) and Botanischer Garten der Universitat Zurich (Z).

*Macro-morphological studies of cypselas*

In cases, where intact cypselas were available, the first and foremost step was to mark the posterior and anterior surface of the cypselas. Then 10 dry and 10 FAA preserved mature cypselas were randomly taken in glass slides and graphed slides and observed under Olympus stereo dissecting microscope (DM) and Olympus binocular microscope (No. 611062). Suitable images were taken using Zeiss Stemi DV4 camera equipped microscope.

Shape and direction of cypselas were noted carefully. Length and width of the cypselas were measured visually by graphed slides, in few cases they were counted by ocular and stage micrometer. The length of the cypselas in the present study is defined as the length of the body of cypselas from basal meristematic zone (carpopodium) up to apical end excluding pappus. The width of the cypselas was measured at the widest part of the cypselar body.

*Micro-morphological studies of cypselas*

Mature cypselas were dipped in 1-5% NaOH solution for 2-7 days depending upon the hardness. Then they were transferred into saturated chloral hydrate solution for few hours, repeatedly washed with water and properly stained in 0.2-0.5% aqueous Safranin solution. After staining, specimens were placed in 70% phenol glycerine solution and dissected carefully for studying different parts of cypselas. Suitable photographs were taken using Olympus C-310 zoom digital camera (3.2 Megapixel) and Zeiss-stereo microscope.

Nature of ribs, types, distribution and orientation of hairs, nature of surface cells, other epidermal structures and carpopodial cells were critically observed. Pappus characters such as nature of pappus bristles, their number, arrangement, length and apex organization were also examined.

**Results and Discussion**

Cypselas of all the studied species (Figs 1&2) are invariably homomorphic with a length ranging from 1.5 to 8.0 mm. Cypselas are generally straight. Beside *Senecio alpinus* and *S. aquaticus* cypselas of other two species bear 7-10, prominent and straight ribs (Figs 1A,F,G, 2A). Pubescent cypselas have been noted except in *S. alpinus* (Fig. 1F). Hairs of all the pubescent species are of twin or duplex type, common in the family (Hess, 1938). They are typically three-celled, with two parallel cells and a smaller basal cell. Occurrence of few myxogenic hairs having mucilaginous properties when soaked in water have been reported by Nordenstam (1977), Konechaya (1981) and Mukherjee (2001) in the members of Senecioneae. Sahu (1983) has mentioned that these "Achenial hairs" has sharply pointed apex. In the contrary, Mukherjee (2001) has pointed out that "tips of the hairs are obtuse or rounded but not sharply pointed", as mentioned by Sahu (*l.c.*). Interestingly, in our observations both the statements are found to be true, as in *Senecio aegyptius* and *S. viscosus* tips of the hairs are sharply pointed (Figs 1C, 2B), while that of *S. aquaticus* are rounded (Fig. 1H).

Well-developed stylopodium with broaden base have been found in *Senecio aquaticus* (Fig. 1G). In other studied species stylopodium is found to be ill-developed or insignificant.

Carpopodium is symmetric, either well-developed ring-like as found in *Senecio viscosus* (Fig. 2A), or ill-developed thickened band-like as in other three investigated species of *Senecio*. Carpopodium in all the studied taxa, is made up of rectangular, thick-walled cells arranged in several tangential rows. Such findings are well supported by Wetter (1983), who mentioned that in carpopodium of different members of Senecioneae "the squarish to rectangular (quadrate) cells were arranged in one to several rows or series. The number of rows which composed of carpopodium was constant in each species." Wetter (*l.c.*) also documented variation in the number of rows of cells among the species. This variation is also evident in the present observation, as the number of rows is 3-4 in *Senecio viscosus* and 1-2 in other three studied species. Haque and Godward (1984) have reported the absence of carpopodium in all four species of *Senecio* studied by them. But the present investigation is not in agreement with the above view, as carpopodium are found to be present in all the *Senecio* species studied.

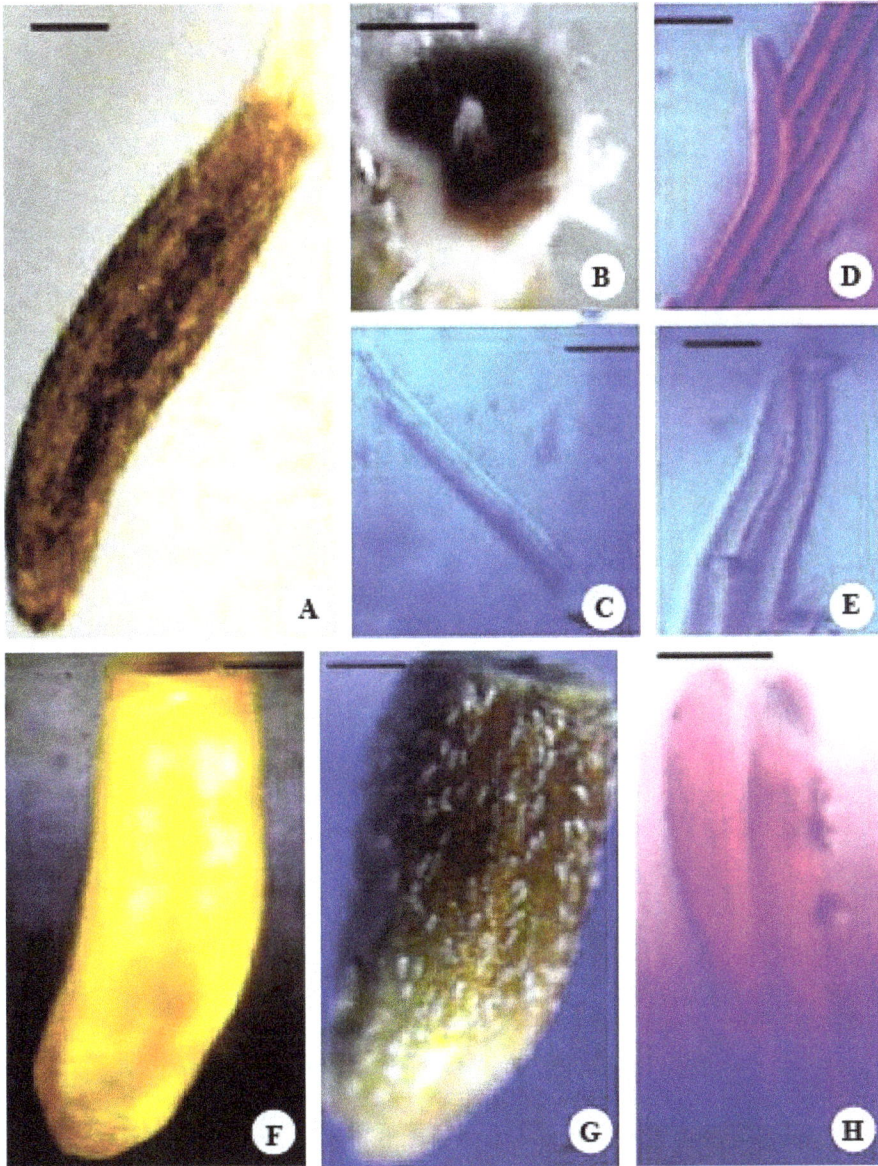

Fig. 1. Cypselar morphology of *Senecio aegyptius* (A-E), *S. alpinus* (F) and *S. aquaticus* (G-H). A, F, G. cypsela; B. apex with stylopodium; C, H. twin hair; D, E. parts of pappus bristle. Bar: 0.2 mm (A,F,G); 0.1 mm (B); 0.02 mm; (C-E); 0.005 mm (H).

Pappus usually represented by many, free, 2-5 mm long, persistent or caducous (as in *Senecio aegyptius*), scabrous or barbellate (as in *S. aegyptius*), biseriate bristles; with unequal, sharply pointed apical cells. Apical cells were two in number in *S. aegyptius* and three in *S. viscosus* (Figs 1E, 2E). Biseriate pappus bristles of the genus *Senecio* have also been noted by Drury and Watson (1965), who mentioned that the outer series contain minute fimbrillae with retrorsely barbed tips called "fluked". However, our observation is not similar with the above view. Multiseriate pappus bristles also have been marked in *Senecio viscosus*. Often pappus bristles is reduced as in *Senecio aquaticus*, where pappus is represented by apical corona. So pappus with all its features can be employed in classification of taxa.

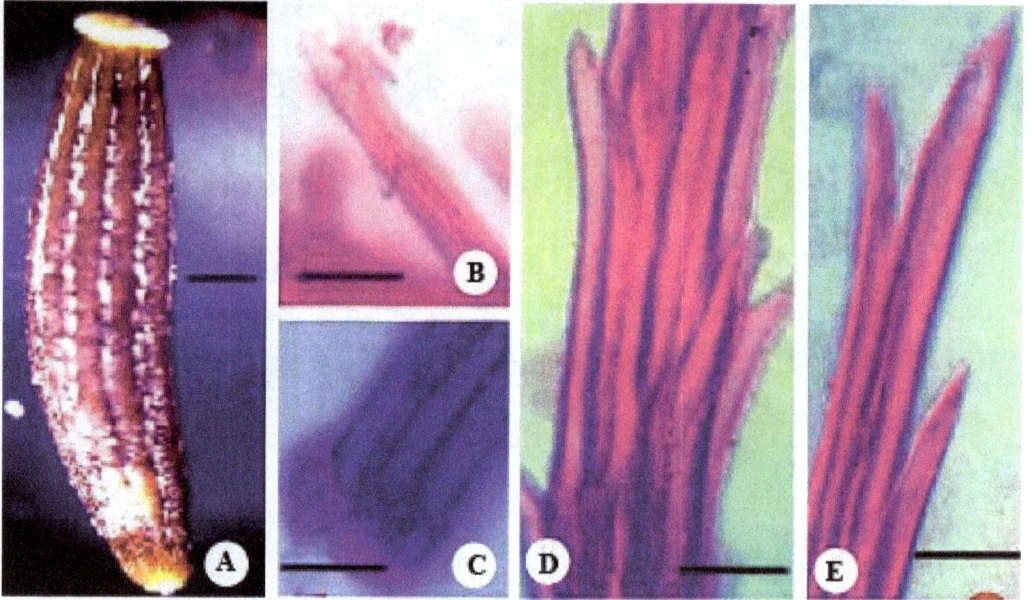

Fig. 2. Cypselar morphology of *Senecio viscosus*. A. cypsela;  B. twin hair; C. base of pappus bristle; D. middle part of pappus bristle; E. apical part of pappus bristle. Bar: 0.2 mm (A); 0.02 mm (B-E).

Considering all these cypselar features, an attempt has been made to construct an artificial key to the species.

*Key to the species of Senecio* L.

1. Cypsela pubescent.                                                                                                   2
- Cypsela glabrous.                                                                                            *S. alpinus*
2. Cypsela truncate at the apex, quadrangular, not ribbed; stylopodium well-developed; insertion of cypsela oblique; pappus of apical corona.                           *S. aquaticus*
- Cypsela rounded at the apex, cylindrical, 9-10 ribbed; stylopodium ill-developed; insertion of cypsela straight; pappus of capillary bristles.                                3
3. Stylopodium conical; carpopodium ill-developed, thickened band-like; pappus caducous, of biseriate barbellate bristles; apex of bristle made of two unequal cells.                                                                                      *S. aegyptius*
- Stylopodium tubular; carpopodium well-developed, symmetric, circular ring-like; pappus persistent, of multiseriate scabrous bristles; apex of bristle made of three unequal cells.                                                                       *S. viscosus*

The present study on detailed macro- and micro-morphological features of cypselas of four species of *Senecio* L. is a preliminary attempt to assess the usefulness of cypsela as species delimiting factor. The analysis clearly indicates that in comparison to size and shape of cypsela, nature of carpopodium, presence or absence of rib, trichome tip, pappus features like arrangement of pappus bristle, number of apical cells in bristle etc. are much more reliable characters for inter-specific grouping or separation.

## Acknowledgement

We extend our special thanks to Dr. Hans Vilhelm Hansen, Curator, Denmark and to Dr. Peter Enz, Curator, Zurich for their active assistance in despatching the identified mature cypselas for our studies.

## References

Bremer, K. 1994. Asteraceae. Cladistics and Classification. Timber Press, Portland.

Calabria, L.M., Emerenciano, V.P., Scott, M.T. and Mabry, T.J. (Eds). 2009. Secondary Chemistry of Compositae. *In*: Funk, V., Susanna, A., Stuessy, T.F. and Bayer, R.J. (Eds), Systematics, Evolution, and Biogeography of Compositae. Smithsonian Institution, Washington, DC., USA, pp. 369-383.

Cassini, H. 1816. *In*: King, R.M. and Dawson, H.W. (Eds), Cassini on Compositae, collected from the Dictionnaire des Sciences Naturelles. New York, Oriole Editions, pp. 535-602.

Drury, D.G. and Watson, L. 1965. Anatomy and the taxonomic significance of gross vegetative morphology in *Senecio*. New Phytol. **64**: 307-314.

Haque, M.Z. and Godward, M.B.E. 1984. New records of the carpopodium in Compositae and its taxonomic use. Bot. J. Linn.Soc. **89**: 321-340.

Hess, R. 1938. Vergleichende Untersuchungen uber die Zwillingshaare der Compositen. Bot. Jahrb. Syst. **68**: 435-496.

Heywood, V.H., Harborne, J.B. and Turner, B.L. 1977. An overview to the Compositae. *In*: Heywood, V.H., Harborne, J.B. and Turner, B.L. (Eds), The Biology and Chemistry of the Compositae. Vol. **1**. Academic Press, London, pp.780-802.

Konechaya, G.Y. 1981. Carpological and anatomical characters of species of the genus *Senecio* (Asteraceae) with reference to their taxonomy. Bot. Zh. (Leningr.) **66**(6): 834-842.

Mukherjee, S.K. 2001. Cypselar features in nineteen taxa of the tribe Senecioneae (Asteraceae) and their taxonomic significance. *In*: Maheshwari, J.K. (Ed.), Recent Researches in Plant Anatomy and Morphology. Scientific Publishers, Jodhpur, India, pp. 253-274.

Nordenstam, B. 1977. Senecioneae and Liabeae - systematic review. *In*: Heywood, V.H., Harborne, J.B. and Turner, B.L. (Eds), The Biology and Chemistry of the Compositae. Vol. **II**. Academic Press, London, pp. 799-830.

Nordenstam, B. 2007 Senecioneae. *In*: Kadereit, J.W. and Jeffrey, C. (Eds), The Families and Genera of Vascular Plants. Vol. **8**. Flowering Plants, Eudicots, Asterales. Springer, Berlin, pp. 208-241.

Sahu, T.R. 1983. Trichome studies in *Senecio* Linn: Structure, distribution and taxonomic significance. J. Indian Bot. Soc. **62**: 84-89.

Small, J. 1919. The origin and development of the Compositae. V. The pappus. New Phytol. **11**: 98-123.

Wetter, M.A. 1983. Micromorphological characters and generic delimitation of some New World Senecioneae (Asteraceae). Brittonia **35**: 1-22.

# 9

# TYPIFICATION OF TEN SPECIES OF *LITSEA* LAM. (LAURACEAE) ENDEMIC TO INDIA

RAJEEV KUMAR SINGH, ARTI GARG[1] AND PARAMJIT SINGH[2]

*Botanical Survey of India (BSI), Central Regional Centre (CRC), 10-Chatham Lines, Allahabad 211 002, Uttar Pradesh, India*

*Keywords*: Lauraceae; *Litsea*; Type; India.

## Abstract

This paper deals with the lectotypification of eight binomials of seven recognized species of *Litsea* Lam. endemic to India, namely *Litsea assamica* (Meisn.) Hook. f., *L. coriacea* (B. Heyne *ex* Nees) Hook. f., *L. membranifolia* Hook. f., *L. oleoides* (Meisn.) Hook. f., *L. stocksii* (Meisn.) Hook. f., *L. venulosa* (Meisn.) Hook. f., and *L. wightiana* (Nees) Hook. f. Types of three other endemic species, viz. *L. beddomei* Hook. f., *L. mishmiensis* Hook. f. and *L. oreophila* Hook. f. are also specified.

## Introduction

The family Lauraceae Juss. comprises 52 genera with about 2,550 species, distributed mainly in the tropical and warm regions of southeastern Asia and Brazil (Mabberley, 2008; Bhuinya *et al.*, 2010). The genus *Litsea* Lam. consists of more than 300 species worldwide, especially in tropical Asia and Australia (Mabberley, 2008; Bhuinya *et al.*, 2010). In India, there are 45 species, occurring in moist deciduous, semi-evergreen, and evergreen forests at 200–3,650 m elevation, with 18 species endemic to different states (Bhuinya *et al.*, 2010; Singh, 2015). As part of the revisionary studies of *Litsea* in India, we realized the necessity for typification of some endemic species; hence, here we lectotypified eight binomials of seven species of the genus. While designating lectotypes, we followed the guidelines of Art. 9.2 of the Melbourne Code (McNeill *et al.*, 2012). Specification of holotype for other three endemic species was done as no holotype was cited in their protologues.

## Taxonomy

1. **Litsea assamica** (Meisn.) Hook. f., Fl. Brit. India 5: 161 (1886). *Tetranthera rangoonensis* Meisn. var. *assamica* Meisn. in De Candolle, Prodr. 15(1): 188 (1864).

*Type*: India. Assam, 1850, *W. Griffith 1190* [*Tetranthera* n. 17, Herb. Hook. f. *et* Thoms.] (lectotype K-000793183!, here designated; isolectotype BM-000951039!).

*Distribution*: India, endemic (Arunachal Pradesh, Assam and Meghalaya).

*Notes*: In protologue of *Tetranthera rangoonensis* var. *assamica*, Meisner cited 'In Assam (Jenkins !)., *Tetranthera* 17. Hook. fil. et Thoms.! Hb. Ind. Or.' Three specimens belonging to the herbarium of Hooker f. & Thomson, *Tetranthera* no. 17 are now extant, which comprised of a single specimen gathered by *T.J. Jenkins s.n.* (K-000357513) and two of *W. Griffith 1190* (K-000793183 and BM-000951039). Of these, K-000793183, is designated here as the lectotype as it agrees well with the protologue and also includes a short description of the flower by Hooker.

[1]Corresponding author. Email: kad_arti396@yahoo.com
[2]Botanical Survey of India, Head Quarter, CGO Complex, Salt Lake City, Kolkata 700 064, West Bengal, India.

From the protologue, it seems that "Jenkins" and "*Tetranthera* 17. Hook. fil. et Thoms" are different collections and two of *W. Griffith 1190* (K-000793183 and BM-000951039) likely belong to "*Tetranthera* 17. Hook. fil. et Thoms".

2. **Litsea coriacea** (B. Heyne *ex* Nees) Hook. f., Fl. Brit. India 5: 166 (1886). *Tetranthera coriacea* B. Heyne *ex* Nees in Wall., Pl. Asiat. Rar. 2: 66 (1831).

*Type*: India. Deccan Peninsula, *s.d., B. Heyne s.n.* [in *Wallich* Number. List no. 2556] (lectotype GZU-000254469!, here designated; isolectotype BM-000793685!).

*Distribution*: India, endemic (Goa, Karnataka, Kerala and Tamil Nadu).

*Notes*: Three specimens are extant, two at GZU (GZU-000254468 and GZU-000254469) and one at BM (BM-000793685). The GZU-000254469 specimen is designated here as the lectotype as it agrees well with the protologue.

3. **Litsea membranifolia** Hook. f., Fl. Brit. India 5: 159 (1886).

*Type*: India. Arunachal Pradesh, Dibang Valley, Mishmi Hills, *s.d., W. Griffith s.n.* [*Kew Distrb. 4310*] (lectotype K-000357530!, here designated; isolectotypes K-000793176!, GH-00415039!).

*Distribution*: India, endemic (Arunachal Pradesh and Nagaland).

*Notes*: Hooker described *Litsea membranifolia* on the basis of specimens from 'Upper Assam; Mishmi Hills, and woods at Yen, *Griffith* (*Kew Distrib.* 4310)', but no specific specimen was indicated as the holotype. Furthermore, for this species, three specimens of Griffith's collections are extant, two at K (K-000357530 and K-000793176) and one at GH (GH-00415039). The K-000357530 specimen is designated here as the lectotype as it agrees well with the protologue.

Ngernsaengsaruay *et al.* (2011: 72) cited the following: "Type: India, East Bengal, *Griffith 4310* (holotype K!)". Although they cited K as housing the "holotype," their citation of "holotype" is not corrected to lectotype. Because, from 2001, an act of lecto-, neo-, and epitypification needs the citation of the phrase, "here designated" or its equivalent. The critical scrutiny of description provided in Ngernsaengsaruay *et al.* (*l.c.*) shows that it is best matches with *L. glutinosa* (Lour.) C.B. Rob., and thus the occurrence of this species in Thailand is uncertain and needs further confirmation based on fresh collections.

4. **Litsea oleoides** (Meisn.) Hook. f., Fl. Brit. India 5: 175 (1886). *Tetranthera oleoides* Meisn. in DC., Prodr. 15(1): 195 (1864).

*Type*: India. Kerala, Sispauray (Sispara), April 1846, *R. Wight 2530* (lectotype K-000357533!, here designated).

*Syntype*: India. Kerala, *s.d., R. Wight 54* (NY-00355989!).

*Distribution*: India, endemic (Kerala and Tamil Nadu).

*Notes*: Meisner described *Tetranthera oleoides* on the basis of specimen(s) from 'In Penins. Indiae Or.? (Wight!)'. The type data lack Wight's collection number and date, and information on the number of specimens used by Meisner to describe this species. Currently, Meisner's main herbarium is at New York (NY), which has one fragmentary specimen of this species (NY-00355989); the specimen shows "*Tetranthera oleoides* Meisn. In Hb. Hook., "Wight. Hb. N. 54!" and an illegible writing of what seems to be the locality. In contrast, the K specimen (K-000357533) is complete and fits well with the protologue. Therefore, the K-000357533 specimen is designated here as the lectotype as it agrees well with the protologue.

5. **Litsea stocksii** (Meisn.) Hook. f., Fl. Brit. India 5: 176 (1886). *Cylicodaphne oblonga* Nees var. *stocksii* Meisn. in DC., Prodr. 15(1): 205 (1864).

*Type*: India. Canara (Karnataka), *s.d.*, *J.E. Stocks s.n.* (lectotype K-000357539!, here designated; isolectotype K-000793237!).

*Litsea stocksii* (Meisn.) Hook. f. var. *glabrescens* (Meisn.) Hook. f., Fl. Brit. India 5: 176 (1886). *Cylicodaphne wightiana* Meisn. var. *glabrescens* Meisn. in DC., Prodr. 15(1): 201 (1864).

*Type*: India. Canara (Karnataka), *s.d.*, *J.E. Stocks s.n.* (lectotype K-000793240!, here designated; isolectotype K-000357538!).

*Litsea josephii* S. M. Almeida, Fl. Savantwadi 1: 364 (1990), nom. illeg. *et* superfl. for *L. stocksii*

*Litsea vartakii* M. R. Almeida, J. Bombay Nat. Hist. Soc. 86(2): 180 (1989), nom. illeg. *et* superfl. for *L. stocksii*

*Distribution*: India, endemic (Goa, Karnataka, Kerala, Maharashtra and Tamil Nadu).

*Notes*: Hooker (*l.c.*) based his *Litsea stocksii* on *Cylicodaphne oblonga* var. *stocksii* Meisn.; at the same time, he also included "*Tetranthera lancaefolia* (sensu) Graham (Cat. Pl. Bombay. 174. 1839, non Roxb. 1832)" with a query sign. Because of the doubtful inclusion, the citation does not cause superfluity to the name *L. stocksii*.

Meisner described *Cylicodaphne oblonga* var. *stocksii* on the basis of specimens from 'In Canara (Stocks!).' Two specimens collected by *Stocks* from Canara are now extant at K (K-000357539 and K-000793237). The K-000357539 specimen is designated here as the lectotype as it agrees well with the protologue.

Meisner described *Cylicodaphne wightiana* var. *glabrescens* Meisner on the basis of specimens from 'Wight!, Perrott. n. 451!, 452!, 1843!, Hohenack. n. 1335! Stocks! Gardn.! Law!', but no specific herbarium sheet was designated as the holotype. Pertaining to this specification, only two specimens collected by *Stocks* from Canara are now extant at K (K-000793240 and K-000357538). The best one, K-000793240, is designated here as the lectotype as it agrees well with the protologue.

6. **Litsea venulosa** (Meisn.) Hook. f., Fl. Brit. India 5: 161 (1886). *Tetranthera venulosa* Meisn. in DC., Prodr. 15(1): 187 (1864).

*Type*: India. Peninsula Indiae Orientalis, *s.d.*, *R. Wight s.n.* (lectotype K-000793184!, here designated).

*Syntypes*: India. Peninsula Indiae Orientalis, *s.d.*, *R. Wight 2532a* (L-0037108!); Tamil Nadu, Courtallam, Sep 1835, *R. Wight 710b* (L-0037109!); Penins. Ind. Or. (Peninsula Indiae Orientalis), *s.d.*, *R. Wight 30* (NY-00356000!).

*Distribution*: India, endemic (Kerala and Tamil Nadu).

*Notes*: Meisner described *Tetranthera venulosa* on the basis of specimens from 'In Penins.? Indiae (Wight!).' Four specimens belonging to Peninsula Indiae Orientalis of the Wight Herbarium are extant (K-000793184, L-0037108, L-0037109, and NY-00356000). The K-000793184 specimen is designated here as the lectotype as it agrees well with the protologue.

7. **Litsea wightiana** (Nees) Hook. f., Fl. Brit. India 5: 177 (1886). *Cylicodaphne wightiana* Nees in Wall., Pl. Asiat. Rar. 2: 68 (1831).

*Type*: India. Peninsula Indiae Orientalis, *s.d.*, *R. Wight 2232* [= Wall. Cat. n. 2557 A] (lectotype P-02003106!, here designated; isolectotypes, BR-0000013053000!, K-000357542!).

*Distribution*: India, endemic (Maharashtra, Goa, Karnataka, Kerala and Tamil Nadu).

*Notes*: In the protologue of *Cylicodaphne wightiana*, Nees cited 'Wall. Cat. n. 2557, A, B. Habitat in sylvis Nilghiry (Deenhutty in schedis). (E. Noton.) Vidi etiam in Hb. Wight.' The "Wall. Cat. n. 2557 A" specimen belongs to the Wight Herbarium, whereas the "Wall. Cat. n. 2557 B" specimen belongs to collections from Nilghiry by E. Noton. Three specimens belonging to the Wight herbarium (Wall. Cat. n. 2557 A) are extant (BR-0000013053000, K-000357542, and P-02003106). Of the choice of specimens, P-02003106! is designated here as the lectotype as it agrees well with the protologue.

**Types of three endemic species are also specified below as in their protologues no holotypes were cited.**

1. **Litsea beddomei** Hook. f., Fl. Brit. India. 5: 177 (1886).

*Type*: India. Tamil Nadu, Tinnevelly (Tirunelveli), Dec 1880, *R.H. Beddome 15* (Lectotype, here designated (or perhaps holotype): K-000357515!).

*Distribution*: India, endemic (Kerala and Tamil Nadu).

*Notes*: For his new species *Litsea beddomei*, Hooker cited the type collection information as follows: "SOUTH DECCAN; Tinnevelly Hills, Beddome." The protologue lacks the collection date and number. Furthermore, neither a holotype nor the name of the herbarium housing the type was mentioned. A single specimen collected by Beddome from Tirunelveli in December 1880 bearing Hooker's annotation '*L. beddomei*' is now extant at K (K-000357515). Because of the uncertainty whether Hooker's description was based on the K single specimen alone or on additional specimens, it is concluded here that the single specimen at K may serve as the holotype or a lectotype.

2. **Litsea mishmiensis** Hook. f., Fl. Brit. India 5: 161 (1886).

*Type*: India. Arunachal Pradesh, Dibang Valley, Laim-planj-thaya, Mishmi Hills, *s.d.*, *W. Griffith 4317* (Lectotype, here designated (or perhaps holotype): K-000357531!).

*Distribution*: India, endemic (Arunachal Pradesh).

*Notes*: Within the protologue of *Litsea mishmiensis*, Hooker mentioned the following: 'Upper Assam; Mishmi Hills at Laim-planj-thaya, *Griffith*'. The collection date and number were not cited. A single type specimen is now extant at K (K-000357531). Hooker, however, did not cite a holotype or mention the herbarium housing the type. It is uncertain whether Hooker's description was based on the K single specimen alone or on additional specimens. Therefore, it is concluded here that the single specimen at K may serve as the holotype or a lectotype.

3. **Litsea oreophila** Hook. f., Fl. Brit. India 5: 156 (1886). *Lindera hookeri* Meisn. in DC., Prodr. 15(1): 245 (1864).

*Type*: India. Sikkim, Lachoong (Lachung), 10,000–11,000 ft., 29 Aug 1849, *J.D. Hooker s.n.* [*Tetranthera* n. 4, Herb. Hook. f. *et* Thoms.] (holotype K-000357534!).

*Distribution*: India, endemic (Sikkim).

*Notes*: Within the protologue of *Lindera hookeri*, Meisner mentioned: 'In Sikkim (Hook. fil. et Thoms. hb. Ind. or. *Tetranthera* 4!)' and "(v. s. in hb. Hook.)" (= vidi siccam in Hooker herbarium (K)]. Only one specimen is extant at K (K-000357534), which is therefore the obligatory holotype.

## Acknowledgements

The authors are grateful to the curators of BM, BR, GH, GZU, K, L, NY and P herbaria for information and images of type specimens. We gratefully acknowledge late Prof. James L. Reveal, Cornell University, Ithaca, USA for his critical comments and suggestions on the manuscript.

## References

Bhuinya, T., Singh, P. and Mukherjee, S.K. 2010. An account of the species of *Litsea* Lam. (Lauraceae) endemic to India. Bangladesh J. Pl. Taxon. **17**(2): 183–191.

Mabberley, D.J. 2008. Mabberley's Plant-Book: A portable dictionary of plants, their classification and uses. 3rd edition. Cambridge University Press, Cambridge, 1020 pp.

McNeill, J., Barrie, F.R., Buck, W.R., Demoulin, V., Greuter, W., Hawksworth, D.L., Herendeen, P.S., Knapp, S., Marhold, K., Prado, J., Proud'homme van Reine, W.F., Smith, G.F., Wiersema, J.H. and Turland, N.J. (Eds) 2012. International Code of Nomenclature for algae, fungi and plants (Melbourne Code): Adopted by the Eighteenth International Botanical Congress, Melbourne, Australia, July 2011. Regnum Vegetabile **154**: 1–274.

Ngernsaengsaruay, C., Middleton, D.J. and Chayamariat, K. 2011. A revision of the genus *Litsea* Lam. (Lauraceae) in Thailand. Thai Forest Bull. Bot. **39**: 40–119.

Singh, R.K. 2015. Typification of seven species of *Litsea* (Lauraceae) endemic to India. Phytotaxa **201**(4): 278–286.

# 10

# REPRODUCTIVE BIOLOGY OF SEVEN TAXA OF *MAGNOLIA* L. IN THE SOUTH OF RUSSIAN FAR EAST

Lyubov A. Kameneva and Inna M. Koksheeva[1]

*Botanical Garden-Institute, Far Eastern Branch of Russian Academy of Sciences (BGI FEB RAS), Vladivostok, Russia.*

*Keywords*: *Magnolia* L.; Pollen; Seed germination; Stratification; Seed productivity.

## Abstract

This paper presents the phenology of seven taxa of the genus *Magnolia* L., pollen biology on germination and storage conditions, seed productivity, germinating ability of seeds and its dependence on stratification and germination conditions. It has been found that *M. kobus* DC, *M. kobus* var. *boreales* Sarg., *M. obovata* Thurb., *M. officinalis* Rehd. *et* Wils., *M. salicifolia* (Sieb. *et* Zucc.) Maxim., *M. sieboldii* K. Koch. and *M. tripetala* L. in cultivated condition produce pollen at a low viability rate (9.4 - 31.7%). Real seed productivity of the taxa being studied is less than their potential productivity. However, *M. obovata, M. officinalis* and *M. tripetala* are characterized by high seed germinating ability, up to 94%. Optimal germination conditions for seeds of *M. tripetala, M. officinalis* and *M. obovata* require protected ground (greenhouse) and stratification at 4°C during 30 days. The high germinating ability of *M. sieboldii* seeds in the open ground is explained by stratification duration and temperature required for this species. Sarcotesta effects on higher seed germinating ability have been observed in *M. obovata* only.

## Introduction

*Magnolia*, belonging to the family Magnoliaceae Juss. includes over 80 species, and is distributed in Southeast Asia, North and Central America (Low, 1996). The only *Magnolia* species that occurs in Russia is *Magnolia obovata* Thunb., growing in Kunashir Islands (Barkalov, 2009). Primorsky Krai has favourable climatic conditions for the cultivation of representatives of the genus *Magnolia* (Turkenya, 1991).This fact has made it possible to bring some other magnolia species under cultivation in this area.

The collection of *Magnolia* in the Botanical Garden-Institute, Far Eastern Branch of Russian Academy of Sciences (BGI FEB RAS), Vladivostok (Russia, Far East, Primorsky Krai), dates back to 1972 and at that time some seeds of *M. sieboldii* K. Koch. were brought from the Pyongyang Botanical Garden (North Korea). Forty years later, there are 20 magnolia species successfully cultivated in the Botanical Garden Institute FEB RAS.

Prospects for bringing plants under cultivation are assessed by correspondence of plant phenology with climatic conditions in which they grow. Knowledge of pollen quality, pollination and fertilization issues and production of quality seed material has both theoretical and practical importance. Pollen quality is important for prediction of seed productivity of plants and for hybridization studies (Termena, 1972; Koksheeva, 2004; Pshennikova, 2007; Liza *et al.*, 2010). Production of quality seed material is an indicator of reproductive capacity of plants which depends on a number of factors: morphogenetic, genetic, physiological and ecological.

---

[1] Corresponding author. Email: koksheeva@yandex.ru

Individual questions of reproductive biology of some species of magnolia in different climatic conditions have been studied by several authors (Minchenko and Korshuk, 1987; Kikuzawa and Mizui, 1990; Ishida, 1996; Grigorenko, 1998; Hirayama and Ishida, 2005; Korshuk and Palagecha, 2007; Setsuko *et al.*, 2008; Han and Long, 2010; Wang, 2010). The present work deals with comprehensive study of the genus *Magnolia* in the Russian Far East, including the study of phenology, pollen germination, seed productivity and seed germination.

## Materials and Methods

Seven taxa of *Magnolia* growing in the Botanical Garden Institute in Vladivostok (Russian Far East, Primorsky Krai) are appended in Table 1. Observations on phenology have been made using the methodology of Lapin (1967).

**Table 1. List of taxa of *Magnolia* L. used in the present study.**

| Taxa | Locality | Origin material | h/d (m) | Beginning | |
| --- | --- | --- | --- | --- | --- |
| | | | | flowering (year) | fruiting (year) |
| *M. kobus* | Central and northern Japan, southern part of the Korean Peninsula | Ukraine, Kiev, Institute of Botanical Gardens, 1984, seedlings | 4.5/5.0 | 9 | 9 |
| *M. kobus* var. *borealis* | Japan, Hokkaido | Ukraine, Kiev, Institute of Botanical Gardens, 1984, seedlings | 05.0/5.0 | 14 | 26 |
| *M. obovata* | Japan, Kuril Islands, Russia | Ukraine, Kiev, Institute of Botanical Gardens, 1986, seedlings | 3.0/3.0 | 13 | 24 |
| *M. officinalis* | Central China | Ukraine, Kiev, Institute of Botanical Gardens, 1989, seedlings | 4.5/5.0 | 16 | 20 |
| *M. salicifolia* | Central and northern Japan | Czech Republic, 1996, seeds | 63.5/3.5 | 13 | 15 |
| *M. sieboldii* | Japan, China, the Korean Peninsula | North Korea, Pyongyang, the Botanical Gardens, 1974, seeds | 04.0/6.0 | 9 | 14 |
| *M. tripetala* | The southern part of North America | Ukraine, Kiev, Institute of Botanical Gardens, 1988, seedlings | 4.0/2.0 | 12 | 22 |

h = height of stem,  d = diameter of crown of tree.

Pollen was collected during mass flowering period. Three growth medium varieties were used for study of pollen germination: 5%, 10% and 15% glucose solutions, with distilled water used for control purposes (Golubinsky, 1974). Pollen was germinated in a thermostat at 24°C and laboratory temperature 18°C to 20°C. Specimens were observed through a microscope through 24h after pollen was sown. The number of germinated pollen grains was counted in five microscope fields of view for each specimen. The length of pollen tubes was measured as an indicator of pollen viability. An optimal growth medium was assumed in which germinated pollen percentage was at its highest and corresponded to maximum length of pollen tubes.

Three storage methods were used to determine optimal conditions of pollen storage: storage in laboratory conditions at 18°C to 20°C, storage in a household refrigerator at 4°C and storage in a freezing cabinet at –18°C. Pollen was stored for 7 days in tight sealed test tubes.

Seed productivity and productivity rate were determined using the method of Rabotnov (1960). Potential seed productivity (PSP) is number of seed buds per one fruit aggregate. Real seed productivity (RSP) is number of mature seeds per one fruit aggregate. Productivity rate (PR) is the ratio of potential and real seed productivity expressed in per cent. Seed productivity studies were based on 60 fruit aggregates taken for each plant. Seed and fruit parameters were also measured: length, width and weight (weight of 1,000 seeds with/without sarcotesta). Seed germinating ability was determined by sowing seeds in the open ground and in the protected ground (greenhouse) and further natural lengthy (mean air temperature in winter varying from –5°C to –27.2°C) stratification and artificial stratification (at 4°C during 30 days). Also, seeds were germinated with and without sarcotesta. Studies were based on 90 seeds of each species taken for each experimental condition.

## Results and Discussion

*Phenology of plants:*

Vegetative period of *Magnolia* species begins with swelling of generative buds (20 April 2012) at temperature around 5.4°C. Vegetative buds swell later (5-28 May 2012). For all the seven *Magnolia* taxa studied, flowering period normally begins 10-20 days after commencement of generative bud swelling and continues for 17 to 40 days (Fig. 1A-C). The flowering of early flowering *Magnolia* species occurs before leaf unfolding period and mass flowering occurs in mid-May at 9.7°C. Mass flowering of late flowering *Magnolia* species, which flower after leaf unfolding, occurs in mid-June at 14.5°C. In the south of Primorsky Krai, seeds ripen in late September or in first ten-day period of October (16.5°C). Vegetation period duration for the species being studied is 169-179 days (Table 2).

**Table 2. Phenology of the genus *Magnolia* L. in cultivated condition.**

| Taxa | Genera-tive bud swelling | Vegeta-tive buds swelling | Leaf unfold-ing | Bud-ding | Flowering | | | Fruit forma-tion | Fruit ripen-ing | Defo-liation | Vegeta-tion period (days) |
|---|---|---|---|---|---|---|---|---|---|---|---|
| | | | | | Start | Mass | Stop | | | | |
| | | | | | Date | | | | | | |
| *M. kobus* | 20.04 | 16.05 | 27.05 | 29.04 | 05.05 | 15.05 | 22.05 | 02.10 | 16.10 | 03.11 | 179 |
| *M. kobus* var. *borealis* | 23.04 | 16.05 | 27.05 | 30.04 | 03.05 | 18.05 | 25.05 | 01.10 | 18.10 | 28.10 | 173 |
| *M. obovata* | 9.05 | 22.05 | 28.05 | 5.06 | 16.06 | 18.06 | 9.07 | 3.10 | 12.10 | 26.10 | 172 |
| *M. officinalis* | 10.05 | 28.05 | 29.05 | 6.06 | 14.06 | 17.06 | 9.07 | 8.10 | 13.10 | 26.10 | 169 |
| *M. salicifolia* | 21.04 | 05.05 | 10.05 | 28.04 | 03.05 | 19.05 | 03.06 | 02.10 | 13.10 | 25.10 | 175 |
| *M. sieboldii* | 29.04 | 08.05 | 20.05 | 24.05 | 6.06 | 20.06 | 18.07 | 27.09 | 08.10 | 25.10 | 169 |
| *M. tripetala* | 7.05 | 19.05 | 5.05 | 13.06 | 19.06 | 20.06 | 12.07 | 27.09 | 9.10 | 27.10 | 175 |

*Pollen germination biology:*

Seed productivity of plants is known to be largely dependent on pollen viability. Pollen quality is governed by many factors, *viz.* species particulars, climatic conditions for growing and pollen maturity. Pollen germination results for seven taxa of *Magnolia* genus on various growth mediums are shown in Table 3. Pollen germination in laboratory conditions at temperature

18-20°C revealed that its viability does not exceed 4.8%. Because of that, further pollen germination studies were continued in a thermostat at 24°C. This temperature increase resulted in a higher percentage of pollen viability and longer pollen tubes (Fig. 1L).

Fig. 1. Development stages of the genus *Mognolia: Magnolia sieboldii* (A. budding; F. flower; H. fruit); *Magnolia officinalis* (B,C. budding; E,G. flowers; K. seeds); *Magnolia kobus* (D. flower); *Magnolia kobus* var. *borealis* (I. fruit; L. pollen); *Magnolia obovata* (J. fruit).

Pollen germination results for *Magnolia* taxa in a thermostat at 24°C on various growth mediums showed that the optimal medium is a 5% glucose solution in which the percentage of germinated pollen was at its highest and corresponded to the maximum length of pollen tubes. In general, the taxa being studied are characterized by low pollen viability from 9.4% to 31.7%. Among these species, highest viability data were registered for pollen of *M. kobus* var. *borealis* (31.7%) with a pollen tube length of 8.2 µm. Low pollen viability data were observed for *M. kobus* (9.4%), with its flowering period occurring in early May (9.7°C). These data are supported by Minchenko and Korshuk (1987) showing that the main reason for poor pollen viability (in a cultivated condition in Kiev, Ukraine) may be low temperatures during flowering period preventing complete maturation of pollen.

**Table 3. *Magnolia* pollen viability in different growth medium.**

| Taxa | Glucose concentration (%) | | |
|---|---|---|---|
| | 5 | 10 | 15 |
| *M. kobus* | 9.4±1.2 / 46±2.9 | 6.1±0.8 / 5.5±1.9 | 3.5±0.6 / 5.8±1.9 |
| *M. kobus* var. *borealis* | 31.7±2.8 / 34.3±2.6 | 10.2±2.6 / 5.8±1.8 | 5.2±0.6 / 4.7±1 |
| *M. obovata* | 11.3±1.4 / – | 6.1±0.9 / – | 3.4±0.6 / – |
| *M. officinalis* | 13.8±1.7 / – | 11.7±1.3 / – | 7±0.7 / – |
| *M. salicifolia* | 10.4±1.2 / 62±1.2 | 7.06±1.09 / 5.3±1.3 | 1.3±0.4 / 5.02±1.1 |
| *M. sieboldii* | 19.6±1.4 / 8.2±1.5 | 14.2±1.1 / 3.9±0.09 | 5.6±0.8 / 2.2±0.3 |
| *M. tripetalla* | - | - | - |

In the numerator - pollen viability (%), denominator - the length of pollen tubes (µm); «-» not examined.

Pollen ability to be stored during a long time is an important feature for hybridization studies. Results of pollen storage of *Magnolia* for different temperature conditions showed that pollen of all taxa being studied rapidly loses its viability at 18-20°C (Fig. 2). Highest pollen viability data were observed for *M. officinalis* (8%) when stored in a freezing cabinet at –18°C. However, this temperature of –18°C adversely affected pollen viability during storage for the rest taxa being studied. Therefore, an optimal pollen storage condition for the majority of species is a lowered above-zero temperature of 4°C. Pollen storage results for *Magnolia* species are confirmed by data of Minchenko and Korshuk (1987) who indicated that even five-day-long storage of magnolia pollen reduces its viability twice and more and that such pollen cannot be used for hybridization.

*Seed productivity:*
Results of seed productivity studies for seven taxa of *Magnolia* in a cultivated condition in the south of the Russian Far East are presented in Table 4. Potential seed productivity of the species being studied is defined by the number of seed buds per one fruit aggregate, being an upper limit of a species' seed productivity, and characterizes its potential that is little dependent on environmental conditions. Potential seed productivity varies depending on species and is equal to 48-210 ovules per one fruit aggregate. Real seed productivity was found by the number of seeds beginning to develop in a fruit. It amounted up to 55 seeds per one fruit aggregate which is much lower than potential seed productivity. Despite low real seed productivity, productivity rate for *M. sieboldii* and *M. tripetala* is above 50%. The lowest productivity rate was registered for *M. salicifolia* (1.7%).

**Table 4. Seed productivity of seven taxa of *Magnolia* in cultivated condition.**

| Taxa | PSP | | | RSP | | | PR (%) | | |
|---|---|---|---|---|---|---|---|---|---|
| | Min | Max | Mean | Min | Max | Mean | Min | Max | Mean |
| *M. kobus* | 26 | 66 | 74 | 9 | 31 | 11 | 15.4 | 61.3 | 14,8 |
| *M. kobus* var. *borealis* | 50 | 78 | 62 | 2 | 32 | 15 | 10 | 55.1 | 24,2 |
| *M. obovata* | 36 | 158 | 120 | 9 | 69 | 47 | 11.2 | 47.8 | 39.1 |
| *M. officinalis* | 166 | 240 | 210 | 6 | 65 | 31 | 8.4 | 25.3 | 14.7 |
| *M. salicifolia* | 40 | 55 | 60 | - | 1 | - | - | 1.7 | - |
| *M. sieboldii* | 26 | 52 | 48 | 6 | 39 | 28 | 13.4 | 64.2 | 58.2 |
| *M. tripetala* | 74 | 112 | 96 | 4 | 86 | 55 | 4.1 | 86.8 | 57.5 |

PSP = potential seed productivity, RSP = real seed productivity, PR = productivity rate.

Characteristics of seeds and fruits of *Magnolia* taxa are presented in Table 5. Large fruit aggregates (8.7-15.0 cm) and seeds (0.8-1.3 cm) are typical in *M. officinalis, M. obovata* and *M. tripetala*, while small ones (3.2 cm) are observed in *M. salicifolia* (Fig. 1H-K).

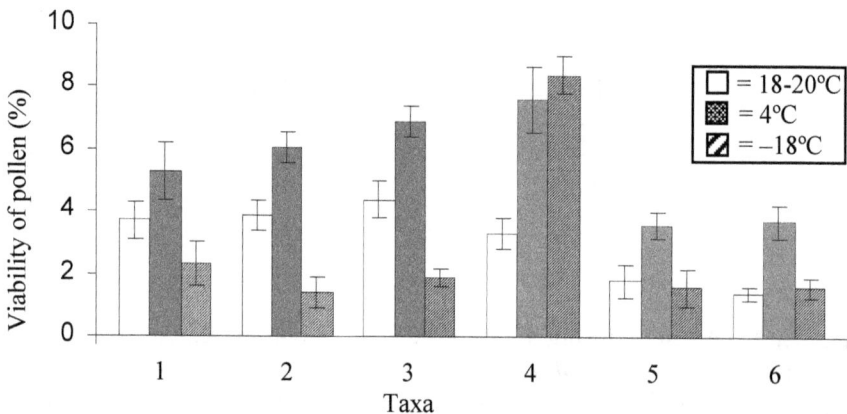

Fig. 2. Viability of *Magnolia* pollen after storage at different temperatures. 1. *M. kobus*, 2. *M. kobus* var. *borealis*, 3. M. *obovata*, 4. *M. officinalis*, 5. *M. salicifolia*, 6. *M. sieboldii*.

*Seed germination:*

Seed germination of *Magnolia* taxa in a cultivated condition in the south of Russian Far East showed that high germinating ability (67-94%) is typical in *M. tripetala, M. officinalis* and *M. obovata* (Table 6). Despite a high productivity rate of *M. kobus*, germinating ability of its seeds is low, 3.0% to 4.4%. Seed germination in the open ground and in the protected ground showed that optimal conditions for seeds of *M. tripetala, M. officinalis* and *M. obovata* are artificial stratification at 4°C during 30 days and further germination in the protected ground. Optimal conditions for *M. sieboldii* seeds are in the open ground (long-time stratification at 0°C to −27°C).

The present study revealed that a favourable effect on seed germinating ability was observed for *M. obovata* only 57.3% without sarcotesta and 94.3% with sarcotesta. No sarcotesta effects on seed germination were registered for other species.

**Table 5. Morphometric characteristics of the fruits and seeds of seven taxa of *Magnolia*.**

| Taxa | Fruit | | | Seed | | | |
|---|---|---|---|---|---|---|---|
| | Length (cm) | Width (cm) | Weight (g) | Length (cm) | Width (cm) | Weight (S) 1000 (g) | Weight (WS) 1000 (g) |
| *M. kobus* | 7.2 ± 0.8 | 2.1 ± 0.06 | 9.5 ± 0.7 | 0.9 ± 0.01 | 0.8 ± 0.02 | 323 | 169 |
| *M. kobus* var. *borealis* | 4.5 ± 0.1 | 2 ± 0.1 | 8.9 ± 0.9 | 0.8 ± 0.02 | 0.7 ± 0.02 | 226.9 | 122 |
| *M. obovata* | 8.7 ± 0.3 | 4.6 ± 0.12 | 43.3 ± 3.3 | 1.1 ± 0.02 | 0.9 ± 0.01 | 262.5 | 152.6 |
| *M. officinalis* | 15 ± 0.5 | 4.9 ± 0.07 | 108.8 ± 8.5 | 1.3 ± 0.01 | 1.1 ± 0.02 | 528.2 | 399.3 |
| *M. salicifolia* | 3.2 ± 0.7 | 1.2 ± 0.6 | 2.2 ± 0.4 | 0.8 ± 0.01 | 0.7 ± 0.01 | – | – |
| *M. sieboldii* | 5.3 ± 0.9 | 2.07 ± 0.02 | 3.4 ± 0.1 | 0.5 ± 0.02 | 0.48 ± 0 | 52.7 | 39.5 |
| *M. tripetala* | 7.2 ± 0.2 | 3.9 ± 0.09 | 40.3 ± 2.2 | 0.8 ± 0.02 | 0.7 ± 0 | 151.98 | 99.5 |

S = seeds with sarkotesta, WS = seeds without sarkotesta, – = not examined.

**Table 6. Seed germination of seven taxa of *Magnolia* in different condition.**

| Taxa | % of seeds germination in greenhouse | | % of seed germination in the open ground | |
|---|---|---|---|---|
| | with sarcotesta | without sarcotesta | with sarcotesta | without sarcotesta |
| *M. kobus* | 3 | 0 | 4.4 ± 0.1 | 0 |
| *M. kobus* var. *borealis* | 0 | 0 | 0 | 0 |
| *M. obovata* | 94.3 ± 5.7 | 57.3 ± 4.7 | 59.8 ± 12.6 | 45.5 ± 7.8 |
| *M. officinalis* | 70.1 ± 11,8 | 47.3 ± 1,3 | 61±5.9 | 45.5 ± 6.1 |
| *M. salicifolia* | – | – | – | – |
| *M. sieboldii* | 37.7 ± 2.3 | 56.7 ± 5.7 | 42.4 ± 3.9 | 52 ± 6.1 |
| *M. tripetala* | 67.3 ± 11.6 | 56 ± 5.8 | 66.6 ± 15.04 | 47.8 ± 9.4 |

'0' denotes not germinated, '–' denotes not examined.

The present study addresses phenology of seven taxa of the genus *Magnolia* L., pollen biology of germination and storage conditions, seed productivity, germinating ability of seeds and its dependence on stratification and germination conditions. It was found that the flowering period of the taxa being studied occurs in May – June at a time of low air temperatures varying from 9°C to 14.5°C, which has adverse effects on pollen viability. Due to this circumstance, *M. kobus, M. obovata, M. officinalis, M. tripetala, M. salicifolia* and *M. sieboldii* are characterized by low pollen fertility (9.4-31.7%) in the south of the Russian Far East which affects their seed productivity. Real seed productivity of these taxa is considerably lower (1-55 seeds per one fruit aggregate) than their potential seed productivity (48-210 ovules per one fruit aggregate). Despite their low pollen viability and productivity rate, *M. tripetala, M. officinalis* and *M. obovata* in a cultivated condition produce seeds with high germinating ability (67-94%). At the same time, *M. sieboldii* features a high productivity rate but low seed germinating ability. Seed germination results showed that optimal conditions for species with high germinating ability such as *M. tripetala, M. officinalis* and *M. obovata* are artificial stratification, presence of sarcotesta and protected ground conditions. We believe that low germinating ability of *M. sieboldii* and *M. kobus* in the protected ground can be explained by stratification duration and temperature. Sarcotesta effects on higher seed geminating ability were observed for *M. obovata* only.

## References

Barkalov, V.Y. 2009. Flora of the Kuril Islands. Dalnauka, Vladivostok. 54 pp.

Golubinsky, I.N. 1974. Biology of germination of pollen. Naukova Dumka, Kiev. 368 pp.

Grigorenko, I.V. 1998. Flowering ecology of the family Magnoliaceae Juss. in the industrial city of the South-east of Ukraine. Problems Dendrology, floriculture, horticulture. Abstracts of the VI Confer. Yalta. pp. 12-17.

Han, C.-Y. and Long, C.-L. 2010. Seed dormancy, germination and storage behavior of *Magnolia wilsonii* (Magnoliaceae), an endangered plant in China. Acta Bot. Yun. **32**(1): 47-52.

Hirayama, K. and Ishida, K. 2005. Effect of pollen shortage and self-pollination on seed production of an endangered tree *Magnolia stellata*. Ann. Bot. **95**(6): 1009-1015.

Ishida, K. 1996. Beetle pollination of *Magnolia praecocissima* var. *borealis*. Plant Species Biology **11**: 199-206.

Kikuzawa, K. and Mizui, N. 1990. Flowering and fruiting phenology *Magnolia hypoleuca*. Plant Species Biology **5**: 255-261.

Koksheeva, I.M. 2004. On the methods of determining the viability of the pollen in the genus *Rhododendron* L. (Ericaceae). Bot. Journal **89**(6): 147-150.

Korshuk, T.P. and Palagecha, R.M. 2007. *Magnolia* L. Kiev University. Ukraine. 207 pp.

Lapin, P.I. 1967. Seasonal rhythm of woody plants and its significance for the introduction. Bull. GBS. **65**:13-18.

Law, Y.W. 1996. Magnoliaceae. Flora Reipublicae Popul256 Sinicae **30**(1): 151-194.

Liza, S.A., Rahman, M.O., Uddin, M.Z., Hassan, M.A. and Begum, M. 2010. Reproductive biology of three medicinal plants. Bangladesh J. Plant Taxon. **17**(1): 69-78.

Minchenko, N.F. and Korshuk, T.P. 1987. *Magnolias* in Ukraine. Nauk. Dumka, Kiev. 184 pp.

Pshennikova, L.M. 2007. Lilac, cultivated in the Botanical Garden-Institute FEB RAS. Dalnauka, Vladivostok. 113 pp.

Rabotnov, T.A. 1960. Methods of study of seed reproduction of herbaceous plants in the communities. Geobotany field. Publishing House of the Academy of Sciences of the USSR, Moscow-Leningrad **2**: 20-40.

Setsuko, S., Tamaki, I., Ishida, K., and Torami, N. 2008. Relationships between flowering phenology and female reproductive success in the Japanese tree species *Magnolia stellata*. Botany **86**: 248-258.

Termena, B.K. 1972. About flowers and fruits of *Magnolia soulangia* in Bukovina. Bull. GBS. **84**: 82-84.

Turkenya, V.G. 1991. Biological aspects of the monsoon climate zone of the Far East. Vladivostok: Far Eastern Branch of the Academy of Sciences of the USSR. 203 pp.

Wang, R. 2010. Flowering and pollination patterns of *Magnolia denudata* with emphasis on anatomical changes in ovule end seed development. Flora **205**: 269-265.

# 11

# DIVERSITY OF STOMATA AND TRICHOMES IN *EUPHORBIA* L. - I

N. Sarojini Devi, Y. Padma, C.L. Narasimhudu and R.R. Venkata Raju[1]

*Biosystematics and Phytomedicine Division, Department of Botany,*
*Sri Krishnadevaraya University, Anantapur - 515 055, Andhra Pradesh, India*

*Keywords:* Euphorbiaceae; *Euphorbia*; Epidermal studies; Classification.

## Abstract

Foliar epidermal features of 18 species of *Euphorbia* L. *s.l.* (Euphorbiaceae) are studied. While the anisocytic and anamocytic stomata are common in herbaceous members (*Euphorbia* subg. *Chamaesyce*), the paracytic type is predominant in succulent species (*Euphorbia* proper). The stomatal types, index and frequency, and the types of trichomes are explored on vegetative as well as floral parts to evaluate their possible taxonomic importance.

## Introduction

Sufficient interest seems to have been revived during the past two decades on the role of internal organization of the individual organs of plants. Leaves occupy a prominent position in this regard and their various features such as venation, stomata and trichomes were found useful in solving taxonomic and phylogenetic issues. The utility of foliar epidermal features in distinguishing taxonomic groups was clearly established (Stace, 1965, 1984; Dilcher, 1974; Raju, 1981; Rao and Raju,1985, 1988; Mohan, 1994; Manohari, 2004).

In view of the above considerations, it was thought worthwhile to investigate the epidermology of the important genus *Euphorbia* L. *s.l.,* with 84 species occurring in India (Binojkumar and Balakrishnan, 2007, 2010). The great diversity in habit and adaptation exhibited by the species of the genus *Euphorbia* provide the added impetus for undertaking the present study. The structural diversity and distribution of trichomes are significant for taxonomic analysis, especially in tropical plants (Stace, 1965; Dilcher, 1974; Rao and Raju, 1985). The distribution and structure of trichomes and stomata are genetically controlled and consistent. The useful epidermal characters in systematics are the distribution of stomata over the two surfaces of the leaf, stomatal index and frequency, the nature of anticlinal walls of the epidermal cells and the types and distribution of trichomes. To fill the gaps in our knowledge in this regard concerning the genus *Euphorbia* sensu Linnaeus in India, the present study was undertaken. It is also intended to view whether these data support the realignments in the genus *Euphorbia* by Yang *et al.* (2012) with regard to the traditional subgenera *Eremophyton* and *Poinsettia*.

## Materials and Methods

In the present investigation, 18 species of *Euphorbia* L. *s.l.* were studied for the organography of epidermal features (Table 1). The plant materials used in the present study were mostly shade dried and of pressed specimens. Whole plants were boiled in water with a few pellets (1 g) of NaOH at 30-40°C. The peels were thoroughly washed and stained with safranin or acetocarmine and mounted in glycerin. The epidermal features such as cuticle, epidermal cells, stomata, and trichomes were observed on all the organs and measured using Nikon Eclipse, E-400 microscope.

[1]Corresponding author. Email: rrvenkataraju@yahoo.com

Ocular micrometer was used for measurement. The terminology adopted for the epidermal features is after Raju and Rao (1977, 1987) and Raju (1981). The species of *Euphorbia* studied were identified with the help of standard regional floras and experts (Ellis, 1990; Venkataraju and Pullaiah, 1995; Babu, 1995; Binojkumar and Balakrishnan, 2010), and the voucher specimens were deposited in the Sri Krishnadevaraya University Herbarium (SKU), Anantapur.

## Results

The epidermis in *Euphorbia* bears a variety of trichomes and stomata dispersed all over the plant surface in a consistent pattern. The extent of variations in the shape of epidermal cells, nature of anticlinal walls, types of stomata, number of stomata per unit area and stomatal indices, and the types of trichomes (glandular and eglandular) observed in 18 species are documented in Tables 1 and 2.

Eleven species of *Euphorbia* bear trichomes on vegetative and floral parts and the remaining seven species are glabrous. The epidermal appendages vary in structure, form and distribution. The trichome types presently recorded in the members of *Euphorbia* are mostly multicellular. However, *Euphorbia agowensis* (Fig. 1A) bears unicellular cylindrical trichomes on vegetative as well as floral parts. A noteworthy feature of *Euphorbia* species currently recorded is the presence of glandular trichomes at nodes and bases of stipules. These can be found in *E. cristata* (Fig. 3A), *E. elegans* (Fig. 3B), *E. hirta* (Fig. 3E), *E. indica* (Fig. 3G), and *E. prostrata* (Fig. 3D) of subg. *Chamaesyce* and *E. heterophylla* (Fig. 3C) of subg. *Poinsettia,* which are often segregated as distinct genera. These glandular trichomes are multicellular, stalked and found to be species specific in regard to their size, shape and number per unit area. The stem of *E. hirta* (Fig. 2G) possesses two types of trichomes; they are multicellular forked with cuticular ornamentations and multicellular uniseriate osteolate ones. However, its involucre bears strictly two-celled cylindrical trichomes (Fig. 2H).

The epidermal cells may be rectangular and polygonal in outline. Depending upon the location on the leaf, i.e., the midrib, margin and apex, their shapes tend to vary. The anticlinal walls are straight to variously arcuate.

The anticlinal walls are straight in *E. dracunculoides, E. deccanensis* var. *nallamalayana* (Fig. 1M), *E. nivulia* (Fig.1H), *E. perbracteata* (Fig. 1L, adaxial) and *E. tirucalli* (Fig. 1K). Most of the stomatal types noted for Dicotyledonae (Magnoliopsida) are met within the genus, with anamocytic, anisocytic and paracytic being most preponderant or basic (Sehgal and Paliwal, 1974; Raju and Rao 1977, 1987). Stomata of more than one types have been encountered on the same leaf surface in *E. hyssopifolia, E. longistyla* (Fig. 1I) and *E. thymifolia* (Fig. 1J), as reported earlier by Raju and Rao (1977, 1987) for the other species of *Euphorbia*. In *E. caducifolia* (Fig. 1B) at places, the stomata are just represented by persistent stomatal initials (due to arrested stomatal development c.f. Raju and Rao, 1977). In *E. tirucalli,* sometimes, the stomatal complex has single guard cell. With regard to the position of stomata to the level of epidermis, different depths of sunkenness have been observed in various succulent *Euphorbia* species (Fig. 1B, H). Obviously, these epidermal features can be usefully employed for diagnostic purposes in *Euphorbia*.

## Discussion

The foliar epidermis offers a number of noteworthy taxonomic characters. The biosystematic and taxonomic studies of a number of families established the importence of leaf epidermis (Baranova, 1972; Raju, 1981; Stace, 1984). Although the taxonomists realized lately the importance of micromorphology of the epidermis, the taxonomic monographs are now considered

**Table 1. Analysis of foliar stomatal diversity in 18 taxa of *Euphorbia* L.**

| No. | Taxon | | Stomatal types | Percentage of richness | Stomatal index | | Stomatal frequency | | Source and voucher specimen number |
|---|---|---|---|---|---|---|---|---|---|
| | | | | | Adaxial | Abaxial | Adaxial ($\mu m^2$) | Abaxial ($\mu m^2$) | |
| | Subg. **Esula** | | | | | | | | |
| 1 | *E. dracunculoides* Lam. | | Anamocytic | 54% | 11.11 | 16.96 | 0.004 | 0.006 | Sanjamula, Kurnool (31412) |
| | | | Anisocytic | 32% | | | | | |
| | | | Paracytic | 14% | | | | | |
| 2 | *E. perbracteata* Gage | Vegetative leaf | Anamocytic | 66% | 20.312 | 17.30 | 0.009 | 0.006 | Sanjamula, Kurnool (31411) |
| | | | Anisocytic | 44% | | | | | |
| | | Floral leaf | Anisocytic | 58% | 18.00 | 18.35 | 0.006 | 0.007 | |
| | | | Anamocytic | 42% | | | | | |
| | Subg. **Euphorbia** L. | | | | | | | | |
| 3 | *E. caducifolia* Haines | | Paracytic | 100% | 8.571 | 7.691 | 0.082 | 0.692 | Rachakunta palle, Kadapa (31416) |
| 4 | *E. nivulia* Buch.-Ham. | | Paracytic | 100% | 8.00 | 11.11 | 0.002 | 0.003 | Srisailam, Kurnool (31417) |
| 5 | *E. tirucalli* L. | | Paracytic | 100% | 9.836 | 13.432 | 0.826 | 0.123 | Ramagiri, Anantapur (31418) |
| | Subg. **Chamaesyce** Raf. | | | | | | | | |
| | Sect. **Scatorhizae** Y. Yang & P.E. Berry | | | | | | | | |
| 6 | *E. agowensis* Hochst *ex* Boiss. | | Paracytic | 82% | 21.608 | 22.608 | 0.130 | 0.179 | Srisailam, Kurnool (31410) |
| | | | Anisocytic | 18% | | | | | |
| | Sect. **Anisophyllum** Roep. | | | | | | | | |
| 7 | *E. deccanensis* V.S. Raju var. nallamalayana (Ellis) V.S. Raju | | Anamocytic | 73% | 15.789 | 18.852 | 0.061 | 0.096 | Nallamalais, Kurnool (31402) |
| | | | Anisocytic | 27% | | | | | |
| 8 | *E. corrigioloides* Boiss. | | Anisocytic | 66% | 17.89 | 18.584 | 0.006 | 0.007 | Vemula, Kadapa (31415) |
| | | | Anamocytic | 44% | | | | | |
| 9 | *E. cristata* Heyne *ex* Roth | | Anisocytic | 81% | 18.584 | 14.893 | 0.144 | 0.096 | Gooty hills, Anantapur (31409) |
| | | | Anamocytic | 19% | | | | | |
| 10 | *E. elegans* Spreng. | | Anisocytic | 57% | 10.344 | 11.578 | 0.061 | 0.075 | Chelama, Kurnool (31404) |
| | | | Anamocytic | 43% | | | | | |

**Table 1 contd.**

| No. | Taxon | Stomatal types | Percentage of richness | Stomatal index | | Stomatal frequency | | Source and voucher specimen number |
|---|---|---|---|---|---|---|---|---|
| | | | | Adaxial | Abaxial | Adaxial ($\mu m^2$) | Abaxial ($\mu m^2$) | |
| 11 | *E. hirta* L. | Anisocytic<br>Anamocytic | 60%<br>40% | 21.568 | 22.707 | 0.227 | 0.289 | Anantapur (31407) |
| 12 | *E. hyssopifolia* L | Anisocytic<br>Anamocytic<br>Paracytic | 79%<br>13%<br>8% | 8.108 | 9.333 | 0.061 | 0.096 | Palamaneru, Chittoor (31407) |
| 13 | *E. indica* Lam. | Anisocytic<br>Anamocytic | 88%<br>12% | 15.714 | 21.787 | 0.151 | 0.268 | Sanjamula, Kurnool (31420) |
| 14 | *E. longistyla* (Boiss.) Croizat | Anisocytic<br>Anamocytic<br>Diacytic | 92%<br>5%<br>3% | 10.769 | 12.50 | 0.048 | 0.061 | Palakonda hills, Kadapa (31406) |
| 15 | *E. prostrata* Aiton | Anisocytic<br>Anamocytic | 58%<br>42% | 22.535 | 20.37 | 0.220 | 0.151 | Batrepalle, Anantapur (31401) |
| 16 | *E. serpens* Kunth | Anisocytic<br>Anamocytic<br>Paracytic | 66%<br>23%<br>11% | 19.727 | 20.43 | 0.199 | 0.261 | Kalasamudram, Anantapur (31421) |
| 17 | *E. thymifolia* L. | Anisocytic<br>Diacytic<br>Paracytic | 58%<br>22%<br>20% | 20.987 | 17.62 | 0.234 | 0.144 | S.K. University campus, Anantapur (31422) |
| 18 | Sect. **Poinsettia** (Graham) Baill.<br>*E. heterophylla* L. | Anamocytic<br>Anisocytic | 72%<br>28% | 16.097 | 19.923 | 0.022 | 0.358 | S.K. University campus, Anantapur (31423) |

Infrageneric classification after Binojkumar and Balakrishnan (2010), and Yang *et al.* (2012).

**Table 2. Diversity and distribution of trichomes of 18 taxa of *Euphorbia* L.**

| Taxon | Plant part | Function of Trichomes | Trichome type | Size |
|---|---|---|---|---|
| **Subg. Esula** | | | | |
| 1. *E. dracunculoides* | | absent on all the organs | | |
| 2. *E. perbracteata* | Floral leaf adaxial | Eglandular trichomes | Multicellular uniseriate cylindrical | 0.4 mm |
| | Floral leaf abaxial | " | " | 0.6 mm |
| | Capsule | " | Unicellular cylindrical | 0.8-0.9 mm |
| **Subg. Euphorbia** | | | | |
| 3. *E. caducifolia* | | Absent on all the organs | | |
| 4. *E. nivulia* | | Absent on all the organs | | |
| 5. *E. tirucalli* | Capsule | Eglandular trichomes | Multicellular uniseriate cylindrical | 1.5-1.9 mm |
| **Subg. Chamaesyce** | | | | |
| **Sect. Scatorhizae** | Stem | Eglandular | Unicellular cylindrical | 189-191 μm |
| | Leaf abaxial | " | " | 190-200 μm |
| 6. *E. agowensis* | Involucre | " | " | 183-184.5μm |
| | Capsule | " | " | 172.8-176.2 μm |
| **Sect. Anisophyllum** | | | | |
| 7. *E. deccanensis* var. *nallamalayana* | | Absent on all the organs | | |
| 8. *E. corrigioloides* | Stem | Eglandular trichomes | Multicellular cylindrical | 75.6 μm |
| | Leaf abaxial | " | Unicellular cylindrical | 56.7 μm |
| | Involucre | " | " | 59.4 μm |
| | Capsule | " | " | 62.1 μm |
| 9. *E. cristata* | Stem | Eglandular | Multicellular uniseriate cylindrical | 1.2 mm |
| | Stipular base | Glandular | Multicellular stalked | 0.8-1.0 mm |
| | Leaf adaxial | Eglandular | Multicellular osteolate | 1.0-1.2 mm |
| | Leaf abaxial | " | " | 1.1-1.3 mm |
| | Involucre | " | Multicellular cylindrical | 0.9 mm |
| | Capsule | " | " | 1.0-1.1 mm |
| 10. *E. elegans* | Stem | Eglandular | Multicellular uniseriate cylindrical | 2 mm |
| | Stipular base | Glandular | Multicellular stalked | 0.4 mm |
| | Leaf abaxial | Eglandular | Multicellular uniseriate cylindrical | 0.9-1.0 mm |
| | Involucre | " | " | 0.6-0.8 mm |
| | Capsule | " | " | 0.8-1.2 mm |

**Table 2 contd.**

| Taxon | Plant part | Function of Trichomes | Trichome type | Size |
|---|---|---|---|---|
| 11. *E. hirta* | Stem | Eglandular trichomes | Multicellular uniseriate osteolate and multicellular biforked osteolate trichomes with cuticular ornamentation | 0.7-1.2 mm & 0.8-1.3 mm |
| | Stipule | " | Multicellular uniseriate osteolate | 0.4 mm |
| | Leaf adaxial | " | " | 0.4-0.9 mm |
| | Leaf abaxial | " | " | 0.4-0.9 mm |
| | Involucre | " | Two celled uniseriate cylindrical | 73-78 μm |
| | Capsule | " | " | 60 μm |
| | Stipular base | Glandular trichomes | Multicellular stalked | 0.6-0.8 mm |
| 12. *E. hyssopifolia* | | | Absent on all the organs | |
| 13. *E. indica* | Stem | Eglandular | Multicellular uniseriate osteolate | 0.2-0.3 mm |
| | Leaf abaxial | " | " | 0.3 mm |
| | Capsule | " | " | 0.4 mm |
| | Involucre | " | " | 0.5 mm |
| | Stipular base | Glandular | Multicellular stalked | 0.4-1.2 mm |
| 14. *E. longistyla* | Petiole | Eglandular | Multicellular uniseriate cylindrical | 0.1 mm |
| 15. *E. prostrata* | Stem | Eglandular | Multicellular uniseriate cylindrical | 0.2-0.4 mm |
| | Capsule | " | " | 81-88 μm |
| | Leaf abaxial | " | Multicellular uniseriate osteolate | 0.7-0.9 mm |
| | Stipule | " | " | 1.0-1.2 mm |
| | Involucre | " | " | 86.4-87.0 μm |
| | Stipular base | Glandular | Multicellular stalked | 161 μm |
| 16. *E. serpens* | | | Absent on all the organs | |
| 17. *E. thymifolia* | Stem | Eglandular | Multicellular uniseriate cylindrical | 0.8-1.0 mm |
| | Leaf abaxial | " | " | 0.8 mm |
| | Involucre | " | " | 0.6 mm |
| | Capsule | " | " | 0.9 mm |
| Sect. *Poinsettia* | | | | |
| 18. *E. heterophylla* | Stem | Eglandular | Multicellular uniseriate cylindrical | 0.3-0.6 mm |
| | Petiole | " | Multicellular uniseriate osteolate | 0.6-0.8 mm |
| | Leaf adaxial | " | " | 0.8-1.0 mm |
| | Leaf abaxial | " | " | 0.9-1.0 mm |
| | Node | Glandular | Multicellular stalked | 0.4 mm |

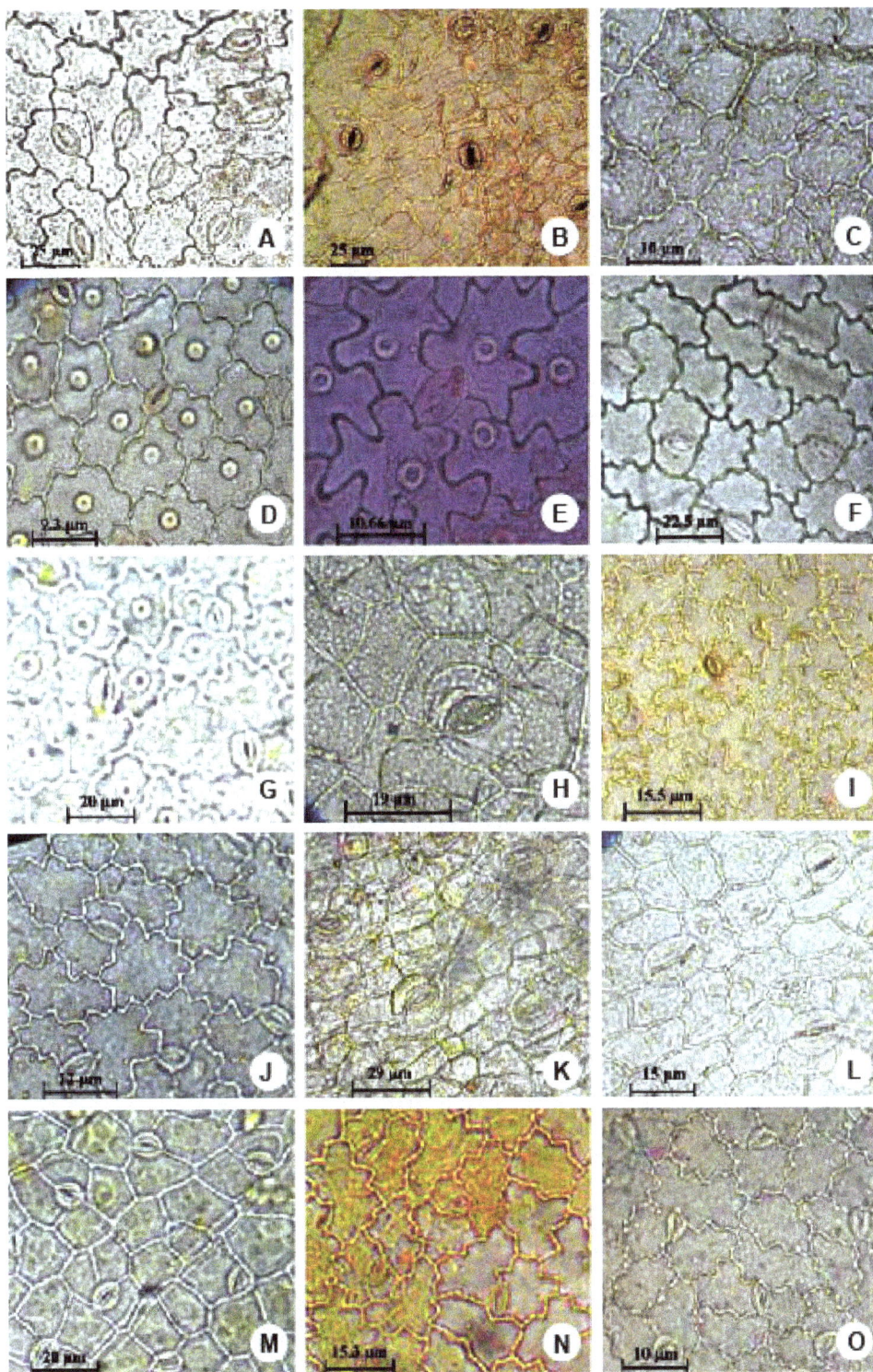

Fig. 1. Organography of epidermal structures in *Euphorbia*. A) *E. agowensis*, B) *E. caducifolia*, C) *E. corrigioloides,* D) *E. cristata*, E) *E. elegans*, F) *E. heterophylla*, G) *E. indica*, H) *E. nivulia*, I) *E. longistyla*, J) *E. thymifolia*, K) *E. tirucalli*, L) *E. perbracteata*, M) *E. deccanensis* var. *nallamalayana*, N) *E. prostrata*, O) *E. serpens*.

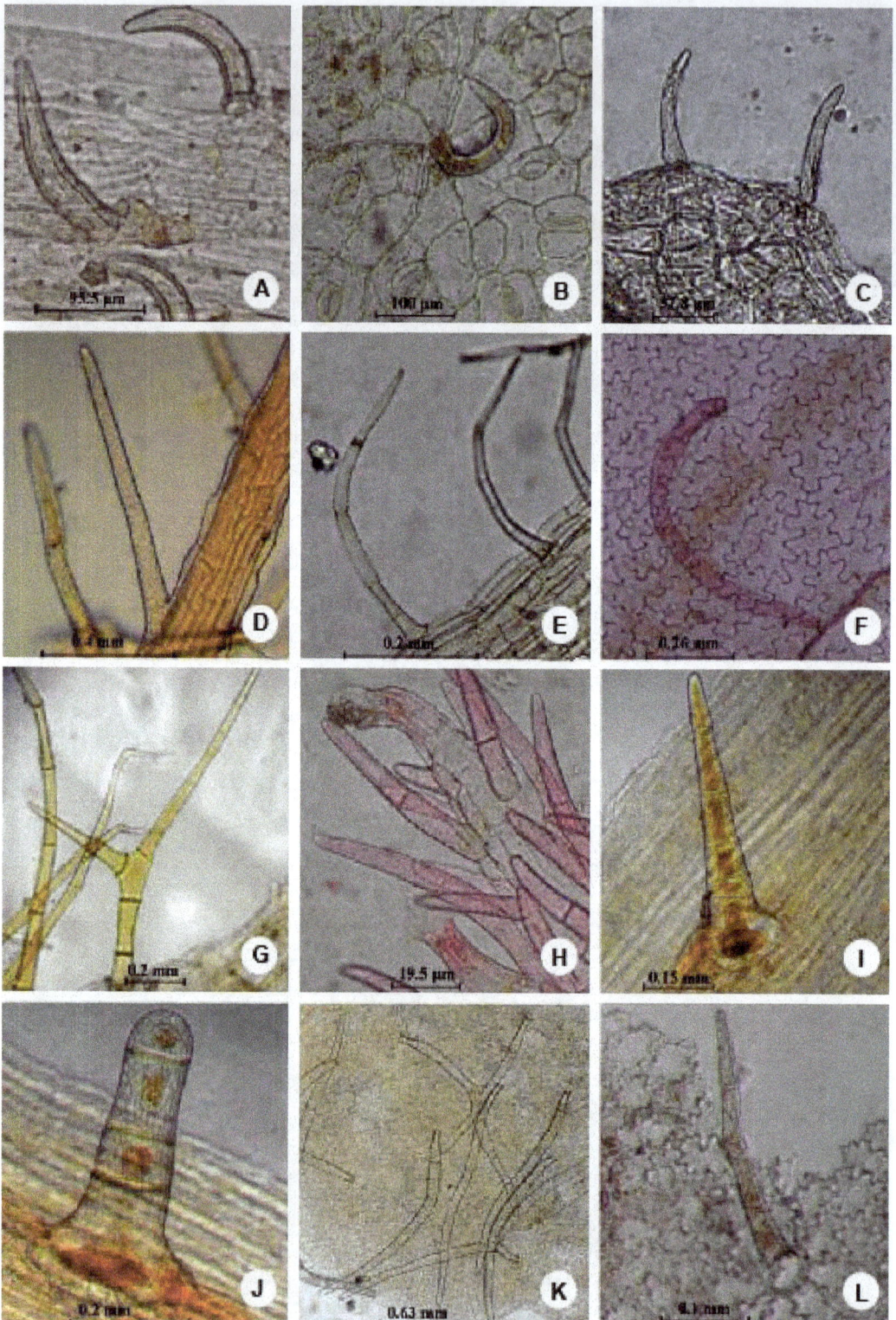

Fig. 2. Trichome diversity in *Euphorbia*. A) & B) *E. agowensis* (stem & leaf), C) *E. corrigioloides* (stem), D) *E. cristata* (appendages), E) *E. prostrata* (stem), F) *E. elegans* (leaf abaxial), G) & H) *E. hirta* (stem & involucre),  I) & J) *E. heterophylla* (stem), K) *E. tirucalli* (capsule), L) *E. indica* (leaf abaxial).

incomplete without it (Rejdali, 1991). The diversity and distributional pattern of stomata and trichomes can be viewed from different perspectives and used as a model system for investigations into developmental biology, ecology, physiology, morphology and evolution.

The work done on the stomata and trichomes was well-documented by Metcalfe and Chalk (1950) and reviewed by Raju and Rao (1977) and Rao and Raju (1985, 1988). The present study fills up the gaps in our knowledge of the other species of the genus *Euphorbia* after Sehgal and Paliwal (1974), Raju and Rao (1977) and Raju (1981) in India. Different species of *Euphorbia* have been found to possess anamocytic, anisocytic and paracytic types of stomata, indicating that Linnean *Euphorbia* is heterogeneous. Therefore, this diversity is of use for infrageneric delimitation. The species of *Euphorbia* subg. *Chamaesyce* such as *E. hyssopifolia*, *E. longistyla* and *E. thymifolia* showed combinations of two or more types of stomata on the same leaf surface. Despite the variation, a single stomatal type is preponderant in any particular Euphorbiaceous taxon (Raju and Rao, 1977). In the case of *Euphorbia* subg. *Chamaesyce*, it is the trilabrate anisomesogenous type. Anisocytic stomata are dominant (Table 1) in the foliar epidermis of *E. longistyla* (92%) followed by *E. indica* (88%), *E. cristata* (81%) and *E. prostrata* (73%). While anamocytic stomata are predominantly found in *E. deccanensis* var. *nallamalayana* (73%), *E. heterophylla* (72%), *E. perbracteata* (66%) and *E. dracunculoides* (54%), paracytic stomata are preponderant in tree species and shrubs like *E. caducifolia*, *E. nivulia* and *E. tirucalli* (Table 1).

Papillate epidermal cells were found in the abaxial surface of leaves of *Euphorbia* subg. *Chamaesyce*, as seen in *E. cristata*, *E. elegans* (Fig. 1D, E) and *E. indica* (Fig. 1G). In *E. perbracteata*, the anticlinal walls are straight in the adaxial foliar epidermis while they are undulate to highly wavy abaxially, as noted in the other Euphorbiaceae (Raju and Rao, 1977). Similarly, more than seven types of eglandular trichomes and three types of glandular trichomes are found on vegetative and floral parts of the Linnean *Euphorbia*.

## Taxonomic Treatment

The diversity in stomata and trichomes is useful for infrageneric distinctions. However, their importance as taxonomic criteria will be greatly enhanced if the information can be interpreted with supportive evidence.

Binojkumar and Balakrishnan (2010) recognised 10 subgenera under *Euphorbia* for the Indian species. The species of *Euphorbia* studied now belong to five subgenera, *viz.*, *Chamaesyce*, *Eremophyton*, *Esula*, *Euphorbia* and *Poinsettia*. *Euphorbia* subg. *Chamaesyce* exibits distinct taxonomic features like varied forms of glandular trichomes at stipular bases, more than four types of eglandular trichomes and four types of stomata, whereas the subg. *Eremophyton* is distinct from the other groups by bearing unicellular trichomes on all vegetative and floral parts, and the predominant paracytic stomata. The subg. *Esula* shows two stomatal types, the abaxial anticlinal walls wavy and adaxial ones straight, while the subg. *Euphorbia* is characterized by paracytic and sunken stomata. The subg. *Poinsettia* exhibits two types of trichomes (eglandular and glandular) and the predominant advanced anamocytic stomata. Yang *et al.* (2012) based on molecular evidence, re-circumscribed the genus *Euphorbia* subg. *Chamaesyce*. They reduced the traditional subgenera *Eremophyton* and *Poinsettia* as sections under *Euphorbia* subg. *Chamaesyce*. *E. agowensis* was placed under subg. *Euphorbia* sect. *Scatorhizae*. However, *E. agowensis* is not allied to the core sect. *Anisophyllum* in its basic stomata and trichome types besides being ecarunculate and non-kranz species (Tables 1 & 2). As a section, *Poinsettia* (*E. heterophylla*) also makes the subg. *Chamaesyce* heterogeneous with its species bearing coloured floral bracts, basic anamocytic stomata (Table 1) and two types of trichomes on the stem (Table 2). Therefore, the micromorphological evidence supplemented with other morphological data are not in agreement with the re-alignment made for these two sections by Yang *et al.* (2012), and instead, the data are

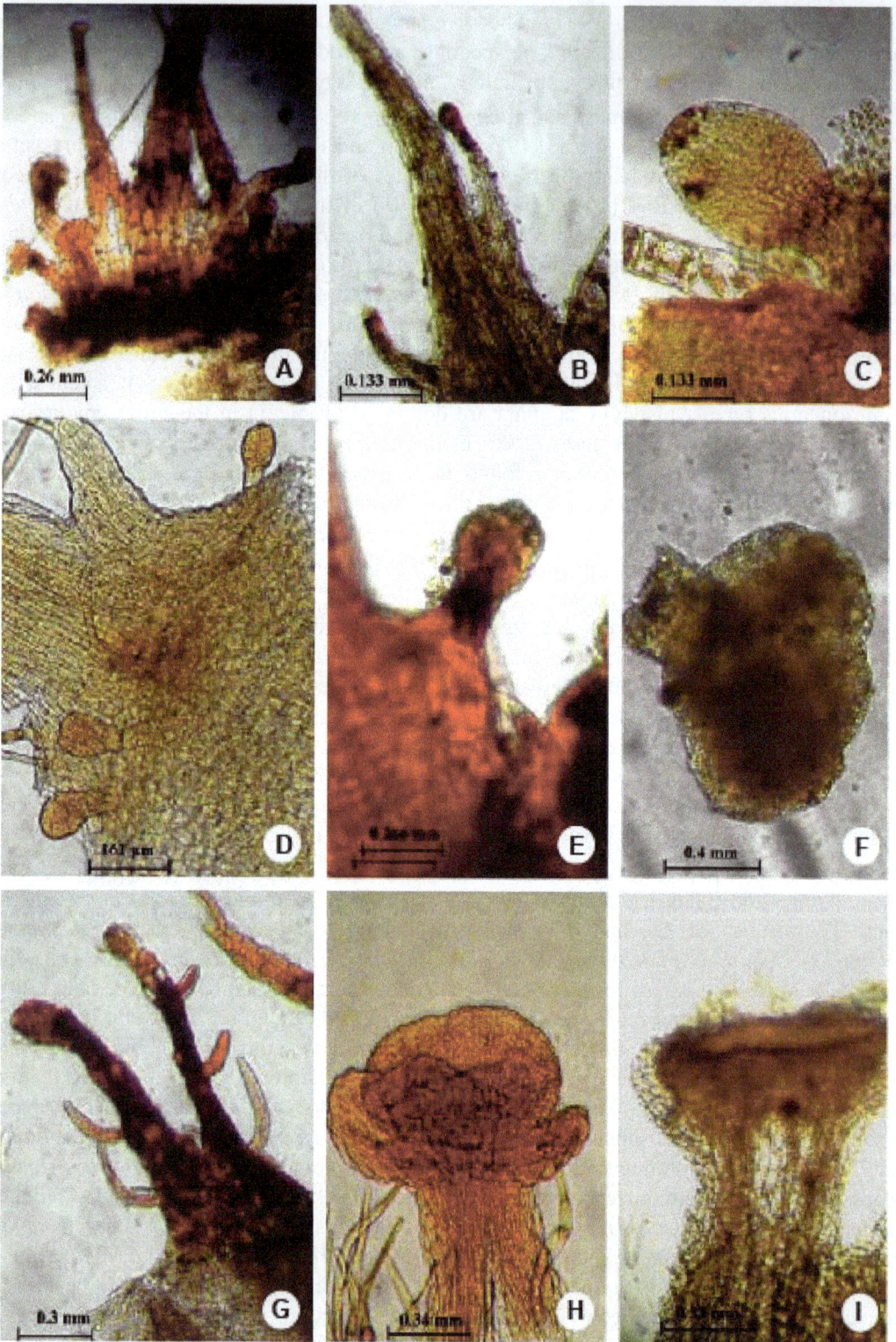

Fig. 3. Vegetative and floral glands in *Euphorbia*. A) *E. cristata*, B) *E. elegans*, C) *E. heterophylla*, D) *E. prostrata*, E) *E. hirta*, F) *E. agowensis* (leaf), G) *E. indica*, H) *E. prostrata* (involucral gland), I) *E. serpens* (involucral gland).

compatible with the traditional treatment adopted by Binojkumar and Balakrishanan (2010), for the species examined.

The present study reveals that the epidermal characters are of taxonomic significance in the members of the *Euphorbia* examined. Despite the fact that the epidermis is being influenced by environmental factors, the traits employed are stable with regard to the mature stomatal type and distribution on different organs. Therefore, the stomata, trichomes and epidermal cells can be effectively used to identify and distinguish different plant species and draw parallels or convergence with the molecular evidence.

## Acknowledgements

The authors are grateful to Prof. Vatsavaya S. Raju, Retired Professor, Department of Botany, Kakatiya University, Warangal, Andhra Pradesh, for his encouragement and inputs and the Forest Officials of Andhra Pradesh for their cooperation and help during the field work. The first author is grateful to the University Grants Commission, New Delhi for financial assistance.

## References

Babu, P.S.P. 1995. Euphorbiaceae of Andhra Pradesh, India. Ph.D. thesis. Sri Krishnadevaraya University, Anantapur, India.

Baranova, M. 1972. Systematic anatomy of the leaf epidermis in the Magnoliaceae and some related families. Taxon **21**: 447-469.

Binojkumar, M.S. and Balakrishnan, N.P. 2007. *Euphorbia. In*: Balakrishnan, N.P. and Chakrabarty, T. (Eds), The Family Euphorbiaceae in India, A Synopsis of its Profile, Taxonomy and Bibliography. Bishen Singh Mahendra Pal Singh, Dehra Dun, India. pp. 238-281.

Binojkumar, M.S. and Balakrishnan, N.P. 2010. The genus *Euphorbia* L. (Euphorbiaceae) in India: A Taxonomic Revision. Bishen Singh Mahendra Pal Singh, Dehra Dun, India. pp. 1-430.

Dilcher, D.L. 1974. Approaches to the identification of angiosperm leaf remains. Bot. Rev. **40**: 1-157.

Ellis, J.L. 1990. Flora of Nallamalais. Vol. **2**. Botanical Survey of India, Calcutta. pp. 351-357.

Manohari, A.L.S. 2004. Studies on the Foliar Epidermology, Architecture and Anatomy of some Apocynaceae. Ph.D Thesis, Andhra University, Visakhapatnam, India.

Metcalf, C.R. and Chalk, L. 1950. Anatomy of the Dicotyledons. Vol. **2**. Clarendon Press, Oxford.

Mohan, A.C. 1994. Foliar Epidermology and Venation Pattern of Amaranthaceae in Relation to its Systematics. Ph.D. Thesis. Kakatiya University, Warangal, India.

Raju, V.S. 1981. Leaf Architecture as an Aid to the Systematics of the Order Euphorbiales. Ph.D. thesis. Nagarjuna University, Nagarjunanagar, Guntur, India.

Raju, V.S. and Rao, P.N. 1977. Variation in the structure and development of foliar stomata in the Euphorbiaceae. Bot. J. Linn. Soc.**75**: 69-97.

Raju, V.S. and Rao, P.N. 1987. The taxonomic use of the basic stomatal type in the generic delimitation of *Chamaesyce* (Euphorbiaceae). Feddes Repert. **98**: 137-141.

Rao, P.N. and Raju, V.S. 1985. Foliar trichomes in the family Euphorbiaceae. *In*: Govil, C.M. and Kumar, V. (Eds), Trends in Plant Research. Prof. Y.S. Murthy Commemorative Volume. Bishen Singh Mahendra Pal Singh, Dehra Dun, India, pp. 128-136.

Rao, P.N. and Raju, V.S. 1988. On the distribution of cuticular markings on the foliar epidermis of the Euphorbiales. J. Econ.Taxon. Bot. **12**: 135-137.

Rejdali, M. 1991. Leaf micromorphology and taxonomy of North African species of *Sideritis* L. (Lamiaceae). Bot. J. Linn. Soc. **107**: 67-77.

Sehgal, L.S. and Paliwal, G.S. 1974. Studies on the leaf anatomy of *Euphorbia* VII. General conclusions and systematic considerations. Phytomorphology **24**: 141-151.

Stace, C.A. 1965. Cuticular studies as an aid to plant taxonomy. Bull. Br. Mus. Nat. Hist. **4**: 1-78.

Stace, C.A. 1984. The taxonomic importance of the leaf surface. *In*: Herwood, V.H. and Moore, D.M. (eds.), Current Concepts in Plant Taxonomy. Systematic Association Special Vol. **25**, Academic Press, London. pp. 67-94.

Venkataraju, R.R. and Pullaiah, T. 1995. Flora of Kurnool (Andhra Pradesh). Bishen Singh Mahendra Pal Singh, Dehra Dun, India. pp. 412-418.

Yang, Y., Riina, R., Morawetz, J.J., Haevermans, T., Aubriot, X. and Berry, P.E. 2012. Molecular phylogenetics and classification of *Euphorbia* subgenus *Chamaesyce* (Euphorbiaceae). Taxon **61**(4): 764-789.

# 12

# NUMERICAL TAXONOMY OF *ABELMOSCHUS* MEDIK. (MALVACEAE) IN INDIA

PRAVIN PATIL[1], SHRIKANT SUTAR[2], SURENDRA KUMAR MALIK, JOSEPH JOHN[3], SHRIRANG YADAV[2] AND KANGILA VENKATARAMAN BHAT

*National Bureau of Plant Genetic Resources, Pusa Campus, New Delhi 110012, India*

*Keywords*: *Abelmoschus*; Phenetics; Species relationships; India.

## Abstract

In the present study, numerical taxonomy approach has been used for the first time to access the taxonomy and species relationships of *Abelmoschus*. Sixteen *Abelmoschus* taxa were subjected to cluster analysis using 52 diagnostic characters related to root, stem, leaf, flower, fruit and seed. In this analysis, the first six principal components (PCs) accounted for the total variance of 79.22%. Similarity values for all 17 operational taxonomic units (OTUs) ranged from 0.18 to 0.82 with an average of 0.34. *Abelmoschus sagittifolius* showed maximum similarity value of 0.82 with *A. moschatus* subsp. *tuberosus*. On the other hand, minimum similarity values (0.18) were observed between *A. tuberculatus* and *A. moschatus* subsp. *tuberosus*, *A. tuberculatus* and *A. sagittifolius*, *A. palianus* and *A. moschatus* subsp. *tuberosus*, and *A. palianus* and *A. sagittifolius*. Neighbour joining (NJ) cluster analysis clearly discriminated 17 OTUs into four major clusters. The present study also validates the utility of morphometric analysis of *Abelmoschus* with respect to the taxonomy and species relationships.

## Introduction

Over the years, taxonomy has found important practical applications in various fields of science such as theoretical and applied biology including agriculture, evolutionary study, forestry, public health, biodiversity management and environmental issues (Godfray, 2002). Adanson (1763) proposed that classification should be based on characters covering all aspects of plant such as leaf, flower, fruit, seed, and each character should be given equal importance. As a consequence, a mathematical approach has been established by taxonomists called Numerical Taxonomy (Sokal and Sneath, 1963). Morphological data are considered significant in systematics because variation exhibited by morphological traits is supposed to be categorized by gaps between taxa which reflect their evolutionary arrangement emerged through morphological changes (Otte and Endler, 1989).

Taxonomy of *Abelmoschus* Medik. has a complex history with uncertainty in the generic status and composition of the genus as well as the species concept applied within the genus. The taxonomic treatment for some species of *Abelmoschus* is not consistent. *Abelmoschus manihot* (L.) Medik. and *A. moschatus* Medik. are the most polymorphic species (Hamon and Charrier, 1983). Hochreutiner (1924) described 14 species of *Abelmoschus*, in which *A. moschatus* and *A. manihot* constitute several varieties. However, Sivarajan and Pradeep (1996) did not consider infra-specific classification of *A. manihot* produced by Van Borssum-Waalkes (1966). Paul and Nayar (1988) and Paul (1993) therefore treated *A. manihot* as a single species without any infra-specific classification. Bates (1968) also suggested that all subspecies and varieties of *A. manihot* should

[1]Corresponding author. Email: pravin.patil99@gmail.com
[2]Botany Department, Shivaji University, Kolhapur 416004, India.
[3]National Bureau of Plant Genetic Resources Regional Station, KAU PO, Thrissur 680656, India.

be compressed in one group. Later, Vredebregt (1991) pointed out that *A. manihot* subsp. *manihot*, *A. manihot* subsp. *tetraphyllus* var. *tetraphyllus* and *A. manihot* subsp. *tetraphyllus* var. *pungens* complex lack discrete species boundaries among them, which further contradicts Hochreutiner (1900), Van Borssum-Waalkes (1966), Paul and Nayar (1988) and Paul (1993). Infraspecific taxonomy of *A. moschatus* is also a matter of debate as many subspecies and varieties have been recognized by Masters (1874), Hochreutiner (1900) and Van Borssum-Waalkes (1966).

Wild species of *Abelmoschus* comprise still larger unexplored variability, not even 2–3% of them have been studied beyond recognizing them as valuable reservoirs of untagged genes of agronomically useful traits (Sandhu *et al.*, 1974). Therefore, a thorough and robust hypothesis is urgently needed on morphological variation and species relationships among all taxonomically valid species of *Abelmoschus* which may provide the species-wise perspective that will be used in okra [*A. esculentus* (L.) Moench.] breeding strategies and effective germplasm management. The main objective of this study were to examine the morphological variation related to root, stem, leaf, flower, fruit and seed characters of *Abelmoschus* taxa by means of numerical taxonomy in order to resolve their relationships.

## Materials and Methods

### Taxon sampling and taxonomic treatment

On the basis of distribution data obtained from literature survey, several field trips were undertaken during 2010–2012 to collect and study different taxa of *Abelmoschus* occurring in India (Table 1). Confirmation of collected specimens was ensured with the help of information gathered from floras, published reports (Van Borssum-Waalkes, 1966; Paul and Nayar, 1988; Sivarajan and Pradeep, 1996) and the herbarium specimens. A standard procedure of using herbarium material was applied (Edlley *et al.*, 2012). Morphological characters of plants related to root, stem, leaf, flower, fruit and seed were described from their natural habitats, to avoid any ambiguity in the characters due to environmental effect. Seed related characters were taken from Patil *et al.* (2015).

### Character selection and data analysis

Species and in some cases their populations were used as operational taxonomic units (OTUs) for numerical taxonomy based on morphological data. Characters selected for morphological description of *Abelmoschus* species were those reported by Bisht *et al.* (1993, 1995), Sivarajan and Pradeep (1996) and based on field observations. A total 52 diagnostic characters related to habit, stem, leaf, flower, fruit and seed were chosen and scored for each OTU (Table 2).

The characters were converted into binary states and multi-states (interval) code. Standardization to $\mu = 0$ and $\sigma = 1$ of morphological data were done based on YBAR option with the software NTSYSpc ver. 2.10e (Rohlf, 1992). Neighbour joining tree was constructed using euclidean distance with the same software. Principal components (PCs) analysis was performed to analyze non-hierarchical relationship among the OTUs. This analysis was executed by calculating the eigenvectors and eigen values from Eigen programme in the NTSYS software. Morphometric analyses of quantitative data related to leaf, flower and fruit were done using SPSS version 11.5 for Windows.

## Results

### Morphological observations

Morphological evaluation of *Abelmoschus* species demonstrated that characters related to root, pubescent stem, leaf, flower, fruit and seed were significantly different between species. Root of *A. enbeepeegearense* John *et al.*, *A. crinitus* Wall. and *A. sagittifolius* (Kurz.) Merr. *ss.*, was tuberous, while rest of the species was non-tuberous. *Abelmoschus enbeepeegearense, A. crinitus, A. manihot* (L.) Medik. subsp. *tetraphyllus* (Roxb. *ex* Hornem.) Borss. var. *pungens* (Roxb.) Hochr. and *A. moschatus* Medik. subsp. *moschatus*, had conspicuous stem, while rest of the species had glabrous stem. Flowers of *A. ficulneus* and *A. angulosus* var. *angulosus* had white corolla, while *A. angulosus* var. *purpureus* had pink corolla. On the other hand, rest of the species

**Table 1. Studied taxa of *Abelmoschus* along with their codes, accession numbers, places of collection (latitude/longitude) and altitude.**

| Sl. No. | Taxon | Code | Accession/ collector no. | Place of collection (Latitude/Longitude) | Altitude (m) |
|---|---|---|---|---|---|
| 1. | *Abelmoschus esculentus* (L.) Moench. | AES | Var. AA | NA | NA |
| 2. | *A. caillei* (A. Chev.) Stevels | ACA | NMB2924 | N5° 26.860'/E20° 88.221' | 1012 |
| 3. | *A. moschatus* Medik. subsp. *moschatus* (odourless seed) | AMO | EC316073 | NA | NA |
| 4. | *A. moschatus* Medik. subsp. *moschatus* (musk scented seed) | AMM | IC141056 | N8° 38.999'/E77° 03.698' | 124 |
| 5. | *A. moschatus* Medik. subsp. *tuberosus* | ATR | IC324070 | NA | NA |
| 6. | *A. sagittifolius* (Kurz.) Merr. ss. | ASG | W357 | N19° 17.265'/E77° 30.977' | 487 |
| 7. | *A. tuberculatus* Pal & Singh | ATB | IC550656 | N19° 24.909'/E78° 03.337' | 432 |
| 8. | *A. ficulneus* (L.) Wight &Arn. | AFI | IC141001 | N15° 30.040'/E74° 59.587' | 644 |
| 9. | *A. crinitus* Wall. | ACR | N/SS2759 | N19° 43.478'/E78° 17.201' | 470 |
| 10. | *A. manihot* (L.) Medik. subsp. *manihot* | AMN | TCR2305 | N16° 40.857'/E74° 12.759' | 569 |
| 11. | *A. manihot* (L.) Medik subsp. *tetraphyllus* (Roxb. *ex* Hornem.) Borss. Waalk. | AMT | IC141019 | N23°34.630'/E78° 33.261' | 1828 |
| 12. | *A. manihot* (L.) Medik. subsp. *tetraphyllus* (Roxb. *ex* Hornem.) Borss. var. *pungens* (Roxb.) Hochr. | AMP | NMB2933 | N19° 11.795'/E73° 42.307' | 904 |
| 13. | *A. angulosus* var. *grandiflorus* Thwaites | AAG | IC470751 | N12° 26.429'/E75° 39.666' | 694 |
| 14. | *A. angulosus* var. *purpureus* Thwaites | AAP | AP1 | N13° 25.799'/E75° 44.921' | 1606 |
| 15. | *A. angulosus* var. *angulosus* Sivrajan & Pradeep | AAA | AA1 | NA | NA |
| 16. | *A. enbeepeegearense* John *et al.* | AEN | JRN/09/25 | NA | NA |
| 17. | *A. palianus* Sutar *et al.* | APA | SUA54 | NA | NA |

*NA = not available

**Table 2. Description of 52 morphological characters used in the cluster analysis of 16 taxa of *Abelmoschus*.**

| Sl. No. | Character | Code | Description/Value |
|---|---|---|---|
| | **Habit:** | | |
| 1 | Growth habit | GRH | erect (0) medium (1) procumbent (2) |
| | **Root:** | | |
| 2 | Root type | ROT | non-tuberous (0) tuberous (1) |
| | **Stem:** | | |
| 3 | Branching habit | BRH | non-branched (0) branched only at base (1) branched evenly (2) branched only at top (3) |
| 4 | Stem pubescence | STP | glabrous (0) slight (1) conspicuous (2) |
| 5 | Stipule shape | STS | long linear (0) linear lanceolate (1) triangular (2) short linear (3) |
| | **Leaf:** | | |
| 6 | Leaf colour | LEC | green (0) green with red veins (1) dark green (2) light green (3) |
| 7 | Leaf length | LEL | in cm |
| 8 | Leaf width | LEW | in cm |
| 9 | Leaf length : width ratio | LLW | - |
| 10 | No. of lobes | NLN | 5 (0) more than 5 (1) less than 5 (2) |
| 11 | Leaf texture | LTX | glabrous (0) slight (1) conspicuous (2) wooly (3) |
| 12 | Leaf margin | LMR | crenate (0) dentate (1) undulate (2) entire (3) serrate (4) serrulate (5) |
| | **Flower:** | | |
| 13 | Flower stalk | FST | straight (0) drooping (1) |
| 14 | Pedicel length | PDL | in cm |
| 15 | No. of epicalyx segment | NES | in no. |
| 16 | Shape of epicalyx segment | SHE | linear (0) lanceolate (1) triangular (2) ovate (3) broadly lanceolate (4) deltoid (5) |
| 17 | Persistence of epicalyx | PEE | caducous (0) partially persistent (up to seven days) (1) persistent (2) |
| 18 | Flower length | FLL | in cm |
| 19 | Flower diameter | FDM | in cm |
| 20 | Flower length : diameter ratio | FLD | - |
| 21 | Flower length : pedicel length ratio | FLP | - |
| 22 | No. of petals | NPT | 5 (0) more than 5 (1) |
| 23 | Petal colour | PTC | yellow (0) light yellow (1) dark yellow (2) red (3) pink (4) white (5) |
| 24 | Length of style | LST | in cm |
| 25 | No. of stigma lobes | NSL | 5 (0) 6 to 8 (1) |
| 26 | Stigma colour | SCO | red (0) dark red (1) light red (2) white (3) pink (4) |
| | **Fruit:** | | |
| 27 | Fruit colour | FCO | green (0) dark green (1) yellow green (2) |

| Sl. No. | Character | Code | Description/Value |
|---|---|---|---|
| 28 | Fruit shape | FSH | lanceolate (0) ovoid (1) lanceolate-ovoid (2) broadly ovoid (3) widely elliptic (4) |
| 29 | Fruit beak | FBE | non-beaked (0) beaked (1) |
| 30 | Fruit length | FRL | in cm |
| 31 | Fruit width | FRW | in cm |
| 32 | Fruit length : width ratio | FLW | - |
| 33 | Fruit pubescence | FPB | tomentose (0) glandular hairy (1) soft strigulose (2) densely hispid (3) hirsute (4) tuberculate hairy (5) |
| 34 | Fruit tuberculation | FTB | non-tuberculate (0) tuberculate (1) |
| 35 | Fruit dehiscence | FDH | laterally (0) apically (1) |
| | **Seed: a. macro-morphology** | | |
| 36 | Seed odour | SOD | odourless (0) musk scented (1) |
| 37 | Seed size | SDS | large (0) medium (1) small (2) |
| 38 | Seed shape | SSH | obovate (0) globose (1) reniform (2) sub-reniform (3) |
| 39 | Seed colour | SCO | dark brown (0) brown (1) greenish (2) blackish (3) |
| 40 | Seed texture | STX | glabrous (0) pubescent (1) |
| 41 | Hilum position | HLP | terminal (0) sub-terminal (1) |
| 42 | Hilum shape | HLS | ovate (0) broad ovate (1) triangular (2) round (3) |
| | **b. micro-morphology** | | |
| 43 | Trichome | TRC | absent (0) present (1) |
| 44 | Trichome density | TRD | sparse (0) dense (1) |
| 45 | Trichome type | TRT | spiral (0) non-spiral (1) |
| 46 | Seed sculpture | SSC | reticulate (0) reticulate-foveate (1) |
| 47 | Epidermal cell shape | ECS | polygonal (0) tetra-hexagonal (1) pentagonal-hexagonal (2) |
| 48 | Anticlinal wall shape | AWS | undulate (0) striate (1) |
| 49 | Anticlinal wall thickness | AWT | thin (0) thick (1) |
| 50 | Anticlinal wall level | AWL | raised (0) grooved (1) |
| 51 | Periclinal wall level | PWL | convex (0) concave (1) flat (2) |
| 52 | Periclinal wall texture | PWT | tuberculate (0) smooth (1) wavy (2) not noticeable (3) |

had yellow corolla. *Abelmoschus angulosus* var. *grandiflorus* Thwaites, *A. angulosus* var. *angulosus* Thwaites, *A. angulosus* var. *purpureus* Thwaites, *A. ficulneus* (L.) Wight & Arn., and *A. sagittifolius* had ovoid fruits, while *A. palianus* fruits were broadly ovoid. Fruits dehiscence was apically in *A. ficulneus*, *A. tuberculatus* Pal & Singh, *A. manihot*, *A. palianus* Sutar *et al.* and *A. crinitus*, while rest of the species laterally dehiscence. Seeds of *A. moschatus* subsp. *moschatus* had musk scent, and the remaining species were odourless.

Using the seed morphological characters, the studied taxa of the *Abelmoschus* revealed two basic types of seeds i.e., Type I: Seeds with deciduous trichomes and Type II: Seeds with persistent trichomes. *Abelmoschus esculentus*, *A. caillei*, *A. crinitus*, *A. moschatus* subsp. *moschatus*, *A. moschatus* subsp. *tuberosus* and *A. enbeepeegearense* belong to the Type I. In contrast, Type II comprises *A. ficulneus*, *A. tuberculatus* and *A. manihot* subsp. *tetraphyllus* var. *pungens*, *A. manihot* subsp. *manihot*, *A. manihot* subsp. *tetraphyllus* var. *tetraphyllus*, *A. angulosus* var. *grandiflorus*, *A. angulosus* var. *purpureus*, *A. angulosus* var. *angulosus* and *A. palianus*.

*Numerical taxonomic analysis*

The ratio of leaf length to leaf width and flower length to flower diameter did not show variation among the studied OTUs. Pearson's correlation analysis was done to determine the correlation among leaf, flower and fruit characters (Table 3). The highest positive correlation value ($r$p) was observed between FDM to FLL (0.878) followed by LEW to LEL (0.862) and FRL to LEL (0.816) at 0.01 level of significance. On the other hand, the lowest positive correlation value was observed between PDL to LEL (0.041) followed by FLP to LEL (0.052) and FRW to FLD (0.070). However, negative correlation was also observed between LLW to LEW (-0.437), FRW to FLP (-0.538) and FLP to PDL (-0.741).

Analysis of the 52×17 correlation matrix data set resulted in 14 eigenvectors (PCs). Out of 14 PCs, first six PCs were retained because they had eigenvalues of equal or higher than 1. For each PC, a component loading of more than 0.05 was considered as being significant. In this analysis, the first six PCs (PC1 = 23.48%, PC2 = 19.34%, PC3 = 12.13%, PC4 = 10.28%, PC5 = 7.47% and PC6 = 6.52%) accounted for the total variance of 79.22% differentiating the 17 OTUs. The first axis (PC-1) was highly influenced by STS, SHE, NSL, FTB, SOD, SCO and HLS, and defined 23.48% of the overall variance. These characters show considerable significant values of taxonomic importance with respect to the species differentiation. For the second axis (PC-2), the characters contributing to the total variability were BRH, FCO, FLW, SDS, SOD and HLP with 19.34% of variance. In the third axis (PC-3), characters such as GRH, LEC, NLN, LMR, NPT, AWL and PWL showed significant value of taxonomic importance to discriminate the 17 OTUs.

Similarity values of all 17 OTUs ranged from 0.18 to 0.82 (Table 4). *Abelmoschus sagittifolius* showed maximum similarity value of 0.82 with *A. moschatus* subsp. *tuberosus*, whereas minimum similarity value (0.18) was observed between *A. tuberculatus* and *A. moschatus* subsp. *tuberosus*, *A. tuberculatus* and *A. sagittifolius*, *A. palianus* and *A. moschatus* subsp. *tuberosus*, and *A. palianus* and *A. sagittifolius*. Neighbour joining (NJ) cluster analysis clearly discriminated 17 OTUs producing four major clusters (Fig. 1).

Cluster I:     *A. esculentus, A. caillei, A. tuberculatus* and *A. ficulneus*
Cluster II:    *A. moschatus* subsp. *moschatus* (musk scented seed), *A. moschatus* subsp. *moschatus* (odourless seed), *A. moschatus* subsp. *tuberosus*, *A. sagittifolius*, *A. crinitus*, *A. enbeepeegearense* and *A. manihot* subsp. *tetraphyllus* var. *pungens*
Cluster III:   *A. angulosus* var. *grandiflorus*, *A. angulosus* var. *angulosus*, *A. angulosus* var. *purpureus* and *A. palianus*
Cluster IV:   *A. manihot* subsp. *manihot* and *A. manihot* var. *tetraphyllus*

**Discussion**

Plant species have been considered as the central units of ecological and evolutionary studies, and therefore, the identification of boundaries among closely related species is an essential target of current systematic studies (Edlley *et al.*, 2012). In this study, morphological variation based on 52 characters (qualitative and quantitative) related to habit, root, stem, leaf, flower, fruit and seed were analyzed, which gave new insights into their potential taxonomic values for the species differentiation in the genus *Abelmoschus*.

Focusing on the root type in *Abelmoschus* species the present study revealed that there are only three species, which have tuberous root and others are non-tuberous. The characters such as shape of stipule, number of lobes in leaf, leaf margin, shape and nature of epicalyx segment, petal colour, number of stigma lobe, fruit colour, fruit tuberculation, seed odour, seed colour and seed size significantly contributed to separating the studied taxa and have always been central diagnostic characters in the genus *Abelmoschus* (Medikus, 1787; Van Borssum-Waalkes, 1966;

**Table 3.** Correlation coefficients among leaf, flower and fruit variables (quantitative) in *Abelmoschus* taxa. Abbreviations of characters correspond to Table 2. ** indicates significant correlation at the 0.01 level; * indicates significant correlation at the 0.05 level.

| | | Leaf | | | Flower | | | | | | Fruit | | |
|---|---|---|---|---|---|---|---|---|---|---|---|---|---|
| | | LEL | LEW | LLW | PDL | FLL | FDM | FLD | FLP | LST | FRL | FRW | FLW |
| Leaf | LEL | - | | | | | | | | | | | |
| | LEW | 0.862** | - | | | | | | | | | | |
| | LLW | 0.070 | -0.437 | - | | | | | | | | | |
| Flower | PDL | 0.041 | 0.147 | -0.061 | - | | | | | | | | |
| | FLL | 0.158 | 0.324 | -0.264 | 0.490* | - | | | | | | | |
| | FDM | 0.112 | 0.338 | -0.397 | 0.479 | 0.878** | - | | | | | | |
| | FLD | 0.155 | -0.010 | 0.322 | -0.081 | 0.070 | -0.405 | - | | | | | |
| | FLP | 0.052 | -0.110 | 0.206 | -0.741** | 0.085 | -0.038 | 0.235 | - | | | | |
| | LST | -0.248 | -0.151 | -0.035 | 0.599* | 0.762** | 0.637** | 0.097 | -0.039 | - | | | |
| Fruit | FRL | 0.816** | 0.583* | 0.273 | -0.114 | -0.012 | -0.175 | 0.404 | 0.183 | -0.287 | - | | |
| | FRW | 0.224 | 0.420 | -0.293 | 0.773** | 0.652** | 0.557* | 0.070 | -0.538* | 0.455 | 0.082 | - | |
| | FLW | 0.614** | 0.264 | 0.481 | -0.428 | -0.294 | -0.350 | 0.216 | 0.448 | -0.406 | 0.842** | -0.390 | - |

**Table 4. Pairwise similarity matrix of 16 *Abelmoschus* taxa based on simple matching coefficient from the matrix of 52 characters. OTUs codes used correspond to the Table 1. Bold represents minimum and underline represents maximum value of similarity.**

| OTUs | AES | ACA | AMO | AMM | ATR | ASG | ATB | AFI | AMN | AMT | AMP | ACR | AAG | AAA | AAP | AEN | APA |
|---|---|---|---|---|---|---|---|---|---|---|---|---|---|---|---|---|---|
| AES | - | | | | | | | | | | | | | | | | |
| ACA | 0.62 | - | | | | | | | | | | | | | | | |
| AMO | 0.42 | 0.34 | - | | | | | | | | | | | | | | |
| AMM | 0.42 | 0.32 | 0.70 | - | | | | | | | | | | | | | |
| ATR | 0.26 | 0.24 | 0.52 | 0.46 | - | | | | | | | | | | | | |
| ASG | 0.26 | 0.24 | 0.54 | 0.46 | <u>0.82</u> | - | | | | | | | | | | | |
| ATB | 0.36 | 0.32 | 0.30 | 0.30 | **0.18** | **0.18** | - | | | | | | | | | | |
| AFI | 0.28 | 0.28 | 0.20 | 0.20 | 0.20 | 0.20 | 0.36 | - | | | | | | | | | |
| AMN | 0.30 | 0.28 | 0.34 | 0.30 | 0.28 | 0.28 | 0.32 | 0.40 | - | | | | | | | | |
| AMT | 0.28 | 0.28 | 0.30 | 0.26 | 0.30 | 0.30 | 0.36 | 0.42 | 0.70 | - | | | | | | | |
| AMP | 0.22 | 0.20 | 0.34 | 0.30 | 0.30 | 0.30 | 0.34 | 0.36 | 0.34 | 0.38 | - | | | | | | |
| ACR | 0.28 | 0.30 | 0.38 | 0.32 | 0.38 | 0.38 | 0.24 | 0.20 | 0.28 | 0.38 | 0.28 | - | | | | | |
| AAG | 0.34 | 0.34 | 0.32 | 0.30 | 0.26 | 0.26 | 0.36 | 0.36 | 0.54 | 0.48 | 0.36 | 0.26 | - | | | | |
| AAA | 0.30 | 0.28 | 0.32 | 0.30 | 0.22 | 0.22 | 0.32 | 0.30 | 0.46 | 0.42 | 0.36 | 0.28 | 0.62 | - | | | |
| AAP | 0.28 | 0.28 | 0.30 | 0.26 | 0.26 | 0.26 | 0.36 | 0.34 | 0.40 | 0.40 | 0.36 | 0.24 | 0.66 | 0.52 | - | | |
| AEN | 0.38 | 0.30 | 0.48 | 0.48 | 0.44 | 0.44 | 0.28 | 0.22 | 0.24 | 0.24 | 0.24 | 0.48 | 0.26 | 0.28 | 0.28 | - | |
| APA | 0.38 | 0.36 | 0.34 | 0.32 | **0.18** | **0.18** | 0.38 | 0.32 | 0.44 | 0.44 | 0.40 | 0.42 | 0.46 | 0.52 | 0.38 | 0.38 | - |

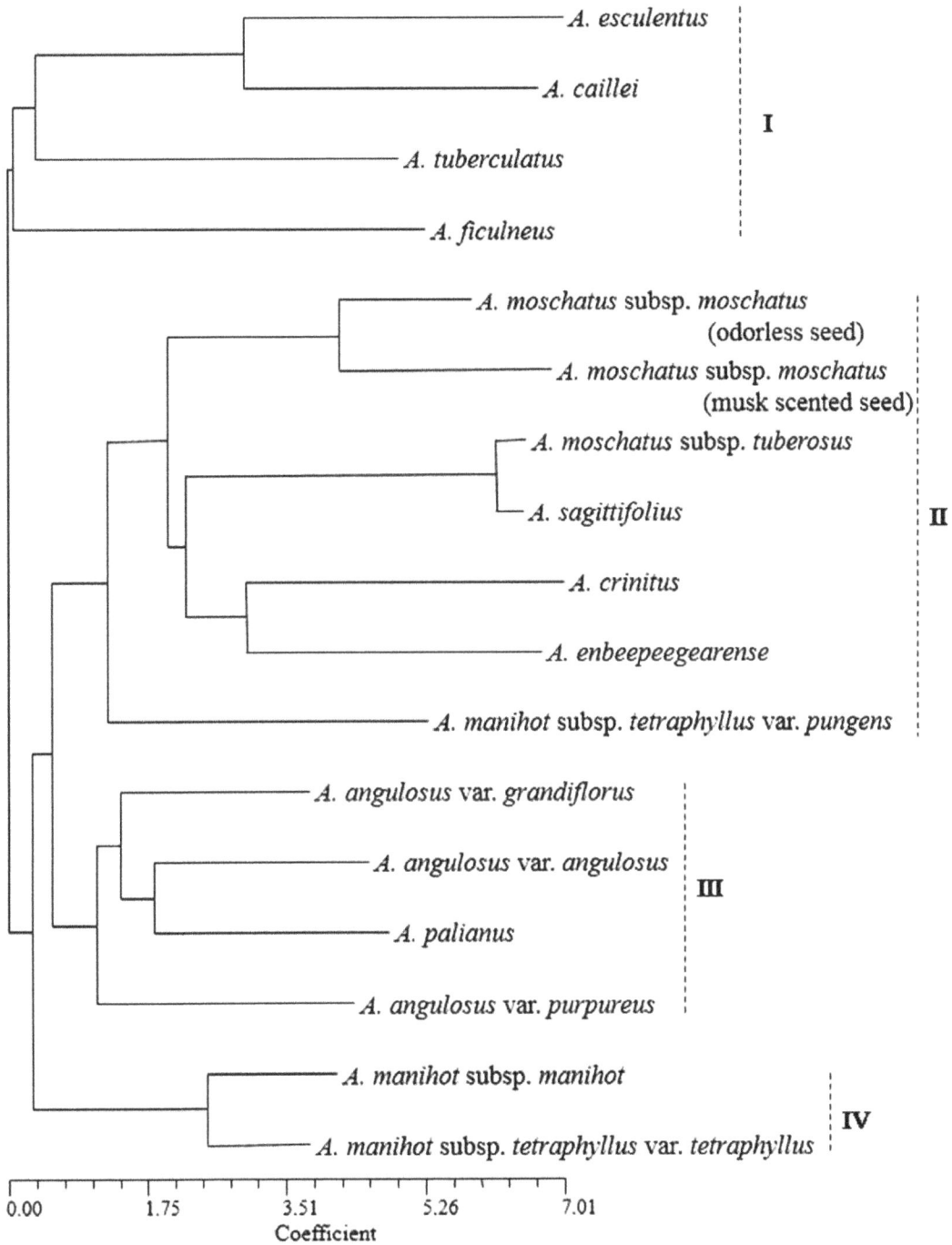

Fig. 1. Dendrogram obtained from neighbour joining (NJ) method showing the relationships of *Abelmoschus* taxa employed in the study.

Paul and Nayar, 1988; Vredebregt, 1991; Sivarajan and Pradeep, 1996; John *et al.*, 2012; Sutar *et al.*, 2013). The large positive correlation value as observed between leaf length and fruit length, pedicel length and fruit width, flower length and flower diameter were found to be most important and can be useful in a combination for more precise identification of *Abelmoschus* species.

In relation to the origin of cultivated okra, *A. tuberculatus* was found to be closely related to the *A. esculentus* in NJ tree, which further supports the hypothesis of Masters (1875) and Joshi *et al.* (1974). On the basis of species relationships as revealed by NJ tree, it is also assumed that *A. ficulneus* might have contributed to the *A. esculentus* genome as a second parent. The conspicuous presence of trichome on the seeds of *A. tuberculatus* is in partial agreement with Van Borssum-Waalkes (1966) who treated it as a wild form of *A. esculentus* since it generally grows along the roadsides and grassy slopes. Among the cultivated okra *A. esculentus* and *A. caillei* have great similarities in reproductive features. These species generally pose challenge for identification. The results of this study confirmed that cultivated species *A. esculentus* (Asian genotype) and *A. caillei* (introduced genotype) are morphologically distinct and easy to recognize. RAPD based characterization (Sunday *et al.*, 2008) revealed significant differences between *A. esculentus* and *A. caillei* accessions which further confirms our findings about their differentiation.

Owing to the close relationships within the species in Cluster II, we observed some common features, such as seed shape and remnants of trichomes on concentric rows in *A. moschatus* subsp. *moschatus*, *A. moschatus* subsp. *tuberosus*, *A. enbeepeegearense* and *A. crinitus*. These characters are confined to these species only indicating their taxonomically diverse nature. Investigations further revealed the remarkable variations in seed coat patterns of two very close taxa, i.e. *A. moschatus* subsp. *tuberosus* and *A. moschatus* subsp. *moschatus* supporting Bates (1968), who proposed to elevate *A. moschatus* subsp. *tuberosus* to the specific rank. Taking only taxonomic treatment into consideration, the present study also assumed that *A. moschatus* subsp. *tuberosus* and *A. sagittifolius* are not two separate entities but same, since both taxa have tuberous root type and yellow flower. Seed odour was found to be distinguishing characters for the correct identification of *A. moschatus* subsp. *moschatus* from other species of *Abelmoschus*. Another interesting new entity *A. enbeepeegearense* recently described by John *et al.* (2012) from the Southern Western Ghats showed intermediate characters (seed shape and seed colour) of *A. moschatus* subsp. *moschatus*, *A. moschatus* subsp. *tuberosus* and *A. crinitus*. However, seed coat features present in this taxon fully support its elevation as a separate species.

Among the species complex in *Abelmoschus*, *A. manihot* has been considered a highly variable taxa. Interestingly, for perennial taxa *A. manihot* subsp. *tetraphyllus* var. *pungens*, the present findings contradict with Hochreutiner (1900), Van Borssum-Waalkes (1966), Paul and Nayar (1988) and Paul (1993) who treat it as a variety of *A. manihot* subsp. *tetraphyllus*. In the NJ dendrogram obtained from 52×17 data matrix, this taxon showed distant position from *A. manihot* subsp. *tetraphyllus* var. *tetraphyllus* and *A. manihot* subsp. *manihot*. Vredebregt (1991) also demonstrated that *A. manihot* subsp. *tetraphyllus* var. *pungens* was not much different from var. *tetraphyllus*. In contrast, *A. manihot* subsp. *tetraphyllus* var. *pungens* was the only taxon which showed triangular hilum when it was rounded in *A. manihot* subsp. *tetraphyllus* var. *tetraphyllus*. Thus, hilum shape played a decisive role in differentiating these two taxa. In view of taxonomic significance, epidermal cell features differentiate *A. manihot* subsp. *tetraphyllus* var. *pungens* from widely distributed *A. manihot* subsp. *tetraphyllus* var. *tetraphyllus* and *A. manihot* subsp. *manihot*. Apart from morphological variability in *A. manihot* complex, species reflected great distinctness in seed micro-morphological characters which implies a need to study the specimens of *A. manihot* subsp. *manihot*, *A. manihot* subsp. *tetraphyllus* var. *tetraphyllus* and *A. manihot* subsp. *tetraphyllus* var. *pungens* using advance molecular markers for precise species differentiation and ranking.

The result obtained confirms the usefulness of seed morphology for identification and categorization of sub-specific taxa of *A. angulosus*. Based on variation in flower color, Sivarajan and Pradeep (1996) defined three varieties of *A. angulosus*, namely *A. angulosus* var. *grandiflorus* (yellow corolla), *A. angulosus* var. *angulosus* (white corolla) and *A. angulosus* var. *purpureus* (pink corolla). The present study significantly provides two more important seed characters which differentiate these varieties: *A. angulosus* var. *grandiflorus* (epidermal cell– tetra or pentagonal, elongate), *A. angulosus* var. *angulosus* (epidermal cell– polygonal) and *A. angulosus* var. *purpureus* (epidermal cell– tetra or pentagonal, not elongate) and therefore, confirm the treatment of Sivarajan and Pradeep (1996). The present study also confirms the uniqueness of recently described *A. palianus* (Sutar *et al.*, 2013). As observed in NJ tree, *A. palianus* was found to be closely related to *A. angulosus*.

In conclusion, classical taxonomy i.e. morphological descriptors, floras, type designations, and identification keys are still important and therefore the present study on *Abelmoschus* provides primary means and promotes further investigations in systematics and genomics.

## Acknowledgments

This work was conducted with funding from National Agriculture Innovation Project of Indian Council of Agriculture Research, Government of India. The facilitation of work by Director, NBPGR, New Delhi is duly acknowledged.

## References

Adanson, M. 1763. Familles des Plantes, Partie I., Paris.

Bates, D.M. 1968. Notes on the cultivated Malvaceae. 2, *Abelmoschus*. Baileya **16**: 99–112.

Bisht, I.S., Patel, D.P., Mahajan, R.K., Koppar, M.N., Thomas, T.A. and Rana, R.S. 1995. Catalogue of wild *Abelmoschus* species germplasm, NBPGR, New Delhi.

Bisht, I.S., Patel, D.P., Mahajan, R.K., Koppar, M.N., Thomas, T.A. and Rana, R.S. 1993. Catalogue on okra [*A. esculentus* (L.) Moench] germplasm. Part III, NBPGR, New Delhi.

Edlley, M.P., Marccus, A., Anderson, A.A., Clarisse, P.S. and Fabio, P. 2012. Integrating different tools to disentangle species complexes: A case study in *Epidendrum* (Orchidaceae). Taxon **61**: 721–734.

Godfray, H.C.J. 2002. Challenges for taxonomy. Nature **417**: 17–19.

Hamon, S. and Charrier, A. 1983. Large variation of okra collected in Benin and Togo. Plant Genetic Resources Newsletter **56**: 52–58.

Hochreutiner, B.P.G. 1900. Revision du genre *Hibiscus*. Conservatoire et Jardin Botaniques Gene've. Annuarie **4**: 23–191.

Hochreutiner, B.P.G. 1924. Genres nouveaux et genres discutes de la famille des Malcacees, Candollea **2**: 79–90.

John, K.J., Scariah, S., Nissar, V.A., Bhat, K.V. and Yadav, S.R. 2012. *Abelmoschus enbeepeegearense* sp. nov. (Malvaceae), an endemic species of okra from Western Ghats, India. Nord. J. Bot. **30**: 1–6.

Joshi, A.B., Gadwal, V.R. and Hardas, M.W. 1974. Evolutionary studies in world crops. *In*: Hutchinson, J.B. (Ed.), Diversity and Change in the Indian Sub–Continent. Cambridge University Press, London, pp. 99–105.

Masters, M.T. 1874. Malvaceae. *In*: Hooker, J.D. (Ed.), Flora of British India, Vol. **1**, London, pp. 317–353.

Masters, M.T. 1875. Flora of British India, Ashford Kent. Hooker, J.D. (Ed.) **1**: 320–348.

Medikus, F.K. 1787. UebereinigekunstlicheGeschlechteraus der Malvenfamilie, den der Klasse der, Monadelphien. 45–46.

Otte, D. and Endler, J.A. 1989. Speciation and its consequences: Sunderland, Massachusetts: Sinauer Associates, pp. 28–59.

Patil, P., Malik, S.K., Sutar, S., John, J., Yadav, S. and Bhat, K.V. 2015. Taxonomic importance of seed macro- and micro-morphology in *Abelmoschus* Medik. (Malvaceae). Nord. J. Bot., DOI: 10.1111/njb.00771.

Paul, T. and Nayar, M.P. 1988. Malvaceae. Flora of India, Fascicle **19**: 61–73.

Paul, T.K. 1993. Malvaceae. *In*: Sharma, B.D. and Sanjappa, M. (Eds), Flora of India, Vol. **3**, Calcutta, pp. 257–394.

Rohlf, F.J. 1992. NTSYS-PC: Numerical taxonomy and multivariate analysis system, version 2.0. Stony Brook: State University of New York.

Sandhu, G.S., Sharma, B.R., Singh, B. and Bhalla, J.S. 1974. Sources of resistance to jassids and white fly in okra germplasm. Crop Impr. **1**: 77–81.

Sivarajan, V.V. and Pradeep, A.K. 1996. Malvaceae of Southern Peninsular India. Daya Pub. House, Delhi, India, pp. 44–73.

Sokal, R. and Sneath, P.H. 1963. Principles of Numerical Taxonomy. W.H. Freeman, San Francisco, 359 pp.

Sunday, E.A., Ariyo, O.J. and Robert, L. 2008. Genetic relationships among West African okra (*Abelmoschus caillei*) and Asian genotypes *(Abelmoschus esculentus)* using RAPD. Afr. J. Biotechnol. **7**: 1426–1431.

Sutar, S.P., Patil, P., Aitawade, M., John, J., Malik, S., Rao, S., Yadav, S., Bhat, K.V. 2013. A new species of *Abelmoschus* Medik. (Malvaceae) from Chhattisgarh, India. Genet. Resour. Crop Evol. **60**: 1953–1958.

Van Borssum–Waalkes, J. 1966. Malesian Malvaceae Revised. Blumea **14**: 89–105.

Vredebregt, J.H. 1991. Taxonomic and ecological observations on species of *Abelmoschus* Medik. *In*: Report of an International workshop on Okra Genetic Resources held at NBPGR, New Delhi, India, 8–12 October 1990, pp. 60–76.

# 13

# POLLEN MORPHOLOGY OF *RHODODENDRON* L. AND RELATED GENERA AND ITS TAXONOMIC SIGNIFICANCE

A.K.M. Golam Sarwar[1] and Hideki Takahashi[2]

*Laboratory of Systematic Botany, Graduate School of Agriculture, Hokkaido University, Japan.*

*Keywords*: Pollen morphology; *Rhododendron*; Infrageneric classification; Generic delimitation; Rhodoreae.

## Abstract

Pollen grains of 40 taxa of *Rhododendron* L. and its closely related genera, *Therorhodion* L. and *Menziesia* Sm., were examined by means of light microscopy and scanning electron microscopy (SEM), or SEM alone. Pollen tetrads of *R. japonicum*, *R. schlippenbachii*, *R. tsusiophyllum* and *M. pentandra* were examined also with transmission electron microscopy. In all the genera studied, 3-colporate, oblate to suboblate pollen grains are arranged in tetrahedral tetrads. The apocolpial pollen wall is composed of the exine - well developed tectum, columellae, foot layer and endexine, and the intine. On the contrary, the septal exine is composed of fragmentary tectum, and the two foot layers of adjacent grains are sometimes connected by columellae and endexine. Among different infrageneric taxa only the subsect. *Ledum* (sect. & subgen. *Rhododendron*) was characterized by small sized pollen tetrads, higher 2f/D value and rugulate exine sculpture. The pollen morphological characteristics overlapped each other in all other taxa. Thus palynological features showed little usefulness in the infrageneric classification of *Rhododendron*, however, they gave additional support to the individual generic status of *Menziesia* and *Tsusiophyllum*, and the sister relationship between *Rhododendron* and *Therorhodion*.

## Introduction

The genus *Rhododendron* L., one of the largest and diverse genera of Ericaceae (Ericoideae, Rhodoreae), comprises over 1000 species (Chamberlain *et al.*, 1996). The centre of diversity of the genus is in the Himalaya, but *Rhododendron* sect. *Vireya* (Blume) Copel. f. is also diverse in Malesia (Sleumer, 1966). Since Linnaeus (1753) established *Rhododendron*, this large genus has posed systematic problems in terms of infrageneric circumscription and rank (for details see Kurashige *et al.*, 2001). *Rhododendron* and closely related genera are included in the tribe Rhodoreae (Kron *et al.*, 2002). The taxonomic history of the Rhodoreae is complex (Gillespie and Kron, 2010), but a brief introduction may help to illustrate the variations in generic composition of this group. According to Stevens (1971), tribe Rhodoreae comprises the genera *Rhododendron*, *Therorhodion*, *Ledum*, *Tsusiophyllum* and *Menziesia*. The genus *Therorhodion* was hypothesized as sister to *Menziesia* + *Tsusiophyllum* + *Rhododendron* (including *Ledum*) (Kron and Judd, 1990). In contrast, Chamberlain *et al.* (1996) recognized only 2 genera, namely *Rhododendron* and *Menziesia*. The recent classification of the Ericaceae (Kron *et al.*, 2002) recognized four genera in this tribe, namely *Diplarche*, *Menziesia*, *Rhododendron* (including *Ledum* and *Tsusiophyllum*) and *Therorhodion*. *Therorhodion* is often placed in *Rhododendron* (Chamberlain *et al.*, 1996; Kurashige *et al.*, 2001; Goetsch *et al.*, 2005); the cladistic analyses of molecular data

[1]Department of Crop Botany, Bangladesh Agricultural University, Mymensingh 2202, Bangladesh. E-mail: drsarwar@bau.edu.bd.
[1]The Hokkaido University Museum, N10 W8, Sapporo 060-0810, Japan.

also support reduction of the genus *Therorhodion* to a subgenus within the genus *Rhododendron* (Kurashige     *et al.*, 2001; Gao *et al.*, 2002a; Goetsch *et al.*, 2005). Recently Craven (2011) has suggested the inclusion of *Diplarche* and *Menziesia* to *Rhododendron*, that will make the tribe Rhodoreae monogeneric.

Pollen morphology has been shown to be useful for taxonomic and phylogenetic analysis of some Ericaceous taxa (Kron *et al.*, 2002; Sarwar, 2007). The pollen of *Rhododendron* has also been studied by many researchers for taxonomic purpose as well as a part of regional flora (Huang, 1972; Vasanthy and Pocock, 1987; Fuhsiung *et al.*, 1995; Mao *et al.*, 2000; Terzioğlu *et al.*, 2001; Gao *et al.*, 2002b, c; Wang *et al.*, 2006; Zhang *et al.*, 2009; Miyoshi *et al.*, 2011). All these results showed that pollen tetrads of *Rhododendron* were diverse in size and exine sculpture; sometime sufficient to differentiate the sections (Gao *et al.*, 2002b; Zhang *et al.*, 2009), but insufficient to differentiate the subgenera (Gao *et al.*, 2002c; Wang *et al.*, 2006). Moreover, our knowledge of pollen morphology and ultrastructure is still very limited for such a large genus as *Rhododendron*. Therefore, the objectives of this study were to clarify the differences of the pollen grains among *Rhododendron* and closely related genera of the tribe Rhodoreae, and to study the systematic significance of the micromorphology of pollen grains for the genus *Rhododendron*.

## Materials and Methods

Pollen morphology of a total of 40 taxa of *Rhododendron* and its closely related genera of the tribe Rhodoreae, *Rhododendron* (34 spp.), *Therorhodion* (2 spp.) and *Menziesia* (4 spp.), was examined by means of light microscopy (LM) and scanning electron microscopy (SEM), or SEM alone (Table 1). The pollen tetrads of *R. japonicum*, *R. schlippenbachii*, *R. tsusiophyllum* and *M. pentandra* were examined with transmission electron microscopy (TEM) to study the exine stratification of respective genera. Polliniferous materials used in this investigation were taken from the dried specimens from the herbaria C, GB, SAPS, SAPT and TUS. Abbreviation of the herbarium names except for SAPT (the Botanic Garden, Hokkaido University, Sapporo) are according to the Index Herbariorum (Holmgren *et al.*, 1990).

**Table 1. List of taxa used in this study along with their voucher specimens.**

| No. | Taxa | Voucher specimens |
|---|---|---|
| 1. | *Rhododendron albrechtii* Maxim. | Japan: Hokkaido, Inaho pass, 22.05.1983, Takahashi 3975 (SAPS) |
| 2. | *R. arborescens* Torr. | USA: no locality, (Herb. Arbor. Harvard Univ.) Fl. 28.06.1892. Unknown *s.n.* (SAPS) |
| 3. | *R. aureum* Gergi. | Japan: Hokkaido, Mt. Daisetsu; Mt. Asahi-dake, 20.06.1982, Takahashi 2512 (SAPS) |
| 4. | *R. brachycarpum* D. Don | Japan: Honshu, Mt. Zao, 08.07.1983, Takahashi *et al.* 40 (SAPS) |
| 5. | *R. dauricum* L. | Japan: Hokkaido, Iburi-shicho, Hobetsu-cho, 11.05.2004, Kanayama *et al.* 04-9050 (SAPS) |
| 6. | *R. davidsonianum* Rehd. & Wils. | Scotland: Royal Botanic Garden, Edinburg, noday, 05.1971, C9180 (GB) |
| 7. | *R. decorum* Franch. | China: Prov. Sze-chuan, Teng-hsiang-ying, 20.05.1922, Smith 2016 (GB) |
| 8. | *R. degronianum* Carr. | Japan: Nagano Pref., Mt. Kimpu-san, 22.06.1975, Iketani 1763 (TUS 129348) |
| 9. | *R. dilatatum* Mig. | Japan: Yamanashi Pref., Minamitsuru-gun, 01.05.1983, Togashi *s.n.* (SAPT) |
| 10. | *R. diversipilosum* (Nakai) Harma | Japan: Prov. Mutsu, Mt. Hakkoda, 30.06.1978, Takahashi 206 (SAPS) |

**Table 1 contd.**

| No. | Taxa | Voucher specimens |
|---|---|---|
| 11. | *R. formosanum* Hemsl. | Taiwan: Taichung Co., Gukan-Chinsan, 16.03.1985, Murata 17561 (TUS) |
| 12. | *R. groenlandicum* (Oeder) Kron & Judd | Greenland: Godthabsfjord, Ilulailik, Igdlorssuit, 17.7.1976, Hansen & Fredskild 1007 (SAPT) |
| 13. | *R. hidakanum* Hara | Japan: Hokkaido, Hidaka, Syoya, 10.5.1977, Tateishi & Togashi *s.n.* (TUS 66107) |
| 14. | *R. indicum* Sw. | Japan: Tokyo, cult., no day.05.1882, Miyabe *s.n.* (SAPS) |
| 15. | *R. japonicum* (A. Gray) Suring. | Japan: Nagano Pref., bet. Shirakaba-ko and Mt. Tateshina-yama, 29.05.1983, Takahashi 3998 (SAPS) |
| 16. | *R. kaempferi* Planch. | Japan: Miyagi Pref., Mts. Abukuma, Wariyama pass, 10.05. 1986, Iketsu *et al.* 95 (SAPT) |
| 17. | *R. keiskei* Miq. | Japan: Kagoshima Pref., Yaku Is., Mt. Tachudake, 10.05.1984, Murata *et al.* 17861 (TUS) |
| 18. | *R. lapponicum* (L.) Wahlenb. | Canada: Manitoba, Churchill, 26.06.1984, Gillett 1835 (C) |
| 19. | *R. macrosepalum* Maxim. | Japan: Shokoku Isl., Kagawa Pref., Kida-gun, 05.05.1982, Takahashi 1033 (SAPS) |
| 20. | *R. macrostemon* Maxim. | Japan: Hondo, Yamamoto in Settsu, cult., 10.05.1953, Togasi 688 (SAPS) |
| 21. | *R. maddeni* Hook. f. | Bhutan: Thimphu–Nimchling–Tanalum Bridge–Bunakha–Chima Khothi, 01.06.1967, Kanai *et al.* 346? (TUS 57346) |
| 22. | *R. mucronulatum* Turcz. var. *ciliatum* Nakai | Korea: Keisho-Nando, 20.05.1039, Yokoyama 299 (SAPS) |
| 23. | *R. nudipes* Nakai. | Japan: Shiga Pref., Mts. Hirasan, 06.05.1981, Murata 10910 (SAPT) |
| 24. | *R. parvifolium* Adams. | Japan: Prov. Nemuro, Ochii-shi, 16.06.1934, Tatewaki 20940 (SAPS) |
| 25. | *R. quinquefolium* Bisset *et* Moore | Japan: Rikuzen, Mt. Funagata, 06.05.1972, Ogura 1637 (TUS 68874) |
| 26. | *R. schlippenbachii* Maxim. | Japan: Hokkaido, Sapporo-shi, Hokkaido University campus, cult., 18.05.2004, Sarwar & Takahashi *s.n.* (SAPS) |
| 27. | *R. semibarbatum* Maxim. | Japan: Kyushu, Mts. Sobo-kutamuki, 07.07.1979, Murata 7987 (TUS 57400) |
| 28. | *R. subarcticum* Harmaja | Japan: Hokkaido, Mt. Taira-yama, 30.06.1982, Takahashi *et al.* 2643 (SAPS) |
| 29. | *R. trinerve* Fr. | Japan: Niigata, Iwafune-gun, Takanosu-yama, 10.07.1974, Togashi *s.n.* (TUS 67214) |
| 30. | *R. tschonoskii* Maxim. | Japan: Honshu, Mt. Zao, 08.07.1983, Takahashi *et al.* 33 (SAPS) |
| 31. | *R. tsusiophyllum* Sugim. | Japan: Hakone, Mt. Koma, 31.07.1926, Sawada *s.n.* (C) Japan: Sagami, Komagatake in Mt. Hakone, 10.08.1927, Asahina & Hisauchi *s.n.* (TUS 4578) |
| 32. | *R. viscistylum* Nakai var. *amakusaense* Tak. *ex* Yam. | Japan: Kumamoto Pref., Mt. Nokogiridake, 30.04.1978, Minamidani 29613 (TUS 100748) |
| 33. | *R. wadanum* Makino | Japan: Prov. Rikuzen, Sendai-shi, Aoba-yama, 29.04.1977, Takahashi 550 (SAPS) |
| 34. | *R. weyrichii* Maxim | Japan: Shikoku, Kagawa Pref., Goshikidai, 28.04.1973, Shimamura *et al. s.n.* (SAPT) |
| 35. | *Therhodion camtschaticum* (Pall.) Small | Japan: Hokkaido, Mt. Chiroro, 07.08.1985, Takahashi *et al.* 5836 (SAPS) |
| 36. | *T. redowskianum* (Maxim.) Hutch. | Russia: South Sakhalin, Poronaysk, 15.7.1937, Yoshimura & Hara *s.n.* (SAPS) |
| 37. | *Menziesia cilicalyx* (Miq.) Maxim. | Japan: Shiga Pref., Mt. Anzouyama, 03.05.1988, Tateishi & Hoshi 13689 (TUS) |
| 38. | *M. goyozanensis* Kikuchi | Japan: Iwate Pref., Mt. Goyozan, Tatamiishi–top, 07.07.1984, Mieno 445 (TUS) |
| 39. | *M. multifora* Maxim. | Japan: Prov. Rikuzen, Miyagi-gun, Izumigatake, 14.06.1978, Takahashi 767 (SAPS) |
| 40. | *M. pentandra* Maxim. | Japan: Hokkaido, Sapporo-shi, Mt. Muine, 06.07.1982, Takahashi 2687 (SAPS) |

Pollen grains were acetolysed following Sarwar and Takahashi (2012a). For LM, the dehydrated (in an ethanol series) pollen was mounted in silicone oil (viscosity 3000 cs), and examined and measured with a Nikon Eclipse E200 microscope. The dimensions "D", "P", "d(E)" and "2f", corresponding to the tetrad diameter, polar length, equatorial length and colpus length of pollen grain were measured, and the D/d, P/E and 2f/D ratio was calculated (Oldfield, 1959). The arithmetic mean, standard deviation and the maximum and minimum values were calculated using the XLSTAT 2009.3 program. Pollen slides of all collections are deposited at the Hokkaido University Museum, Sapporo, Japan. Pollen size and shape classes were used following Erdtman (1986) and descriptive terminology follows Sarwar *et al.* (2006) and Punt *et al.* (2007).

For SEM, the acetolysed pollen samples were dehydrated in an ethanol series, mounted and air dried on aluminum stubs from 70% ethanol, and sputter coated with Platinum-Palladium using a Hitachi E102 ion sputter. Subsequently, these were examined and photographed with a Jeol JSM-5310 LV scanning electron microscope operated at 15 KV. The SEM micrographs of apocolpial exine sculpture from similar positions were used for the purpose of description and comparison.

For TEM, material from herbarium specimens was rehydrated in 3% Aerosol-OT solution for more than one week, and then fixed overnight in 1% osmium tetraoxide solution. Fixed materials were dehydrated through an ethanol series and embedded in Epon 812 epoxy resin. Sections were cut using a Reichert-Jung Ultracut N ultratome, and post-stained with saturated uranyl acetate and lead acetate solution for 23 min (20 min and 3 min, respectively), and observed and photographed using a Hitachi H-800 transmission electron microscope operated at 75 KV.

## Results

*Pollen morphology of Rhododendron:*

Pollen grains are in tetrahedral tetrads, rarely compact or lobed, grains somewhat shrunk in some species (severly in *R. groenlandicum*), rarely with other configurations, sometimes in giant dyads in *R. tsusiophyllum*; viscin threads present; D 30.9-67.1 µm, P 16.3-35.8 µm, E 21.8-47.5 µm, D/d 1.31-1.51, P/E 0.66-0.81, oblate or suboblate; 3-colporate, rarely 4-colporate in *R. kaempferi*, finely demarcated, 2f 14.5-30.4 µm, W 0.7-2.2 µm, 2f/W 6.59-35.43, 2f/D 0.31-0.54, costae present, distinct or indistinct, colpus margin distinct; endocracks present; endoaperture lalongate, 0.6-2.6 µm long, 6.8-15.2 µm wide; apocolpial exine 1.7-3.6 µm thick, septum 0.6-3.6 µm thick; tectate, apocolpial exine sculpture from verrucate to rugulate (Table 2). In SEM, the pollen surface varies from uneven and rugged to flat, primary apocolpial exine sculpture indistinct, secondary sculpture finely (0.1-0.25 µm) to moderate (0.25-0.4 µm) gemmate-pilate (Type GP; Figs. 1F, G, I-O, 2A-D, G-O, 3A-E); or surface rugged to flat, apocolpial exine sculpture coarsely rugulate, grooves distinct (Type R; Figs 2E, F, 3F); or intermediate (Fig. 1H); colpus membrane granulate to granuloid or rarely smooth.

Three species of *Rhododendron*, viz. *R. japonicum*, *R. schlippenbachii* and *R. tsusiophyllum* were studied with TEM. The apocolpial exine is composed of ektexine and endexine (Figs 4A - I). Sexine is c. 1.1-1.3 µm thick, tectum canalized in *R. japonicum* (Fig. 4E), and a total exine is c. 1.8-2.1 µm thick. The septum is c. 0.9-1.9 µm thick. The intine is almost evenly thick around the pollen tetrad, showing lower electron density than the endexine beneath both the apocolpial and septal exine.

*Pollen morphology of Therorhodion:*

Pollen of *T. redowskianum* was studied only with SEM. Pollen grains are in lobed tetrahedral tetrads; viscin threads present; D 50.0 µm, P 26.5 µm, E 35.0 µm, D/d 1.43, P/E 0.76, suboblate;

3-colporate, colpi short and narrow in *T. redowskianum*, 2f 14.8 μm, W 2.9 μm, 2f/W 5.1, 2f/D 0.3, costae present and distinct, colpus margin distinct; endocracks present; endoaperture lalongate, 1.7 μm long, 9.7 μm wide; apocolpial exine 2.2 μm thick, septum 1.4 μm thick; tectate, apocolpial exine sculpture from verrucate to rugulate (Table 2).

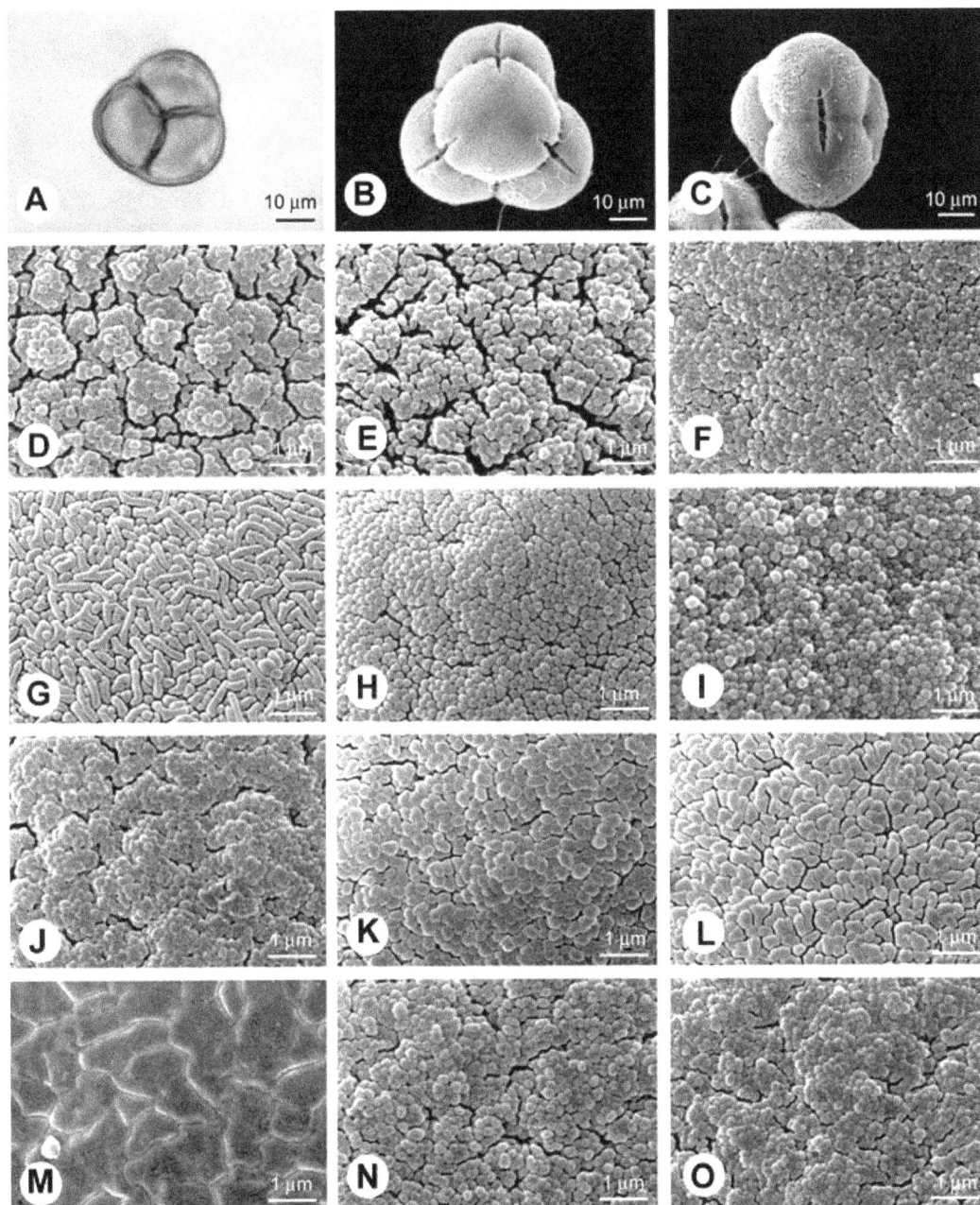

Fig. 1. LM and SEM micrographs of *Rhododendron* pollen. A) *R. formosanum* (Murata 17561); B) *R. maddeni* (Kanai *et al.* 346?). C) *R. tsusiophyllum* (Sawada *s.n.*); D) *R. aureum* (Takahashi 2512); E) *R. kaempferi* (Iketsu *et al.* 95); F) *R. aureum* (Takahashi 2512); G) *R. brachycarpum* (Takahashi *et al.* 40); H) *R. decorum* (Smith 2016); I) *R. degronianum* (Iketani 1763); J) *R. formosanum* (Murata 17561); K) *R. macrostemon* (Togasi 688); L) *R. viscistylum* var. *amakusaense* (Minamidani 29613); M) *R. semibarbatum* (Murata 7987); N) *R. arborescens* (Unknown *s.n.*); O) *R. albrechtii* (Takahashi 3975). Pollen tetrads at polar view (A-D); tetrads with viscin threads (A, B, D); pollen tetrad at equatorial view with viscin threads (E); micrographs with apocolpial exine sculpture details (F-O).

In SEM, pollen surface is uneven and rugged, primary apocolpial exine sculpture indistinct, secondary sculpture finely (0.1-0.25 μm) to moderate (0.25-0.4 μm) gemmate-pilate (Type GP; Figs 5D-E); colpus membrane granuloid or smooth.

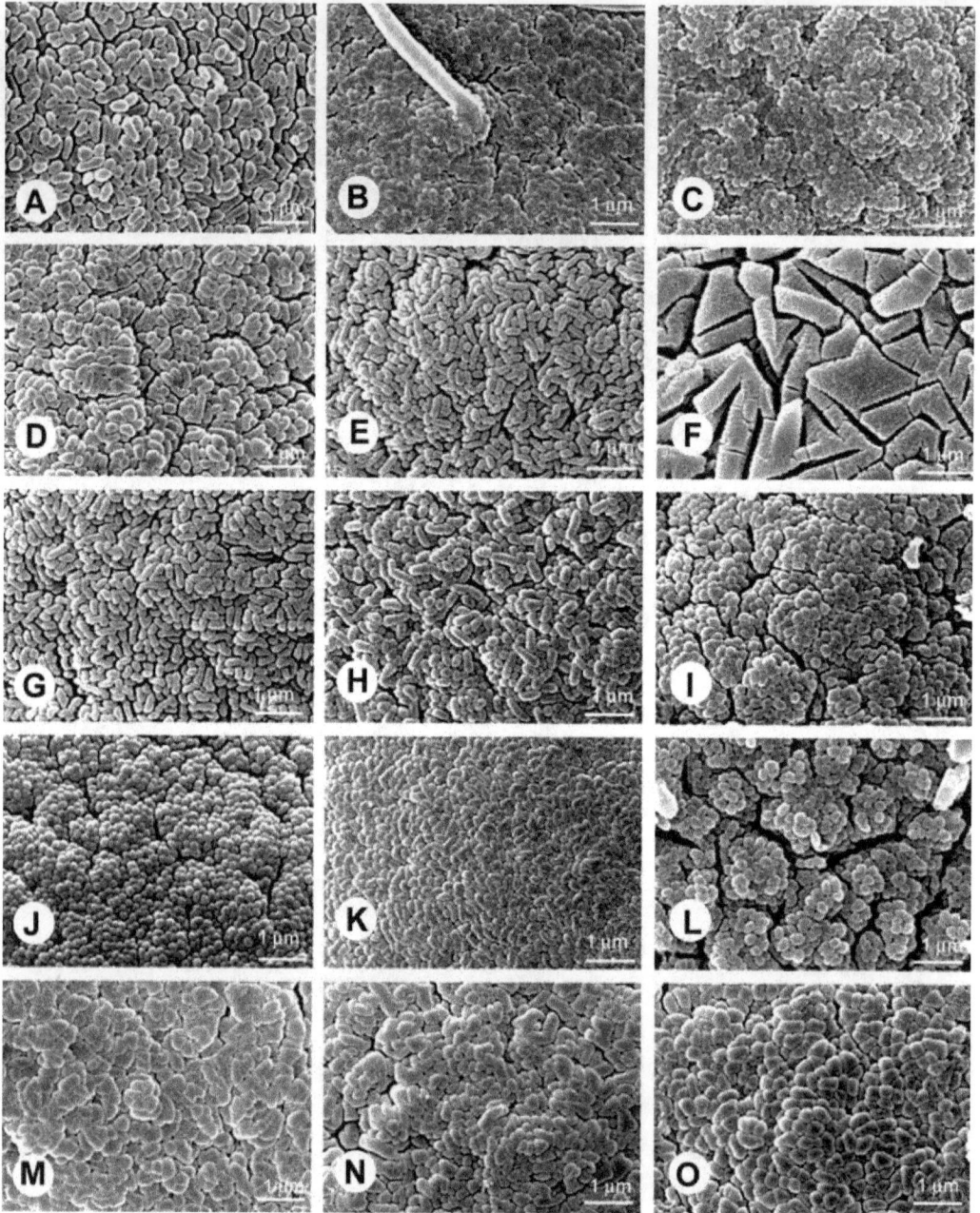

Fig. 2. SEM micrographs of *Rhododendron* pollen. A) *R. quinquefolium* (Ogura 1637); B) *R. schlippenbachii* (Sarwar & Takahashi *s.n.*); C) *R. lapponicum* (Gillett 1835); D) *R. parvifolium* (Tatewaki 20940); E) *R. diversipilosum* (Takahashi 206); F) *R. subarcticum* (Takahashi 2643); G) *R. dauricum* (Kanayama *et al.* 04-9050); H) *R. mucronulatum* var. *ciliatum* (Yokoyama 299); I) *R. davidsoniaum* (C. 9180); J) *R. keiskei* (Murata *et al.* 17861); K) *R. dilatatum* (Togashi *s.n.*); L) *R. hidakanum* (Tateishi & Togashi *s.n.*); M) *R. wadanum* (Takahashi 550); N) *R. weyrichii* (Shimamura *et al. s.n.*); O) *R. indicum* (Miyabe *s.n.*). Micrographs with apocolpial exine sculpture details (A-O); base of viscin threads attached with apocolpial exine (tectum) (A).

*Pollen morphology of Menziesia:*

Pollen of *M. cilicalyx* and *M. goyozanensis* was studied only with SEM. Pollen grains are in tetrahedral tetrads, lobed or compact; viscin threads commonly absent except in *M. pentandra*; D 34.3-36.7 µm, P 17.4-18.7 µm, E 24.0-27.5 µm, D/d 1.33-1.43, P/E 0.68-0.73, oblate; 3-colporate, 2f 15.4-17.6 µm, W 1.1-1.7 µm, 2f/W 9.06-16.0, 2f/D 0.42-0.51, costae present and distinct, colpus margin distinct; endocracks present; endoaperture lalongate, 1.0-1.8 µm long, 6.2-8.7 µm wide; apocolpial exine 1.7 µm thick, septum 1.0-1.2 µm thick, with faint perforations in *M. pentandra*; tectate, apocolpial exine sculpture finely verrucate (Table 2).

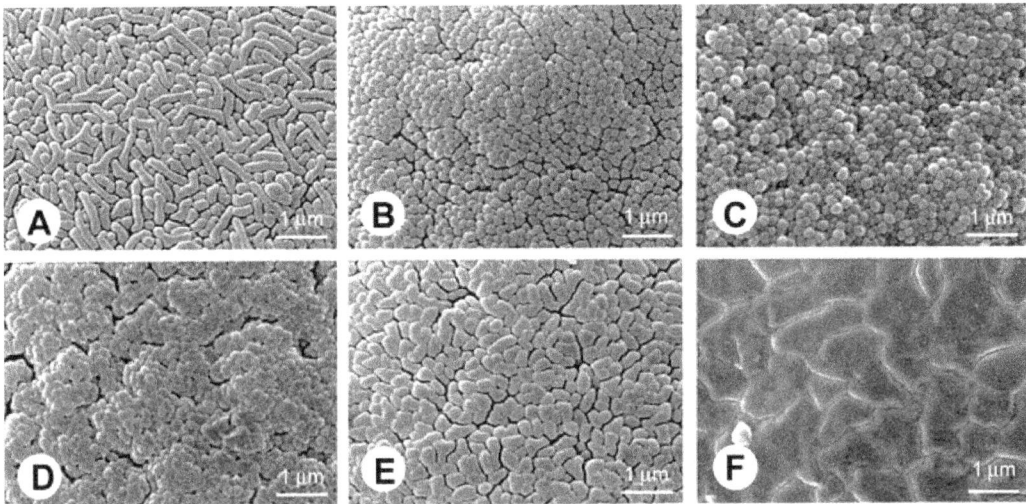

Fig. 3. SEM micrographs of *Rhododendron* pollen. A) *R. japonicum* (Takahashi 3998); B) *R. kaempferi* (Iketsu *et al.* 95); C) *R. macrosepalum* (Takahashi 1033); D) *R. nudipes* (Murata 10910); E) *R. trinerve* (Togashi *s.n.*); F) *R. tsusiophyllum* (Sawada *s.n.*). Micrographs with apocolpial exine sculpture details (A-F).

In SEM, pollen surface is flat, primary apocolpial exine sculpture indistinct, secondary sculpture unit narrowly straight-edged striate (Type NS; Figs 5I, K); or coarsely rugulate, lirae striate (Type R; Fig. 5L); colpus membrane granuloid to smooth.

In TEM of *M. pentandra*, the apocolpial exine is composed of ektexine and endexine (Figs. 5M-O). Sexine is c. 1.0 µm thick, and the total exine is c. 1.9 µm thick (Fig. 5N). The septum is c. 1.3-1.9 µm thick. The intine is almost evenly thick around the pollen tetrad, showing lower electron density than the endexine beneath both the apocolpial and septal exine.

**Discussion**

The genus *Rhododendron* is stenopolynous, having 3-colporate and medium pollen tetrads with viscin threads. A continuous and serial variation was revealed in all quantitative palynological characters within the genus (Tables 2 & 3). The size of *Rhododendron* pollen tetrads varies widely between 30.9 µm and 67.1 µm (Table 2).

No distinct difference in tetrad size was observed among the subfamilies, however, the subgenera *Hymenanthes* and *Pentanthera* produced relatively larger (48 µm) pollen tetrads (Table 3). Variations in ploidy level might be one of the probable causes of this wide variation in pollen size among *Rhododendron* species. In *Rhododendron*, enormous species diversification has

Table 2. Variation in pollen characters of *Rhododendron* and related genera showing mean value in μm and standard deviation. Minimum–maximum value in μm in parenthesis, D= tetrad diameter, P= polar length, d(E)= equatorial diameter, 2f= aperture length, W= aperture width, Apo.= apocolpium, n.d.= not discerned

| Name of taxa | Config-uration[1] | D | P | d(E) | D/d | P/E | 2f | W | 2f/D | Apo. exine thickness | Septum thickness | Orname-ntation[2] | Fig. No. | Re-marks[3] |
|---|---|---|---|---|---|---|---|---|---|---|---|---|---|---|
| *Rhododendron* **Subgenus** *Hymenanthes*    **Section** *Ponticum* | | | | | | | | | | | | | | |
| *R. aureum* | LT(+V) | 57.0±2.5 (51.2-59.4) | 29.1±1.4 (26.7-31.0) | 37.7±1.8 (35.5-40.9) | 1.51 | 0.77 | 20.9±2.4 (18.2-24.8) | 1.0±0.4 (0.5-1.5) | 0.37 | 2.4±0.4 (1.8-3.0) | 1.4±0.3 (0.7-1.8) | GP | 1F | |
| *R. brachycarpum* | T(+V) | 48.5±2.9 (44.2-52.8) | 25.5±2.1 (22.3-28.1) | 35.6±1.7 (34.3-38.0) | 1.36 | 0.72 | 20.2±2.0 (17.3-23.1) | 0.7±0.2 (0.5-1.2) | 0.42 | 3.0±0.5 (1.8-3.3) | 1.1±0.4 (0.5-1.7) | GP | 1G | |
| *R. decorum* | LT(+V) | 65.0±2.3 (62.0-69.3) | 34.9±1.9 (31.0-38.0) | 44.6±0.9 (42.9-45.9) | 1.46 | 0.78 | 24.8±3.5 (18.2-31.4) | 0.7±0.3 (0.5-1.2) | 0.38 | 2.6±0.4 (2.0-3.0) | 2.4±1.0 (1.3-4.1) | NS/RS | 1H | 1 |
| *R. degronianum* | T(+V) | n.d. | n.d. | n.d. | - | - | n.d. | n.d. | - | n.d. | n.d. | GP | 1I | |
| *R. formosanum* | LT(+V) | 65.2±1.9 (62.9-68.4) | 35.8±1.5 (33.1-38.9) | 44.3±1.1 (42.7-46.6) | 1.47 | 0.81 | 27.3±3.1 (22.8-31.9) | 2.2±0.3 (1.9-2.9) | - | 2.7±0.3 (2.4-3.1) | 3.6±0.6 (2.4-4.3) | GP | 1J | |
| *R. macrostemon* | T(+V) | n.d. | n.d. | n.d. | - | - | n.d. | n.d. | - | n.d. | n.d. | GP | 1K | |
| *R. viscistylum* var. amakusaense | T(+V) | n.d. | n.d. | n.d. | - | - | n.d. | n.d. | - | n.d. | n.d. | GP | 1L | |
| **Subgenus** *Mumeazalea* | | | | | | | | | | | | | | |
| *R. semibarbatum* | CT(+V) | n.d. | n.d. | n.d. | - | - | n.d. | n.d. | - | n.d. | n.d. | GP | 1M | |
| **Subgenus** *Pentanthera*    **Section** *Pentanthera* | | | | | | | | | | | | | | |
| *R. arborescens* | T(+V) | n.d. | n.d. | n.d. | - | - | n.d. | n.d. | - | n.d. | n.d. | GP | 1N | |
| **Section** *Sciadorhiodion* | | | | | | | | | | | | | | |
| *R. albrechtii* | T(+V) | 67.1±3.3 (61.1-72.6) | 35.3±3.1 (30.7-41.3) | 47.5±4.3 (41.3-54.0) | 1.41 | 0.74 | 30.4±4.8 (24.8-38.8) | 2.2±0.6 (1.3-3.0) | 0.45 | 2.7±0.5 (1.7-3.1) | 1.6±0.4 (1.2-2.5) | GP | 1O | |
| *R. quinquefolium* | T(+V) | n.d. | n.d. | n.d. | n.d. | n.d. | n.d. | n.d. | n.d. | n.d. | n.d. | GP | 2A | |
| *R. schlippenbachii* | T(+V) | 52.4±2.8 (48.0-57.6) | 28.7±1.9 (25.2-30.7) | 38.8±1.9 (35.0-41.3) | 1.35 | 0.74 | 18.7±1.9 (16.3-21.1) | 2.1±0.4 (1.4-2.6) | 0.36 | 2.6±0.2 (2.4-2.9) | 1.9±0.3 (1.4-2.4) | GP | 2B | |
| **Subgenus** *Rhododendron*    **Section** *Rhododendron*    **Subsection** *Lapponica* | | | | | | | | | | | | | | |
| *R. lapponicum* | T(+V) | 45.7±0.7 (44.9-46.2) | 24.9±0.4 (24.4-25.2) | 32.3±0.9 (31.4-33.0) | 1.41 | 0.77 | 17.7±0.6 (17.3-18.2) | 1.5 | 0.39 | 2.1±0.3 (1.8-2.5) | 0.6±0.3 (0.3-0.8) | GP | 2C | 2, 3 |
| *R. parvifolium* | CT(+V) | n.d. | n.d. | n.d. | - | - | n.d. | n.d. | - | n.d. | n.d. | GP | 2D | |
| **Subsection** *Ledum* | | | | | | | | | | | | | | |
| *R. diversipilosum* | LT(+V) | 30.9±1.5 (29.4-33.5) | 16.3±0.9 (14.9-18.2) | 21.8±0.9 (20.6-23.1) | 1.42 | 0.75 | 16.7±1.7 (13.2-19.0) | 1.4±0.6 (0.3-2.5) | 0.54 | 1.9±0.2 (1.7-2.1) | 1.2±0.3 (0.7-1.8) | R | 2E | |
| *R. subarcticum* | LT(+V) | 31.8±1.1 (30.0-33.2) | 17.3±0.9 (15.8-18.5) | 23.2±0.5 (22.3-23.9) | 1.37 | 0.75 | 16.6±1.2 (14.0-19.0) | 0.8±0.3 (0.5-1.2) | 0.52 | 1.7±0.2 (1.3-2.0) | 0.6±0.3 (0.3-1.2) | R | 2F | 2, 3 |
| **Subsection** *Maddenia* | | | | | | | | | | | | | | |
| *R. maddeni* | LT(+V) | n.d. | n.d. | n.d. | - | - | n.d. | n.d. | - | n.d. | n.d. | GP | - | |
| **Subsection** *Rhodorastra* | | | | | | | | | | | | | | |
| *R. dauricum* | T(+V) | n.d. | n.d. | n.d. | - | - | n.d. | n.d. | - | n.d. | n.d. | GP | 2G | |
| *R. mucronulatum* var. ciliatum | T(+V) | n.d. | n.d. | n.d. | - | - | n.d. | n.d. | - | n.d. | n.d. | GP | 2H | |

**Table 2 contd.**

| Name of taxa | Config-uration[1] | D | P | d(E) | D/d | P/E | 2f | W | 2f/D | Apo. exine thickness | Septum thickness | Orname-ntation[2] | Fig. No. | Re-marks[3] |
|---|---|---|---|---|---|---|---|---|---|---|---|---|---|---|
| **Subsection Triflora** | | | | | | | | | | | | | | |
| R. davidsonianum | T(+V) | 51.2±2.6 (46.7-54.5) | 26.5±1.2 (24.8-28.1) | 37.4±1.2 (35.5-39.6) | 1.37 | 0.71 | 18.6±2.8 (14.9-23.1) | 1.0±0.4 (0.5-1.5) | 0.36 | 3.1±0.3 (2.5-3.5) | 1.2±0.1 (1.0-1.3) | GP | 2I | 3 |
| R. kieskei | T(+V) | n.d. | n.d. | n.d. | - | - | n.d. | n.d. | - | n.d. | n.d. | GP | 2J | |
| **Subgenus Tsutsusi** | | | | | | | | | | | | | | |
| **Section Brachycalyx** | | | | | | | | | | | | | | |
| R. dilatatum | T(+V) | 56.4±1.8 (54.1-59.4) | 29.7±2.1 (26.4-32.7) | 39.6±1.8 (38.0-42.9) | 1.42 | 0.75 | 17.8±2.4 (14.9-23.1) | 0.7±0.3 (0.5-1.3) | 0.32 | 3.4±0.4 (3.0-4.3) | 1.7±0.3 (1.3-2.1) | GP | 2K | |
| R. hidakanum | T(+V) | n.d. | n.d. | n.d. | - | - | n.d. | n.d. | - | n.d. | n.d. | GP | 2L | |
| R. wadanum | T(+V) | 54.1±2.5 (50.8-59.4) | 28.8±1.6 (26.4-31.4) | 37.2±1.5 (34.7-39.6) | 1.45 | 0.77 | 18.6±2.0 (14.9-20.6) | 0.8±0.3 (0.5-1.2) | 0.34 | 3.6±0.3 (3.1-4.3) | 1.5±0.3 (0.8-2.0) | GP | 2M | |
| R. weyrichii | T(+V) | 55.7±1.8 (52.8-57.8) | 29.3±2.1 (27.2-33.7) | 39.4±2.6 (36.3-43.2) | 1.41 | 0.74 | 18.6±3.1 (13.2-23.1) | 1.1±0.4 (0.7-1.7) | 0.33 | 3.3±0.1 (3.1-3.5) | 2.2±0.7 (1.5-3.3) | GP | 2N | |
| **Section Tsutsusi** | | | | | | | | | | | | | | |
| R. indicum | T(+V) | n.d. | n.d. | n.d. | - | - | n.d. | n.d. | - | n.d. | n.d. | GP | 2O | |
| R. japonicum | CT(+V) | 63.2±2.2 (59.7-66.3) | 31.8±1.1 (30.5-33.3) | 48.3±1.6 (45.4-49.5) | 1.31 | 0.66 | 19.6±1.4 (18.2-21.5) | 2.1±0.7 (1.2-3.0) | 0.31 | 3.0±0.2 (2.8-3.3) | 1.5±0.2 (1.3-1.7) | GP/R | 3A | |
| R. kaempferi | CT(+V) | 53.0±3.4 (48.7-58.6) | 27.9±2.8 (23.1-33.3) | 37.4±3.0 (30.0-44.6) | 1.42 | 0.75 | 18.4±2.5 (14.9-22.3) | 1.2±0.5 (0.5-2.1) | 0.35 | 2.6±0.4 (2.2-3.5) | 1.6±0.3 (1.2-2.0) | GP | 3B | 4 |
| R. macrosepalum | T(+V) | 56.7±2.7 (54.8-60.3) | 29.9±2.6 (26.4-32.2) | 40.9±1.6 (39.6-42.9) | 1.39 | 0.73 | 18.6±4.1 (15.7-21.4) | 0.9±0.6 (0.5-1.3) | 0.33 | 3.2±0.2 (3.0-3.3) | 1.7±0.5 (1.3-3.2) | GP | 3C | 4 |
| R. nudipes | T(+V) | 54.6±2.5 (51.2-59.6) | 28.1±1.7 (25.6-30.9) | 36.7±1.7 (34.7-39.6) | 1.49 | 0.77 | 22.6±3.2 (19.8-28.1) | 1.7±1.0 (0.8-3.3) | 0.41 | 3.4±0.3 (3.0-4.0) | 3.2±1.0 (2.2-5.0) | GP | 3D | 3 |
| R. trinerve | T(+V) | n.d. | n.d. | n.d. | - | - | n.d. | n.d. | - | n.d. | n.d. | GP | 3E | |
| R. tschonoskii | T(+V) | n.d. | n.d. | n.d. | - | - | n.d. | n.d. | - | n.d. | n.d. | GP | - | |
| R. tsusiophyllum | T(+V) | 35.0±1.3 (33.0-37.0) | 18.1±1.3 (15.2-20.1) | 26.0±1.6 (23.1-28.1) | 1.35 | 0.70 | 14.5±2.1 (12.4-19.0) | 2.2±0.8 (1.3-3.6) | 0.41 | 2.1±0.2 (2.0-2.5) | 1.8±0.6 (1.3-3.0) | P/R | 3F | 3 |
| Therorhodion camtschaticum | LT(+V) | 50.0±4.2 (43.9-57.8) | 26.5±1.9 (23.3-29.5) | 35.0±3.6 (29.4-41.3) | 1.43 | 0.76 | 14.8±2.0 (12.4-17.3) | 2.9±0.6 (2.1-4.0) | 0.30 | 2.2±0.4 (1.8-3.0) | 1.4±0.1 (1.2-1.5) | GP | 5D | 2, 3 |
| T. redowskianum | T(+V) | n.d. | n.d. | n.d. | - | - | n.d. | n.d. | - | n.d. | n.d. | GP | 5E-F | |
| Menziesia ciliicalyx | T | n.d. | n.d. | n.d. | - | - | n.d. | n.d. | - | n.d. | n.d. | NS | 5I-J | |
| M. goyozanensis | T | n.d. | n.d. | n.d. | - | - | n.d. | n.d. | - | n.d. | n.d. | NS | - | |
| M. multifora | T | 36.7±2.5 (32.8-39.8) | 18.7±1.6 (17.2-21.6) | 27.5±1.3 (25.6-29.7) | 1.33 | 0.68 | 15.4±1.5 (13.2-17.3) | 1.7±0.4 (1.3-2.5) | 0.42 | 1.7±0.2 (1.3-2.0) | 1.0±0.4 (0.3-1.3) | NS | 5K | P |
| M. pentandra | CT(+V) | 34.3±1.5 (31.7-36.6) | 17.4±0.7 (16.7-18.6) | 24.0±1.2 (21.6-25.1) | 1.43 | 0.73 | 17.6±1.0 (16.5-19.8) | 1.1±0.4 (0.5-1.5) | 0.51 | 1.7±0.1 (1.5-2.0) | 1.2±0.1 (1.0-1.5) | NS/R | 5L | 3, P |

[1] T: Tetrahedral tetrad, CT: Compact tetrahedral tetrad, LT: Lobed tetrahedral tetrad, (+V): Viscin threads present.
[2] Exine ornamentation type by SEM. GP: Gemmate-pilate, P: Psilate, R: Rugulate, NS: Narrow straight-edged striate.
[3] 1: Number of endoaperture more than one, 2: Costae indistinct, 3: Endocracks absent/indistinct, 4: Rarely 4-aperturate, P: Perforated septum.

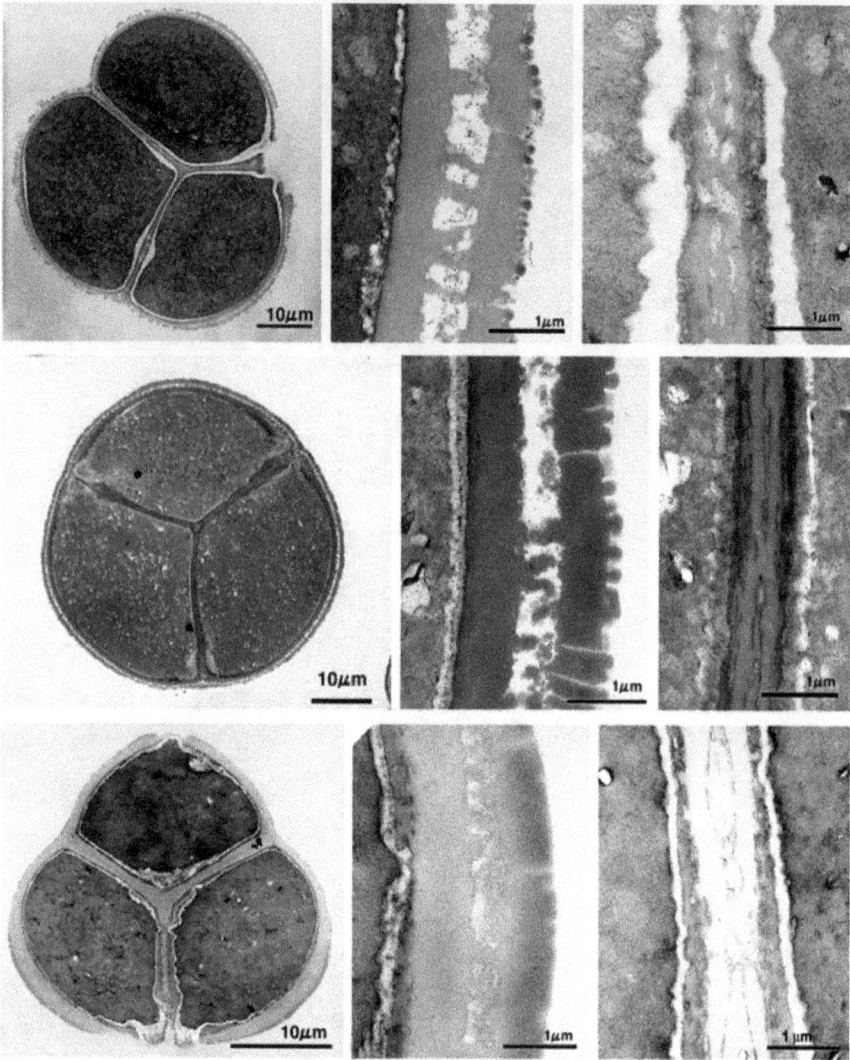

Fig. 4. TEM micrographs of *Rhododendron* pollen. A-C) *Rhododendron schlippenbachii* (Sarwar & Takahashi *s.n.*); D-F) *R. japonicum* (Takahashi 3998); G-I) *R. tsusiophyllum* (Sawada *s.n.*). Whole tetrad (A, D, G); apocolpial exine showing thick canalized tectum with supratectal fine gemmae-pila, thick columellae, thick foot layer and thin endexine (B); in septum, tectum fragmentary, two foot layer of adjacent grains sometimes connected by columellae, endexine thick (C); apocolpial exine showing canalized thick tectum with supratectal fine gemmae-pila, thick columellae, thick foot layer and thin endexine (E); in septum, tectum lacking, two foot layer of adjacent grains connected by thin or rudimentary columellae, endexine thick (F, I); apocolpial exine showing thick tectum, columellae, thick foot layer and endexine with (endo) cracks (H).

clearly occurred at the diploid level (2n=26), and polyploidy occurs among one third of cytologically examined lepidote species, the degree of polyploidy ranging from triploids (2n=3x=39) to dodecaploids (2n=12x=156) (Janaki Ammal, 1950). Aneuploidy (2n=30) has been reported in one case (Jones and Brighton, 1972) and two species namely, *R. wallichii* Hook. f. (as *R. campanulatum* var. *wallichii* Hook. f.) and *R. grande* Wight were reported to have n=12 (i.e. 2n=24) (Mehra, 1976), although the number of *Rhododendron* species having diploid level (2n=24) should be increased after the inclusion of *Therorhodion* (Kron and Judd, 1990). Cockerham and Galletta (1976) reported that the mean pollen diameter was 11% larger in the tetraploids compared to that in the diploids in certain *Vaccinium* species.

Viscin threads occur among the pollen tetrads in *Rhododendron* and *Therorhodion*, and presumably play a role in pollen removal from the anthers and its adhesion to pollinators. Any pollen material with viscin threads points to the highly specialized (entomophilous) pollination mode. It has been suggested that viscin threads increase the efficiency of pollination, and their presence implies highly specific pollinators for accurate delivery of pollen to stigma (Hesse *et al.*, 2000). The viscin threads would also play a role in pollen presentation. According to Skvarla *et al.* (1978) there is significant association between the structure of viscin threads in Onagraceae and the pollen vector: beaded viscin threads associated with birds and moth pollinated taxa whereas smooth ones occur in bee pollinated taxa. No viscin threads were found in *Menziesia* (despite reports to the contrary in Copeland, 1943; Wood, 1961), except in *M. pentandra* (Table 2). Viscin threads are not to be expected in species with urceolate or tubular corollas, since there they might obstruct cross-pollination (Stevens, 1971). Thus they are found in *M. pentandra*, which has broadly urceolate (to campanulate) corollas, but not in other species of the genus with urceolate and/or tubular corollas.

**Table 3. Variation in pollen characters of different subgenera of *Rhododendron* showing minimum-maximum (mean) value in μm. D: tetrad diameter, 2f: aperture length, W: aperture width.**

| Name of subgenera | D | 2f | W | Exine sculpture | Reference |
|---|---|---|---|---|---|
| *Azaleastrum* | 39.18 - 56.93 | 8.89 - 23.92 | 0.86 - 2.36 | GP | Gao *et al.* (2002b) |
| *Candistrum* | 37.97 - 39.95 | 13.80 | 1.29 - 1.38 | GP | Gao *et al.* (2002b) |
| *Hymenanthes* | 48.50 - 65.20 | 16.53 - 20.20 | 0.70 - 2.90 | GP, R | This paper; Gao *et al.* (2002c) |
| *Mumeazalea* | 35.85 - 39.95 | 16.10 | 1.85 - 1.98 | GP | Gao *et al.* (2002b) |
| *Pentanthera* | 49.04 - 67.10 | 18.70 - 30.40 | 1.56 - 3.44 | GP | This paper; Gao *et al.* (2002c); Zhang *et al.* (2009) |
| *Rhododendron* | 30.29 - 54.84 | 9.63 - 20.23 | 0.80 - 1.50 | GP, R | This paper; Gao *et al.* (2002c) |
| *Tsutsusi* | 35. 00 - 63.20 | 13.97 - 22.60 | 0.70 - 2.20 | GP, R | This paper; Gao *et al.* (2002c); Zhang *et al.* (2009) |

Usually, apocolpial exine is thicker than the septal exine, but thinner apocolpial exine has been observed in *R. formosanum* (Table 2). Similar relatively thinner apocolpial exine also has been observed in some taxa of the subfamily Vaccinioideae (Sarwar and Takahashi, 2006; Sarwar *et al.*, 2006) and it may have some taxonomic value in the infrageneric classification of the respected genera. In the lobed tetrads of *R. formosanum*, single pollen grains might be loosely attached together and the septum has not been reduced. A similar cause for comparatively thicker septum has been discussed for tetrads of the family Annonaceae (Le Thomas *et al.*, 1986). However, no significant correlation was found between compactness of tetrad and septum thickness in the present study or published literature (Kim *et al.*, 1988).

The apocolpial exine sculpture can be divided into two distinct groups - pollen surface is uneven and rugged to somewhat flat, apocolpial exine sculpture of Type GP (Figs 1F, G, I-O, 2A-D, G-O, 3A-E); and pollen surface flat or rugged, apocolpial exine sculpture of Type R (Figs 2E, F, 3F). The latter type of exine sculpture characterized *Rhododendron* subsect. *Ledum*, and all other species have almost similar exine sculpture except *R. tsusiophyllum* (Fig. 3F). The subsect. *Ledum* was also characterized by smaller pollen tetrads (30.9-31.8 μm) and a higher value of 2f/D ratio (0.52-0.54) (Table 2). Neither tetrad size nor exine sculpture was able to be used to differentiate among the subgenera and/or sections of *Rhododendron* (Tables 2, 3). Thus, palynological characters showed little usefulness in the infrageneric classification of

*Rhododendron* (Goetsch *et al.*, 2005), but could be used for identification of individual *Rhododendron* species (Table 2; Gao *et al.*, 2002b, c).

Generic delimitation of the tribe Rhodoreae is a subject of dispute until now (Gillespie and Kron, 2010). The phylogenetic analyses of *Rhododendron* based on molecular data did not support the individual generic status of *Menziesia* and *Therorhodion*, or even *Diplarche* (Craven, 2011), but suggested their inclusion within the genus *Rhododendron* (Kurashige *et al.*, 2001; Goetsch *et al.*, 2005). The results of this palynological study added some new points of disagreement within the present generic alignment of this tribe (Gillespie and Kron, 2010). As expected, the quantitative palynological features vary to a large extent in a large genus like *Rhododendron*, and give a little support for the individual generic status of *Menziesia*, *Rhododendron* and *Therorhodion* (Table 2). However, the specialized exine sculpture of Type NS and perforated septum of *Menziesia*, clearly distinguish the genus from other two genera of this tribe, *Rhododendron* and *Therorhodion* (Table 2; Figs 1-3, 5; Gao *et al.*, 2002c; Miyoshi *et al.*, 2011). Both the exine sculpture and septum with perforations have already been identified as taxonomically important characters in different Ericaceous genera (Sarwar, 2007). Along with other morphological and molecular characters (Kron *et al.*, 2002; Gillespie and Kron, 2010), the exceptional exine sculpture may also give additional support to the individual generic status of *Menziesia* (Sarwar and Takahashi, 2012b; Takahashi and Sarwar, 2013). Palynological features of the other two genera, *Rhododendron* and *Therorhodion* are very similar (Table 2; Figs 1-3, 5), and they might support the sister relationship between these two genera as identified by Gillespie and Kron (2010).

The pollen morphological features e.g., tetrad size, exine sculpture, etc. of *R. tsusiophyllum* of sect. *Tsutsusi* are different from those of other members of the same section as well as subgen. *Tsutsusi* (Table 2; Type GP; Figs 2K-O, 3A-E vs. Type R; Fig. 3F). Taking pollen morphology into account *R. tsusiophyllum* might be transferred from the subgen. *Tsutsusi* (Chamberlain *et al.*, 1996) to subsect. *Ledum* of the subgen. *Rhododendron* (Table 2; Type R; Figs 2E, F). Similar transfer of *R. huadingense* from sect. *Brachycalyx* of the subgen. *Tsutsusi* to subgen. *Pentanthera* has also been proposed based on palynological features (Zhang *et al.*, 2009). In TEM, the pollen wall structure of *R. tsusiophyllum* especially the thickness of the columellae and the sexine-nexine ratio also showed a distinct difference (data not shown) compared to the two other taxa of *Rhododendron* (Fig. 4). When considering the differences in the breakdown of the separating wall of the pollen sac, opening of the anther during maturity and the three-locular ovary as well as differences in other morphological characters between *R. tsusiophyllum* and other *Rhododendron* species (Stevens, 1969; Yamazaki, 1991); *R. tsusiophyllum* might be recognized as a separate monotypic genus *Tsusiophyllum*; *T. tanakae* Maxim., which is sister to whole of *Rhododendron* (including *Ledum*) (Kron and Judd, 1990). The recent molecular phylogenetic study of subfamily Ericoideae (Gillespie and Kron, 2010) may also support this supposition. *Rhododendron tsusiophyllum* forms a clade with *Menziesia pilosa*, which is well-supported in Bayesian and Maximum Likelihood analyses, instead of other *Rhododendron* species (Figs 1, 3 in Gillespie and Kron, 2010).

Based solely on molecular data, the classification and evolutionary relationship between plants is not always completely reliable (Stace, 2005), especially in genera like *Rhododendron* where polyploid species are a common phenomenon (Janaki Ammal, 1950). Hörandl (2006) also suggested that clades retrieved by phylogenetic analyses should not be used solely as a basis for classification, but should be regarded primarily as information for a better understanding of relationships. So, detailed phylogenetic analyses, using morphological, palynological and molecular data with larger number of specimens, are necessary to clarify generic circumscription of *Rhododendron* and its relationship with other closely related genera.

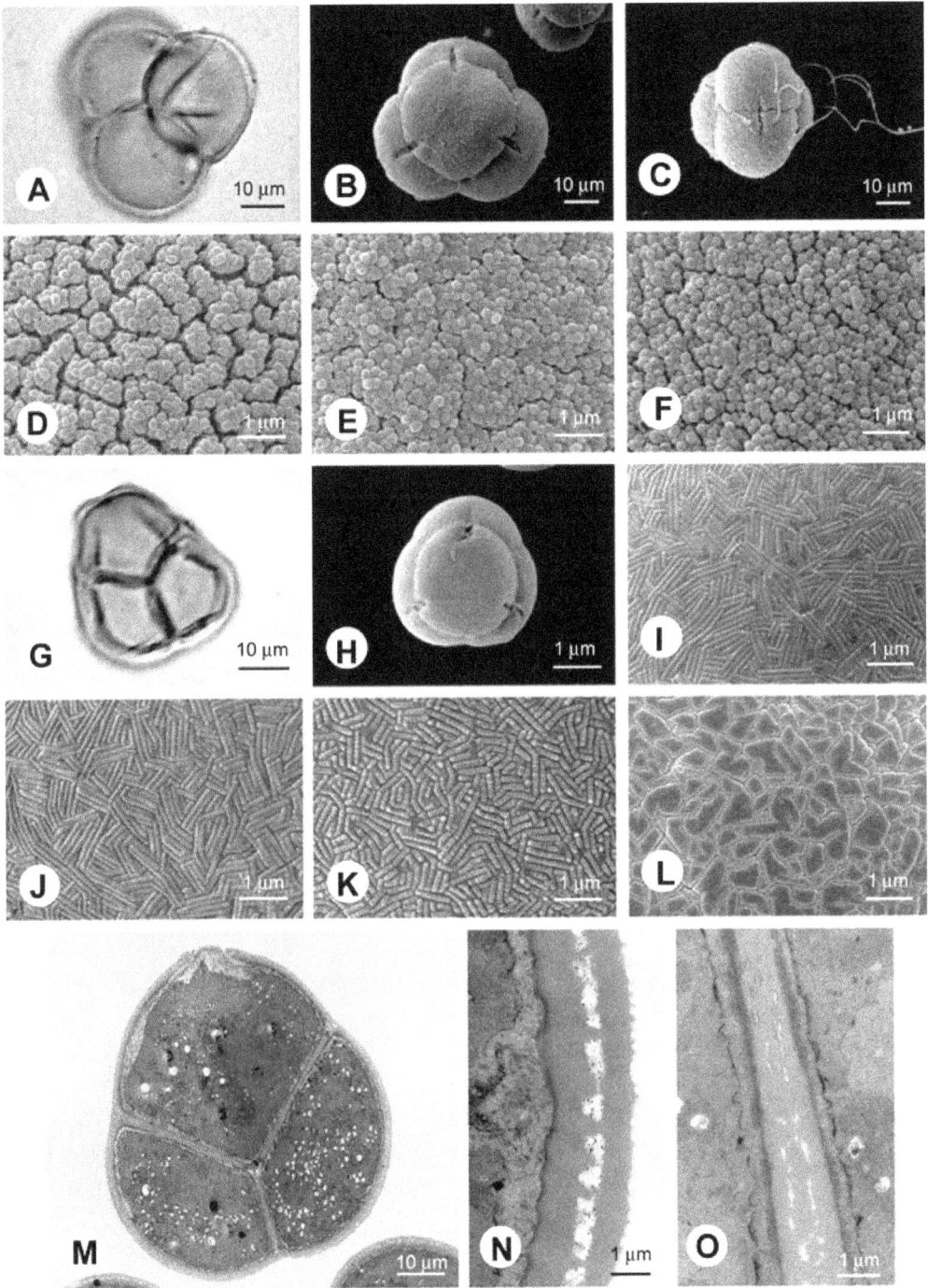

Fig. 5. LM, SEM and TEM pollen micrographs of *Therorhodion* (A-F) and *Menziesia* (G-O). A, B) *Therorhodion camtschaticum* (Takahashi *et al.* 5836); C) *T. redowskianum* (Yoshimura & Hara *s.n.*);  D) *T. camtschaticum* (Takahashi *et al.* 5836); E, F) *T. redowskianum* (Yoshimura & Hara *s.n.*). G) *Menziesia pentandra* (Takahashi 2687); H) *M. multiflora* (Takahashi 767); I, J) *M. cilicalyx* (Tateishi & Hoshi 13689); K) *M. multiflora* (Takahashi 767); L) *M. pentandra* (Takahashi 2687); M-O) *M. pentandra* (Takahashi 2687). Pollen tetrads at polar view (A, B, G, H); pollen tetrads at equatorial view showing aperturate (C); micrographs with apocolpial exine sculpture details (D, E, I, K, L); micrograph with mesocolpial exine sculpture details (F, J); whole tetrad (M); apocolpial exine showing tectum with narrow straight-edged striae, columellae, foot layer and thin undulated endexine (N); septum with tectum, and well defined columellae and foot layer of two adjacent grains (O).

## Acknowledgments

The authors thank the Directors and Curators of the herbaria C, GB, SAPS, SAPT and TUS for allowing them to examine and/or send the specimens on loan and sample polliniferous materials. The first author is thankful to MEXT (Japanese Ministry of Education, Culture, Sports, Science and Technology) Scholarship during the period of this study.

## References

Chamberlain, D.F., Hyam, R., Argent, G., Fairweather, G. and Walter, K.S. 1996. The genus *Rhododendron* – its classification and synonymy. Roy. Bot. Gard. Edinburgh, UK, 192 pp.

Cockerham, L.A. and Galletta, G.J. 1976. A survey of pollen characteristics in certain *Vaccinium* species. J. Amer. Soc. Hort. Sci. **101**: 671-676.

Copland, H.F. 1943. A study, anatomical and taxonomic, of the genera of the Rhododendroideae. Am. Midl. Nat. **30**: 533-625.

Craven, L.A. 2011. *Diplarche* and *Menziesia* transferred to *Rhododendron*. Blumea **56**: 33-35.

Erdtman, G. 1986. Pollen Morphology and Plant Taxonomy - Angiosperms. E. J. Brill, Leiden. 553 pp.

Fuhsiung, W., Nanfen, C., Yulong, Z. and Huiqiu, Y. 1995. Pollen Flora of China. 2nd ed. Institute of Botany, Academia Sinica. 461 pp. + 205 plates (in Chinese).

Gao, L-M., Li, D-Z., Zhang, C-Q. and Yang, J-B. 2002a. Infrageneric and sectional relationships in the genus *Rhododendron* (Ericaceae) inferred from ITS sequence data. Acta Bot. Sin. **44**: 1351-1356.

Gao, L-M., Zhang, C-Q., Li, D-Z. and Wei, Z-X. 2002b. Pollen morphology of *Rhododendron* subgenus *Azaleastrum*. J. Wuhan Bot. Res. **20**: 177-181 (in Chinese with English abstract).

Gao, L-M., Zhang, C-Q., Li, D-Z. and Wei, Z-X. 2002c. Pollen morphology of Rhodoreae (Ericaceae) and its systematic implication. Acta Bot. Yunn. **24**: 471-482 (in Chinese with English abstract).

Gillespie, E. and Kron, K.A. 2010. Molecular phylogenetic relationships and a revised classification of the subfamily Ericoideae (Ericaceae). Mol. Phylogen. Evol. **56**: 343-354.

Goetsch, L., Eckert, A.J. and Hall, B.D. 2005. The molecular systematics of *Rhododendron* (Ericaceae): A phylogeny based on *RPB*2 gene sequences. Syst. Bot. **30**: 616-626.

Hesse, M., Vogel, S. and Halbritter, H. 2000. Thread-forming structures in angiosperm anthers: their diverse role in pollination ecology. Plant Syst. Evol. **222**: 281-292.

Holmgren, P.K., Holmgren, N.H., and Barnett, L.C. (Eds). 1990. Index Herbariorum, Part I: The Herbaria of the World. 8th ed. New York Bot. Gard., Bronx. 704 pp.

Hörandl, E. 2006. Paraphyletic versus monophyletic taxa - evolutionary versus cladistic classification. Taxon **55**: 564-570.

Huang, T.C. 1972. Pollen Flora of Taiwan. National Taiwan University, Bot. Dept. Press. 297 pp.

Janaki Ammal, E.K. 1950. Polyploidy in the genus *Rhododendron*. The Rhododendron Yearbook **5**: 92-96.

Jones, K. and Brighton, C. 1972. Chromosome numbers of tropical *Rhododendrons*. Kew Bull. **26**: 559-561.

Kim, K.H., Nilsson, S. and Praglowski, J. 1988. A note on the pollen morphology of the Empetraceae. Grana **27**: 283-290.

Kron, K.A. and Judd, W.S. 1990. Phylogenetic relationships within the Rhodoreae (Ericaceae) with specific comments on the placement of *Ledum*. Syst. Bot. **15**: 57-68.

Kron, K.A., Judd, W.S., Stevens, P.F., Crayn, D.M., Anderberg, A.A., Gadek, P.A., Quinn, C.J. and Luteyn, J.L. 2002. Phylogenetic classification of Ericaceae: molecular and morphological evidence. Bot. Rev. **68**: 335-423.

Kurashige, Y., Etoh, J.I., Handa, T., Takayanagi, K. and Yukawa, T. 2001. Sectional relationships in the genus *Rhododendron* (Ericaceae): evidence from *mat*K and *trn*K intron sequences. Plant Syst. Evol. **228**: 1-14.

Le Thomas, A., Morawetz, W. and Waha, M. 1986. Pollen of palaeo- and neotropical Annonaceae: definition of the aperture by morphological and functional characters. *In* Blackmore, S. and Ferguson, I.K. (Eds), Pollen and Spores: Form and Function, Academic Press, London. pp. 375-388.

Linnaeus, C. 1753. Species Plantarum. Stockholm.

Mao, Z-J., Yang, Y-F. and Hou, L-J. 2000. A study on pollen morphology of *Rhododendron* in northeast of China. Bull. Bot. Res. (Harbin). **20**: 58-62 (in Chinese with English abstract).

Mehra, P.N. 1976. Cytology of Himalayan Hardwoods. Sree Sarawaty Press, Calcutta, 235 pp.

Miyoshi, N., Fujiki, T. and Kimura, H. 2011. Pollen Flora of Japan. Hokkaido Univ. Press, Sapporo. (in Japanese).

Oldfield, F. 1959. The pollen morphology of some of the West European Ericales - Preliminary descriptions and a tentative key to their identification. Pollen *et* Spores **1**: 19-48.

Punt, W., Hoen, P.P., Blackmore, S., Nilsson, S. and Le Thomas, A. 2007. Glossary of pollen and spore terminology. Rev. Palaeob. Palynol. **143**: 1-81.

Sarwar, A.K.M. Golam. 2007. Pollen morphology and its systematic significance in the Ericaceae. PhD Thesis., Graduate School of Agriculture, Hokkaido Univ., Japan (unpublished), 302 pp.

Sarwar, A.K.M. Golam and Takahashi, H. 2006. The taxonomic significance of pollen morphology in Andromedeae *s.s.*, Gaultherieae, Lyonieae and Oxydendreae (Ericaceae: Vaccinioideae). Jpn. J. Palynol. **52**: 77-96.

Sarwar, A.K.M. Golam and Takahashi, H. 2012a. Pollen morphology of *Kalmia* (Phyllodoceae, Ericaceae) and its taxonomic significance. Bangladesh J. Plant Taxon. **19**: 123-133.

Sarwar, A.K.M. Golam and Takahashi, H. 2012b. Pollen morphology and tribal classification of the subfamily Ericoideae (Ericaceae). Jpn. J. Palynol. **58** (Spec. Issue Abs.: IPC/IOPC 2012): 204.

Sarwar, A.K.M. Golam, Ito, T. and Takahashi, H. 2006. An overview of pollen morphology and its relevance to the sectional classification of *Vaccinium* L. (Ericaceae). Jpn. J. Palynol. **52**: 15-34.

Skvarla, J.J., Raven, P.H., Chissoe, W.F. and Sharp, M. 1978. An ultrastructural study of viscin threads of Onagraceae pollen. Pollen *et* Spores **20**: 5-143.

Sleumer, H. 1966. An account of *Rhododendron* in Malesia. Flora Malesiana. Ser. I, **6**: 674-676.

Stace, C.A. 2005. Plant taxonomy and biosystematics - does DNA provide all the answers? Taxon **54**: 999-1007.

Stevens, P.F. 1969. Taxonomic studies in the Ericaceae. PhD Thesis., University of Edinburgh (unpublished). 678 pp.

Stevens, P.F. 1971. A classification of the Ericaceae: subfamilies and tribes. Bot. J. Linn. Soc. **64**: 1-53.

Takahashi, H. and Sarwar, A.K.M. Golam. 2013. Pollen morphology and plant classification - An example of *Menziesia* (Ericaceae). Abs. 54th Ann. Meet. Palynol. Soc. Japan. 32 p. (in Japanese)

Terzioğlu, S., Merev, N. and Anşin, R. 2001. A study on Turkish *Rhododendron* L. (Ericaceae). Turk. J. Agric. For. **25**: 311-317.

Vasanthy, G. and Pocock, S.A.J. 1987. On the pollen tetrads of the south Indian Ericaceae, *Gaultheria*, *Rhododendron* and *Vaccinium* with special reference to *R. nilagiricum* Zenk. Bull. Jard. Bot. Nat. Belg. **57**: 213-245.

Wang, Y-G., Li, G-Z., Qi, X-X. and Ou, Z-L. 2006. Pollen morphology of *Rhododendron* and its taxonomic implication. Guihaia **26**: 113-119 (in Chinese with English abstract).

Wood, C.E.Jr. 1961. The genera of Ericaceae in the southeastern United States. J. Arnold Arb. **42**: 10-80.

Yamazaki, T. 1991. Morphological structure of the anther of *Tsusiophyllum tanakae* Maxim. J. Jpn. Bot. **66**: 35-38 (in Japanese with English summary).

Zhang, Y-J., Jin, X-F., Ding, B-Y. and Zhu, J-P. 2009. Pollen morphology of *Rhododendron* subgen. *Tsutsusi* and its systematic implication. J. Syst. Evol. **47**: 123-138.

# 14

# NUMERICAL TAXONOMY OF THE GENUS *SENNA* MILL. FROM BANGLADESH

M. Oliur Rahman[1], Md. Zahidur Rahman and Ayesa Begum

*Department of Botany, University of Dhaka, Dhaka 1000, Bangladesh*

*Keywords: Senna*; Cluster analysis; Phenetic relationship; UPGMA.

## Abstract

This study examines the patterns of morphological variation and phenetic relationships among 11 species of *Senna* Mill. from Bangladesh using 32 vegetative and floral characters. The highest similarity is found between *S. obtusifolia* and *S. tora*, while the highest variation is observed between *S. alata* and *S. hirsuta*. UPGMA tree derived from cluster analysis reveals three major clusters, the first of which consists of two species (*S. alata* and *S. auriculata*), the second cluster comprises four species (*S. hirsuta, S. obtusifolia, S. tora* and *S. occidentalis*) and the third one is composed of five species (*S. multiglandulosa, S. sophera, S. siamea, S. timoriensis* and *S. surattensis*). A close relationship is also found between *S. multiglandulosa* and *S. sophera*, and between *S. siamea* and *S. timoriensis*. Results obtained from the present study are found congruent with cytological and anatomical studies showing the significance of numerical analysis for taxonomic relationship in the genus *Senna*.

## Introduction

Numerical taxonomy, also termed as morphometrics deals with grouping by numerical methods of taxonomic units into taxa on the basis of their character state (Sneath and Sokal, 1973). Cluster analysis and principal component analysis are two techniques commonly used in numerical classification. Cluster analysis produces a hierarchical classification of entities (taxa) based on the similarity matrix. It thus provides a logical means of expressing the relationship existing between taxa. Numerical taxonomic studies are important for discovering and documenting new morphological character and character states, and many attempts have been made in this regard for understanding phenetic relationships in different groups of plants (Pinheiro and de Barros, 2007; Mulumba and Kakudidi, 2010; Deshmukh, 2011; Rahman and Rahman, 2012).

The genus *Senna* Mill. (Caesalpiniaceae) is represented by 350 species and is distributed throughout the world (Marazzi *et al.*, 2006). Irwin and Barneby (1982) reports that about 80% of the *Senna* species are found in the American continent, while most of the remaining members occur in tropical Africa, Madagascar and Australia, with only a few species in southeastern Asia and the Pacific Island. *Senna* are characterized by the presence of cylindrical or flattened, irregularly dehiscent pods and longest filaments without sigmoidally curved towards the base and seed surfaces usually with areole. Economically *Senna* species are very important since their bark and oil extract are used for flavouring purposes and in soaps, candy and perfumery (Hill, 1952). Several *Senna* species are reported to have medicinal properties as laxative, expectorant, antimalarial, relaxant and anti-inflammatory (Sadique and Chandra, 1987; Ajagbonna and Mojiminiyi, 2001; Tona and Mesia, 2001).

Studies on the genus *Senna* are very much limited in Bangladesh. Baker (1879) described 18 species of *Cassia s.l.* of which 6 species now included in *Senna* are found in the area of Bangladesh. Prain (1903) listed 7 species of *Senna* from the then Bengal which falls under the

---

[1]Corresponding author. Email: dr_oliur@yahoo.com

territory of present Bangladesh. Although Khan *et al.* (1996) documented 6 species of the *Senna* from Bangladesh, recently Ahmed *et al.* (2008) reported 10 species of the genus from the country. The extensive field surveys through the present study revealed a total of 11 species of *Senna* are now found in Bangladesh. Despite few fragmentary studies are available on *Senna*, numerical approaches have never been tested in this genus to determine species relationships. Therefore, the present study aims at applying numerical method for examining morphological variation and inferring phenetic relationships among *Senna* species occurring in Bangladesh.

## Materials and Methods

*Plant materials:*

Eleven species of *Senna* were used in the present study (Table 1). Both fresh materials collected from different areas of Bangladesh, and herbarium specimens housed at Dhaka University Salar Khan Herbarium (DUSH) and Bangladesh National Herbarium (DACB) were examined for numerical analysis.

**Table 1. List of *Senna* species along with their vouchers used in the present study.**

| No. | Species | Specimens examined |
|---|---|---|
| 1 | *Senna alata* (L.) Roxb. | Bandarban: Chimbuk hills, Mirzapara, 27.11.1983, Khan, Huq, Rahman & Mia K. 6494 (DACB); Dhaka: Dhaka University campus, 23.12.2011, Ayesa 65 (DUSH). |
| 2 | *S. auriculata* L. | Dhaka: Shere-e-Bangla Agricultural University compound, 26.1.2011, Ayesa 07 (DUSH); Sangshad Bhaban, 27.1.2011, Ayesa 08 (DUSH). |
| 3 | *S. hirsuta* (L.) Irwin & Barneby | Bandarban: Lama, 6.12.2007, Bushra, Halib & Mafiz B 609 (DACB); Cox's Bazar: Bamiachara near Chakaria, 2.12.1999, Khan, Mia, Rashid & Islam K. 10177 (DACB); Gazipur: Gazipur, 30.6.2011, Ayesa 40 (DACB). |
| 4 | *S. multiglandulosa* (Jacq.) Irwin & Barneby | No fresh or herbarium specimens available. |
| 5 | *S. obtusifolia* (L.) Irwin & Barneby | Bandarban: Chimbuk hills, 26.11.1983, Khan, Huq, Rahman & Mia K. 6472 (DACB); Cox's Bazar: Teknaf, Mouchuni, 24.4.2011, Ayesa 32 (DUSH); Patuakhali: Islampur, 6.2.2011, Ayesa 17 (DUSH). |
| 6 | *S. occidentalis* Roxb. | Bogra: Mohasthangarh, 22.8.1989, Mia, Rahman, Mahbuba & Rezia M 2117 (DACB); Dhaka: Dhaka University campus, 26.12.2010, Ayesa 02 (DUSH); Gazipur: Rajendrapur forest, 22.12.2011, Ayesa 63 (DUSH). |
| 7 | *S. siamea* (Lamk.) Irwin & Barneby | Chittagong: Chunati range, 10.6.1979, Khan, Huq & Rahman K 5515 (DACB); Dhaka: Tejgaon, Old Airport, 27.12.2011, Ayesa 74 (DUSH). Panchagarh: Fakirhat, 30.6.1998, Mia *et al.* M 3883 (DACB). |
| 8 | *S. sophera* (L.) Roxb. | Chittagong: Sandwip, Rahmatpur 12.2.1988, Mia and Mahfuz M 1590 (DACB); Cox's Bazar: Teknaf, Shilkhali, 30.2. 2011, Ayesa 25 (DUSH); Dhaka: Dhaka University Campus, 30.4.11, Aeysa 33 (DUSH). |
| 9 | *S. surattensis* (Burm. f) Irwin & Barneby | Dhaka: Dhaka University campus, 20.12.2011, Ayesa 47 (DUSH); Div 22, 19.10.1977, M. Naskar 3955 (DUSH). |
| 10 | *S. timoriensis* (DC.) Irwin & Barneby | Bandarban: Ruma bazar, 28.10.1984, Khan, Huq, Rahman & Mia K 6724 (DACB); Chittagong Hill Tracts: Ruma P.S., Changnakra, 25.1.1965, M. S. Khan 1166 (DUSH). |
| 11 | *S. tora* (L.) Roxb. | Dhaka: Dhaka University Botanical garden, 26.12.2011, Ayesa 69 (DUSH); Khulna: Jatrapur Railways line side, 16.6.1982, A. M. Huq 5542 (DACB). |

*Characters:*

Thirty two characters were investigated and used in this analysis. Characters and character states were determined through examination of both living and herbarium specimens housed at DUSH and DACB. Both qualitative and quantitative characters were coded as binary-state. The characters and their binary states used for numerical taxonomic studies are listed in Table 2. Neither herbarium nor living specimens of *Senna multiglandulosa* were available; therefore character states for this species were determined from the relevant literature (Ahmed    *et al.,* 2008).

*Statistical analysis:*

The data matrix was scored using binary matrix. Dissimilarity matrix was prepared based on the data matrix. Cluster analysis was performed using UPGMA (unweighted pair group method with arithmetic mean) and a dendrogram was constructed to show the relationship among the species (Sneath and Sokal, 1973). All analyses were carried out using the program STATISTICA (Version 3.0).

**Result and Discussion**

Thirty two vegetative and reproductive characters have been identified for numerical analysis of *Senna* species (Table 2).

**Table 2. Morphological characters and their state used in the numerical analysis.**

| No. | Characters | Character states |
|-----|------------|------------------|
| 1 | Habit | Herb or undershrub (1), Shrub or tree (0). |
| 2 | Stem | Hairy (1), Glabrous (0). |
| 3 | Leaves | 6-20 pairs (1), 2-5 pairs (0) |
| 4 | Stipules | Persistent (1), Cauducous or subpersistent (0). |
| 5 | Shape of stipules | Deltoid or ovate (1), Linear or cordate (0). |
| 6 | Size of stipules | 1-5 mm (1), 8-20 mm (0). |
| 7 | Leaf attachment | Alternate (1), Opposite (0). |
| 8 | Petiole length | 0.1-0.3 cm (1), > 0.4 cm (0). |
| 9 | Gland | Present (1), Absent (0). |
| 10 | Laminar shape | Oblong or elliptic (1), Ovate or cordate (0). |
| 11 | Base angle | Obtuse (1), Acute (0). |
| 12 | Apex angle | Acute (1), Acumminate or obtuse (0). |
| 13 | Base shape | Rounded or obtuse (1), Oblique or unequal (0). |
| 14 | Apex shape | Rounded or obtuse (1), Acute or acumminate |
| 15 | Margin type | Entire (1), Serrulate (0). |
| 16 | Lobation of vein | Present (1), Absent (0). |
| 17 | Inflorescence | Axillary and terminal (1), Terminal (0). |
| 18 | Sepal | Free (1), United (0). |
| 19 | Bract | Present (1), Absent (0). |
| 20 | Shape of bract | Ovate (1), Linear to lanceolate (0). |
| 21 | Bracteole | Present (1), Absent (0). |
| 22 | Corolla | Free (1), United (0). |
| 23 | Anther | Bilobed (1), Not bilobed (0). |

**Table 2 Contd.**

| No. | Characters | Character states |
|---|---|---|
| 24 | Anther opening | Apical pore (1), Lateral (0). |
| 25 | Ovary | Glabrous (1), Hairy (0). |
| 26 | Stigma | Truncate (1), Punctiform or others (0). |
| 27 | Shape of pod | Linear to curved (1); Oblong (0). |
| 28 | Surface of pod | Pubescent (1), Glabrous (0). |
| 29 | Number of seeds per pod | > 30 (1), 6-12 (0). |
| 30 | Dehiscence of pod | Dehiscent (1), Indehiscent (0). |
| 31 | Areole | Present (1), Absent (0). |
| 32 | Seed shape | Ovoid or oblong (1), Rhomboidal (0). |

The present study reveals that the lowest morphological variation is observed between *S. obtusifolia* and *S. tora* indicating that they are most closely related among all species studied. *S. occidentalis* is also very close to *S. obtusifolia*. The highest variation is found between *S. alata* and *S. hirsuta* (Table 3). A high variation has also been detected between *S. alata* and *S. tora*; and *S. alata* and *S. obtusifolia*.

**Table 3. Morphological variation among 11 species of *Senna* based on Squared Euclidean distance.**

| Species | alat | auri | hirs | mult | obtu | occi | siam | soph | sura | timo | tora |
|---|---|---|---|---|---|---|---|---|---|---|---|
| alat | 0 | | | | | | | | | | |
| auri | 9 | 0 | | | | | | | | | |
| hirs | 18 | 13 | 0 | | | | | | | | |
| mult | 10 | 9 | 14 | 0 | | | | | | | |
| obtu | 15 | 10 | 9 | 13 | 0 | | | | | | |
| occi | 12 | 13 | 8 | 10 | 5 | 0 | | | | | |
| siam | 12 | 13 | 12 | 8 | 13 | 10 | 0 | | | | |
| soph | 12 | 13 | 12 | 6 | 9 | 6 | 10 | 0 | | | |
| sura | 14 | 11 | 12 | 10 | 9 | 12 | 10 | 10 | 0 | | |
| timo | 11 | 10 | 11 | 11 | 12 | 13 | 7 | 13 | 9 | 0 | |
| tora | 16 | 9 | 8 | 12 | 3 | 6 | 12 | 8 | 6 | 11 | 0 |

alat = *Senna alata*, auri = *S. auriculata*, hirs = *S. hirsuta*, mult = *S. multiglandulosa*, obtu = *S. obtusifolia*, occi = *S. occidentalis*, siam = *S. siamea*, soph = *S. sophera*, sura = *S. surattensis*, timo = *S. timoriensis*, tora = *S. tora*

The numerical analysis presents the phenetic relationships among 11 *Senna* species. The UPGMA dendrogram based on cluster analysis reveals three clusters. The first cluster consists of two species, *viz. S. alata* and *S. auriculata*; the second one comprises four species, namely *S. hirsuta*, *S. obtusifolia*, *S. tora* and *S. occidentalis*; while the third cluster includes five species, *viz., S. multiglandulosa*, *S. sophera*, *S. siamea*, *S. timoriensis* and *S. surattensis* (Fig. 1).

In the first cluster *S. alata* is grouped with *S. auriculata* indicating that they are closely allied, and this is evidenced by the presence of their puberulent stem, persistent stipule and linear to oblong pod. A close association between *S. hirsuta*, *S. obtusifolia*, *S. tora* and *S. occidentalis* is evident in the second cluster. The common characters shared by these four species include linear stipule, racemose inflorescence, orbicular to rhomboidal seeds and presence of glands on the rachis. In this cluster the highest similarity has been observed between *S. obtusifolia* and *S. tora* showing that they are most closely related among all the species studied, and this highest affinity is supported by the following shared characters: leaflets obovate, stipules linear, falcate, inflorescence short-racemose, axillary, ovary ribbed, style glabrous, stigma truncate, pod linear or subtetragonous and seeds are 4-5 mm long, with an areole on each face.

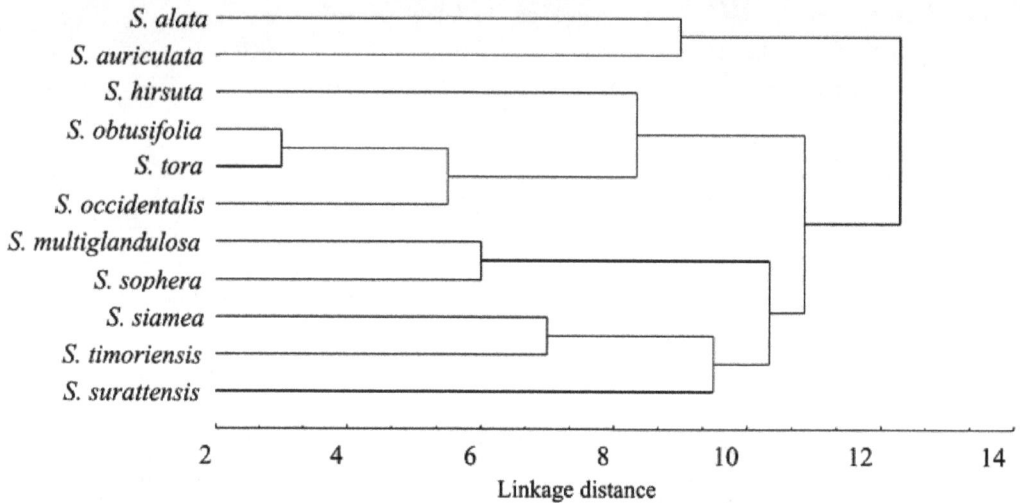

Fig.1. UPGMA dendrogram showing species relationship in *Senna*.

In the third cluster two distinct subclusters are found. The first one consists of *S. multiglandulosa* and *S. sophera*, and they are in the same line by presence of glandular leaves, ovate bracts, caducous stipule, pubescent ovary, and compressed, pointed seeds. The second subcluster contains *S. siamea*, *S. timoriesnsis* and *S. surattensis*. In this subcluster *S. siamea* and *S. timoriensis* are found to be more close to each other than they are to *S. surattensis*. *Senna siamea* and *S. timoriensis* both are evergreen trees and the close affinity between them is supported by their eglandular leaves, linear bracts, puberulent sepals, glossy brown seeds and presence of areoles.

Morphometric studies received considerable attention for species relatedness in different genera (Gomez-Campo *et al.*, 2001; Henderson and Ferreira, 2002; Sonibare *et al.* 2004; Bolourian and Pakravan, 2011). Although such studies were carried out in different legume genera, for example, *Cassia* (Boonkerd *et al.*, 2005), *Indigofera* (Soladoye *et al.*, 2010a), *Daniellia* (de La Estrella *et al.* 2009), however very little is known about the morphometrics in the genus *Senna*. Recently Soladoye *et al.* (2010b) made a morphometric study of eight species of *Senna* from south-western Nigeria and using 13 morphological characters they showed that *S. sophera* is closely related to *S. hirsuta*. Our results suggest that *S. sophera* is closely allied to *S. multiglandulosa* which is incongruent with that of Soladoye *et al.* (2010b). The present study reveals a close association between *S. hirsuta, S. obtusifolia, S. tora* and *S. occidentalis* as they grouped together. Ogundipe *et al.* (2009) have shown that paracytic and anomocytic types of stomata are found both in *S. hirsuta* and *S. occidentalis*. The anticlinal walls are straight and undulate in both these species indicating a close relationship between them. Our result is supported by Ogundipe *et al.* (2009) since a close affinity has been observed between *S. hirsuta, S. obtusifolia, S. tora* and *S. occidentalis*. The close affinity among these species is also evidenced by cytological investigation where the somatic chromosome number 2n=28 was reported for these four species (Irwin and Turner, 1960; Bir and Kumari, 1980). In conclusion, our study shows the significance of numerical analysis for detecting variation and taxonomic relationships among *Senna* species available in Bangladesh as it is attested by previous studies based on cytological and anatomical characters.

## Acknowledgement

The authors are thankful to the Director, Bangladesh National Herbarium, Dhaka for allowing to use herbarium facilities. Thanks are also due to Prof. Md. Abul Hassan and Prof. Momtaz Begum, Department of Botany, University of Dhaka for their help and cooperation during this study.

## References

Ahmed, Z.U., Hassan, M.A. Begum, Z.N.T., Khondker, M., Kabir, S.M.M.H., Ahmed, M., Ahmed, A.T.A., Rahman, A.K.A. and Huque, E.U. (eds). 2008. Encyclopedia of Flora and Fauna of Bangladesh, Vol. 7. Angiosperms: Dicotyledons (Balsaminaceae-Euphorbiaceae). Asiatic Society of Bangladesh, Dhaka, 546 pp.

Ajagbonna, O.P. and Mojiminiyi, F.B.O. 2001. Relaxant effects of the aqueous leaf extract of *Cassia occidentalis* on rat aortic rings. African J. Biomed. Res. **4**(3): 127.

Baker, J.G. 1879. *In*: Hooker, J.D., The Flora of British India, Vol. **2**. Reeve & Co. Ltd. England, pp. 261-267.

Bir, S.S. and Kumari, S. 1980. Cytological evolution of the Leguminous flora of the Punjub plain. *In*: Bir, S.S. (ed.), Recent Researches in Plant Science. Kalyani Publishers, Ludhiana, India, pp. 261-271.

Bolourian, S. and Pakravan, M. 2011. A morphometric study of the annual species of *Alyssum* (Brassicaceae) in Iran based on their macro- and micromorphological characters. Phytologia Balcanica **17**(3): 283-289.

Boonkerd, T., Pechsri, S. and Baum, B.R. 2005. A phenetic study of *Cassia* sansu lato (Leguminosae-Caesalpinioideae: Cassieae: Cassiinae) in Thailand. Plant Syst. Evol. **252**: 153-165.

de La Estrella, M., Aedo, C. and Velayos, M. 2009. A morphometric analysis of *Daniellia* (Fabaceae-Caesalpinioideae). Bot. J. Linn. Soc. **159**: 268-279.

Deshmukh, S.A. 2011. Morphometrics of the genus *Cassia* L. from Kolhapur district. The Bioscan **6**(3): 459-462.

Gomez-Campo, C., Herranz-Sanz, J.M. and Montero-Riquelme, F. 2001. The genus *Coincya* Rouy (Cruciferae) in South-central Spain revisited: A morphometric analysis of population structure. Bot. J. Linn. Soc. **135**: 125-135.

Henderson, A. and Ferreira, E. 2002. A morphometric study of *Synechanthus* (Palmae). Syst. Bot. **27**(4): 693-702.

Hill, F.A. 1952. Economic Botany. McGraw Hill Books Company, New York, 560 pp.

Irwin, H.S. and Barneby, R.C. 1982. The American Cassiinae. Mem. New York Bot. Gard. **35**: 1-918.

Irwin, H.S. and Turner, B.L. 1960. Chromosomal relationships and taxonomic considerations on the genus *Cassia*. Am. J. Bot. **47**(4): 309-318.

Khan, M.S., Khatun, B.M.R. and Rahman, M.M. 1996. A preliminary account of legume diversity in Bangladesh. Bangladesh J. Plant Taxon. **3**(1): 1-33.

Marazzi, B., Endress, K.P., de Queiroz, L.P. and Conti, E. 2006. Phylogenetic relationships within *Senna* (Leguminosae, Cassiinae) based on three chloroplast DNA regions: Patterns in the evolution of floral symmetry and extrafloral nectaries. Am. J. Bot. **93**(2): 288-303.

Mulumba, J.W. and Kakudidi, E. 2010. Numerical taxonomic study of *Acacia senegal* (Fabaceae) in the cattle corridor of Uganda. South African J. Bot. **76**: 272-278.

Ogundipe, O.T., Kadiri, A.B. and Adekanmbi, O.H. 2009. Foliar epidermal morphology of some Nigerian species of *Senna* (Caesalpiniaceae). Indian J. Sci. Technol. **2**(10): 5-9.

Pinheiro, F. and de Barros, F. 2007. Morphometric analysis of *Epidendrum secundum* (Orchidaceae) in southeastern Brazil. Nordic J. Bot. **25**: 129-136.

Prain, D. 1903. Bengal Plants, Vol. **1**, Botanical Survey of India, Calcutta, India, pp.313-315.

Rahman, M.Z. and Rahman, M.O. 2012. A morphometric analysis of *Desmodium* Desv. (Fabaceae) in Bangladesh. Bangladesh J. Bot. **41**(2): 143-148.

Sadique, J. and Chandra, T. 1987. Biochemical modes of action of *Cassia occidentalis* and *Cardiospermum halicacabum* in inflammation. J. Ethnopharmacol. **19**: 201-212.

Sneath, P.H.A. and Sokal, R.R. 1973. Numerical Taxonomy, Freeman and Company, San Francisco, USA, 573 pp.

Soladoye, M.O., Sonibare, M.A. and Chukwuma, E.C. 2010a. Morphometric study of the genus *Indigofera* Linn. (Leguminosae-Papilionoideae) in South-western Nigeria. International J. Bot. **6**(3): 343-350.

Soladoye, M.O., and Onakoya, M.A., Chukwuma, E.C. and Sonibare, M.A. 2010b. Morphometric study of the genus *Senna* Mill. in South-western Nigeria. African J. Plant Science **4**(3): 44-52.

Sonibare, M.A., Jayeola, A.A. and Egunyomi, A. 2004. A morphometric analysis of the genus *Ficus* Linn. (Moraceae). African J. Biotechnol. **3**(4): 229-235.

Tona, L. and Mesia, K. 2001. *In vivo* antimalarial activity of *Cassia occidentalis, Morinda morindoides* and *Phyllanthus niruri.* Ann.Trop. Med. Parasitol. **95**: 47-57.

# 15

# REVISION OF *AMORPHOPHALLUS* BLUME *EX* DECNE. SECT. *AMORPHOPHALLUS* (ARACEAE) IN INDIA

V. Abdul Jaleel[1], M. Sivadasan[2,3], Ahmed H. Alfarhan[2],
Jacob Thomas[2] and A. A. Alatar[2]

*Department of Botany, University of Calicut, Calicut University P. O., 673 635, Kerala, India*

*Keywords: Amorphophallus* sect. *Amorphophallus*; Araceae; Endemics; India.

## Abstract

*Amorphophallus* Blume *ex* Decne. sect. *Amorphophallus* in India is revised. It is the smallest of the three sections in India with five species, viz. *A. hirsutus* Teysm. & Binn., *A. kachinensis* Engl. & Gehrm., *A. longistylus* Kurz, *A. napalensis* (Wall.) Bogner & Mayo and *A. paeoniifolius* (Dennst.) Nicolson. *Amorphophallus paeoniifolius* is the widely distributed species in India with two varieties, viz. *A. paeoniifolius* var. *paeoniifolius*, and var. *campanulatus* (Decne.) Sivad. *Amorphophallus longistylus* is the only species of the section endemic to India.

## Introduction

*Amorphophallus* Blume *ex* Decne. sect. *Amorphophallus* (Araceae) is the smallest of the three sections represented in India. Revisions of the other two sections, viz. *Rhaphiophallus* and *Conophallus* have been carried out recently by Jaleel *et al.* (2011, 2012).

Engler (1911) in his revision of the genus treated *Amorphophallus campanulatus* Decne. [=*A. paeoniifolius* (Dennst.) Nicolson] under the section *Cundarum* Engl. which is a renaming of *Candarum* Rchb. *ex* Schott. According to Engler (1911), composition of the name '*Candarum*' was improper and he renamed it as '*Cundarum*' stating that the name is derived from the Indian name '*Kunda*' (= *Amorphophallus campanulatus*) to which the suffix '*arum*' was added resulting in '*Kundarum*' and written as '*Cundarum*'. He considered it as a new name with his authorship. But as per Art. 60.1 of ICN (McNeill *et al.*, 2012) the original spelling of a name or epithet is to be retained, and hence the name *Cundarum* Engl. is treated as illegitimate. Engler (1911) included *Amorphophallus campanulatus*, the type of the genus *Amorphophallus* under the section *Cundarum* Engl. As per Art. 22.1. of ICN, the name of any subdivision of a genus that includes the type of an adopted, legitimate name of the genus to which it is assigned is to repeat that generic name unaltered as its epithet, and accordingly the correct name of the section of *Amorphophallus* that includes *Amorphophallus campanulatus* [=*A. paeoniifolius*] should have been *Amorphophallus* sect. *Amorphophallus*, and it is used in the present study.

Out of the seventeen species of *Amorphophallus* included in *Flora of British India* by Hooker (1894), only eight were reported to be occurring in India and *Amorphophallus longistylus* Kurz *ex* Hook. f., *A. campanulatus* (=*A. penoniifolius*) and *A. dubius* Blume (=*A. penoniifolius*) are among those belonging to the section *Amorphophallus* as per the present standards adopted for species delimitation. Engler (1911) in his monographic work included *A. longistylus* under the section *Conophallus*, and *A. napalensis* was treated under the genus *Thomsonia* which was later transferred to *Amorphophallus* by Bogner *et al.* (1985). Since Engler's work, several new species have been discovered from various parts of India. Brief accounts on earlier taxonomic work on Indian *Amorphophallus* have been provided by Jaleel *et al.* (2011, 2012). The present article is the third and final part of revision of the genus *Amorphophallus* in India.

[1]Present address: Department of Botany, Sir Syed College, Taliparamba, Kannur-670 142, Kerala, India
[2]Department of Botany & Microbiology, College of Science, King Saud University, P. O. Box 2455, Riyadh-11451, Kingdom of Saudi Arabia
[3]Corresponding author: Email: drmsivadasan@rediffmail.com

The methodology adopted in earlier works (Jaleel *et al.*, 2011, 2012) have been followed in the present work. Extensive and exhaustive field explorations covering all seasons were made all over India for collection and recording relevant data of the specimens.

Indian specimens available at various major herbaria such as ASSAM, BM, BSA, BSD, BSHL, BSI, CAL, CALI, DD, GH, JCB, K, KFRI, L, M, MH, PBL, TBGT and US were consulted, and representative specimens were cited.

**Taxonomic treatment**

**Amorphophallus** Blume *ex* Decne., Nouv. Ann. Mus. Hist. Nat. 3: 366 (1834), *nom. cons.* [Taxon 31: 310 (1982)].

   *Type: Amorphophallus campanulatus* Decne. [= *A. paeoniifolius* (Dennst.) Nicolson].

**Amorphophallus** Blume *ex* Decne. sect. **Amorphophallus**, *emend.* Sivad. *mut. char.* [*Amorphophallus* sect. *Candarum* Blume, Rumphia 1:139 (1835). *Amorphophallus* sect. *Cundarum* Engl., Pflanzenr. IV. 23C (48): 74 (1911), *nom. illegit.*].

   Spathe usually campanulate with a basal convolute tube and an upper horizontally spreading limb or with basal convolute tube and erect, oblong or oblong-ovate or ovate limb; appendix usually conoid, hemispheric or cylindric; style always long, equal to or 2-4 times the height of ovary; stigma lobed.

**Key to the Indian species of *Amorphophallus* sect. *Amorphophallus***

1.  Peduncle 3-8 cm long; spathe broadly campanulate with basal convolute tube and horizontally spreading limb; spadix-appendix sub-globose or conoid.                                              2

-   Peduncle more than 20 cm long; spathe ovate with basal convolute tube and upper erect open limb; spadix-appendix elongate-ovoid or cylindric.                                              3

2.  Spadix-appendix with subglobose base and apical short cylindric truncate column bearing short stiff hairs.                                              *A. hirsutus*

-   Spadix-appendix subglobose or conoid, glabrous.                                              *A. paeoniifolius*

3.  Petiole and peduncle smooth with even surface; spadix shorter than spathe; appendix elongate-ovoid with irregular longitudinal grooves or fissures, or cylindric with warts; style length more or less equal to the height of ovary.                                              4

-   Petiole and peduncle smooth with uneven surface having small bumps; spadix longer than spathe; appendix elongate, cylindric with tapered tip, smooth; style 2-3 times the height of ovary.                                              *A. longistylus*

4.  Spadix stipitate; appendix elongate-ovoid, surface with irregular longitudinal grooves or fissures; style straight; stigma inconspicuously 3-lobed.                                              *A. kachinensis*

-   Spadix sessile; appendix cylindric, surface rough with short prominent protuberances; style bent towards spadix-axis; stigma 4-lobed.                                              *A. napalensis*

**Amorphophallus hirsutus** Teysm. & Binn., Naturk. Tijdschr. Nederl. Ind. XXIV: 332 (1862); Engl., Pflanzenr. IV. 23C (48): 106 (1911); Sivad. & Jaleel, Rheedea 10(2): 143 (2000)    (**Fig. 1**).

   *Type*:  West Sumatra, Soeka Menanti, Ophir, (no date), *Buennemeijer* 1019 (*Neotype:* BO).

   Tubers depressed-globose, 4.5-8.5 cm diam. and 2.5-5.0 cm thick in vegetative phase; c. 11 cm diam. and 7 cm thick in reproductive phase; root scars thickened, annulate. Petiole 73-82 cm long, pale green with small blackish green irregular specks and mottles with minute dark green spots in between, paler towards the tip, extreme base with purplish blue hue; some petioles with pale green background having large irregular blackish green patches with light greyish margin, and minute greenish spots in between, and paler towards the tip. Leaflets sessile, elliptic-lanceolate, large leaflets 11-17 × 3-5 cm, small 6.2-10.5 × 1.8-4.0 cm, acuminate at apex, base unequal, decurrent on rachis, greenish above and pale below. Peduncle short, 3.0-5.5 cm long, greenish in colour. Spathe campanulate, broadly triangular-ovate, c. 17.5 × 26.0 cm, basal tube separated from limb by a constriction, tip acute, margin undulate; tube c. 7.5 cm diam. and 5.5 cm

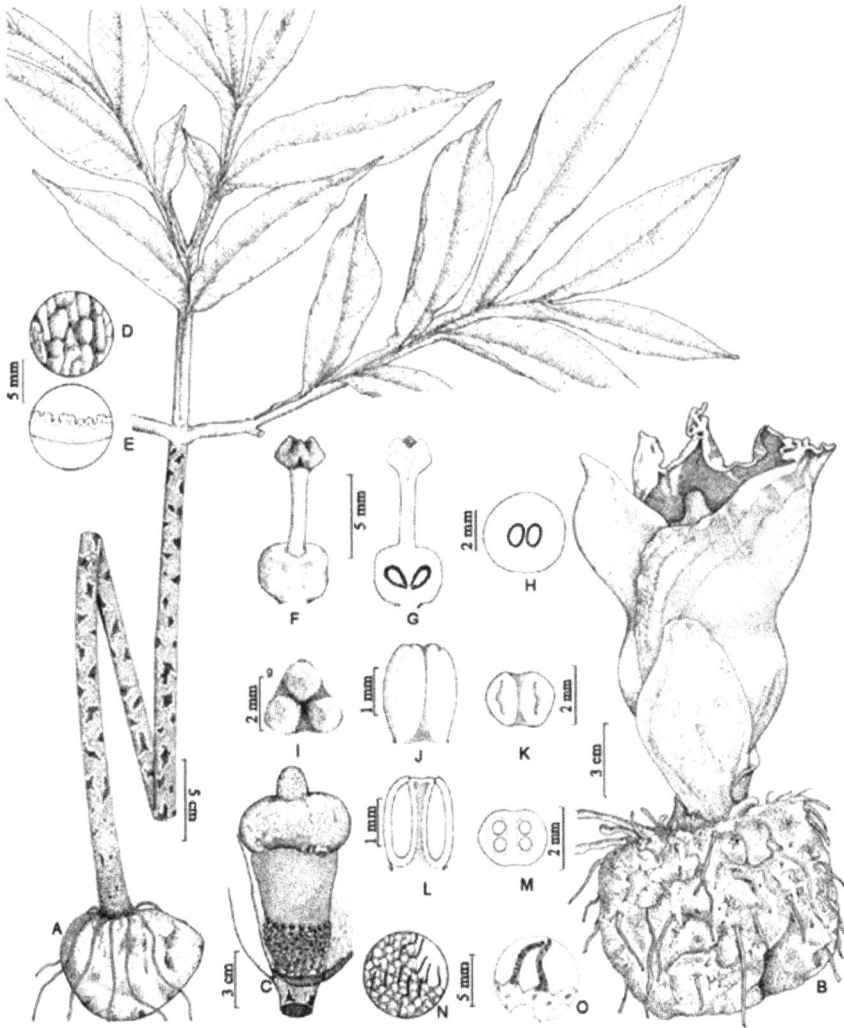

Fig. 1. *Amorphophallus hirsutus* Teysm. & Binn. A. Tuber with leaf; B. Tuber with inflorescence; C. Inflorescence - spathe partially removed showing spadix; D. Small basal inside portion of spathe; E. A small basal portion of spathe - c.s.; F. Female flower; G. Female flower - l.s.; H. Ovary - c.s.; I. Stigma - view from top; J. Male flower - view from broad side; K. Male flower - view from top; L. Male flower - l.s.; M. Male flower - c.s.; N. A small portion of apical portion of spadix-appendix showing papillae; O. Appendix papillae - enlarged.

high, greenish outside with few small white mottling, smooth, pale greenish inside, purplish-orange or maroon at extreme base, rough, irregularly and longitudinally rugose and verrucate; limb c. 9.5 cm long spreading, up to c. 12 cm diam., purplish outside and inside. Spadix shorter than spathe, c. 12.5 cm long; sessile, female zone c. 3.3 cm long; male zone c. 3.8 cm long; appendix c. 3 cm high and c. 5.5 cm diam. Female flowers dense, each flower c. 10 mm high; ovary subglobose, pale greenish, c. 4 mm diam. and 3 mm high, 2-3-locular, each locule with a single basal anatropous ovule; style c. 5 mm long, purplish; stigma 2-3-lobed, c. 2.5 mm diam., pale yellowish. Male flowers dense, pale yellowish with purplish tinge at top of connectives; each c. 2 mm high, sessile, inconspicuously 2-lobed. Spadix-appendix subglobose, light purplish yellow, rough, abruptly narrowed to a cylindric truncate column of c. 1.8 cm high and 0.9 cm diam., light purplish yellow, rough; cylindric column and its basal surrounding portion covered with stiff slender bulbous-based papillae; papillae on cylindrical column smaller compared to that of basal neighbouring portion; appendix becomes unevenly bullate after anthesis.

*Phenology:* Flowering: May; fruiting: Fruiting specimens could not be collected.

*Representative specimens examined*: **Andaman and Nicobar Islands:** Great Nicobar Island, On the way to East-West Road, 17.5.1999, *Abdul Jaleel* RIA 350 (Infl.) (CALI); *Ibid.*, 17.5.1999, *Abdul Jaleel* RIA 351 (tuber and leaf) (CALI); *Ibid.*, 20.5.1999, *Abdul Jaleel* RIA 355 (tuber and leaf) (CALI). **Kerala:** Calicut University Botanical Garden, 17.4.2000, *Abdul Jaleel* RIA 382 (Infl.) (CALI) (originally collected from East West Road, Great Nicobar Island, and introduced and flowered in the Calicut University Botanical Garden).

*Notes*: *Amorphophallus hirsutus* resembles *A. paeoniifolius* and *A. prainii* in general vegetative morphology and inflorescence, especially during the early stages. It differs from the latter two by having a subglobose appendix with a cylindric apical column covered with short stiff papillae. There is no other Indian species with hairs on the spadix-appendix.

*Distribution*: Originally collected from Java and Hetterscheid and Ittenbach (1996) reported its occurrence in Western Sumatra. Sivadasan and Jaleel (2000) first reported it from Great Nicobar islands of India where it is rare.

**Amorphophallus kachinensis** Engl. & Gehrm. in Engler, Pflanzenr. IV. 23C (48): 91 (1911); Hett. & Itten., Aroideana 19: 87 (1996).                                               (**Fig. 2**).

*Type*: Upper Burma, Kachin Hills, 20.5.1898, *Shaik Mokim*, s.n. (*Holotype:* CAL).

Tubers depressed-globose, 5-30 cm diam. and 3-5 cm thick, skin brownish; produce offsets. Petiole c. 20 cm long, smooth, dirty white background with green to reddish brown spots. Leaflets elliptic, 6-9 × 2-3 cm, tip acute-acuminate. Peduncle 24-80 cm long. Spathe more or less boat-shaped, slightly convolute at base, 8-29 × 7-14 cm, tip rounded or truncate, green or greenish brown outside with green spots or purplish stripes and spots; light purplish within, with scattered, shallow, punctiform warts at base. Spadix much shorter than spathe, 6.5-18.0 cm long, stipitate, stipe 0.2-1.0 cm long; female zone c. 5 cm long; male zone c. 7.5 cm long; spadix-appendix c. 15 cm high. Female flowers dense, each c. 2.5 mm high; ovary sub-globose, c. 1.5 mm high, unilocular with basal anatropous ovule; style c. 0.7 mm long; stigma inconspicuously 3-lobed. Male flowers dense, each c. 2 mm broad, sessile, inconspicuously 2-lobed. Spadix-appendix ellipsoid or ovoid with several irregular longitudinal grooves or fissures.

*Phenology:* Flowering: April - May.

*Representative specimens examined*: **Arunachal Pradesh:** Ziro, 10.4.2006, *Abdul Jaleel* RIA 424 (infl.) (CALI); *Ibid.*, 7.8.2006, *Abdul Jaleel* RIA 430 (leaf) (CALI).

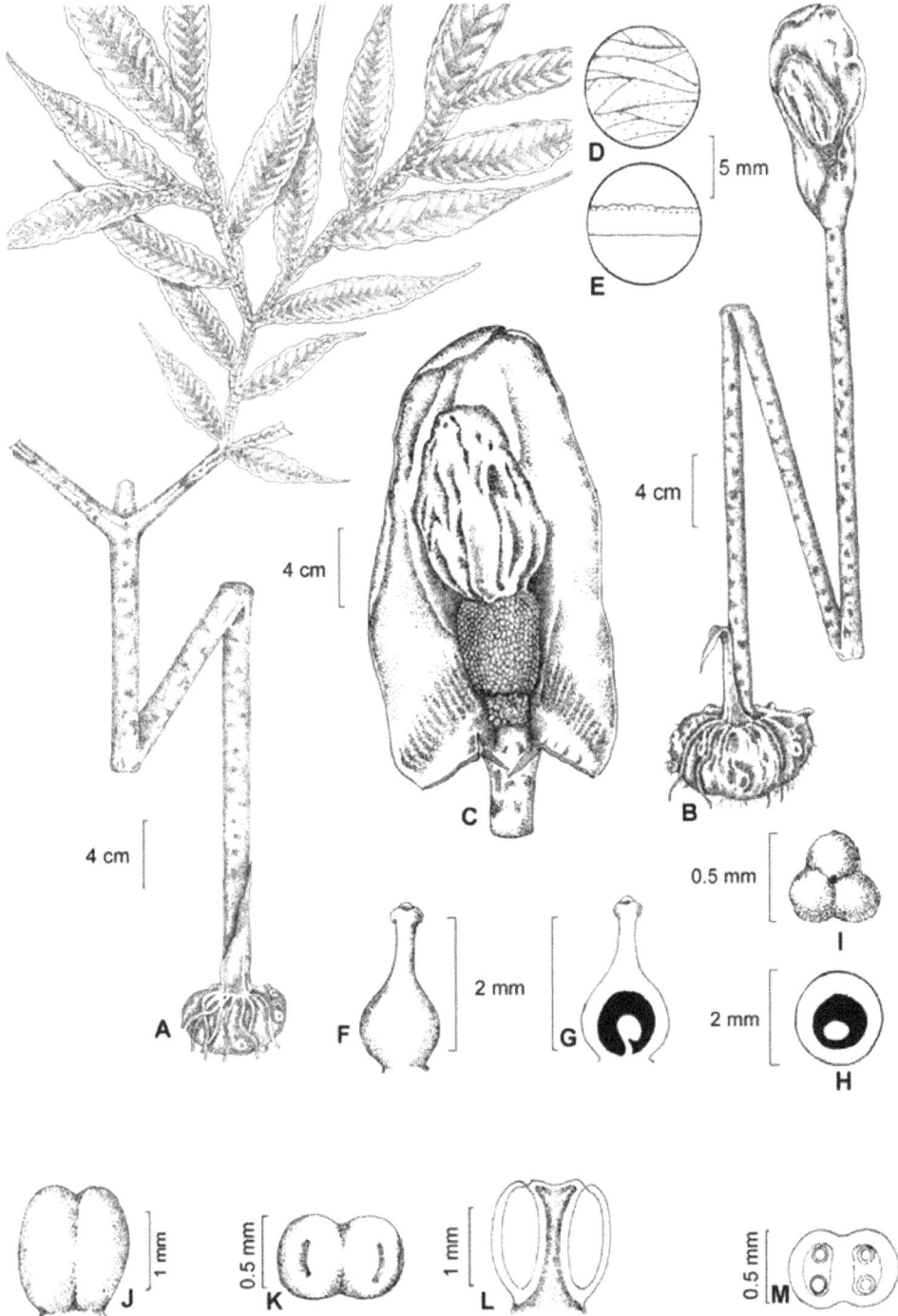

Fig. 2. *Amorphophallus kachinensis* Engl. & Gehrm. A. Tuber with leaf; B. Tuber with inflorescence; C. Inflorescence - spathe cut-opened showing spadix; D. A small basal portion of spathe; E. A small basal inside portion of spathe - c.s.; F. Female flower; G. Female flower - l.s.; H. Ovary - c.s.; I. Stigma - view from top; J. Male flower - view from broad side; K. Male flower - view from top; L. Male flower - l.s.; M. Male flower - c.s.

*Notes*: *Amorphophallus kachinensis* resembles *A. corrugatus* N. E. Br. and *A. yunnanensis* Engl. (1911) in general morphology but differs in the nature of spadix-appendix by having longitudinal irregular grooves. It also differs from *A. yunnanensis* in having female flowers with long style and smaller stigma.

*Distribution*: Northern Myanmar (Kachin state), Northern Thailand, Laos, China (Yunnan) and India. In India, it was collected from Arunachal Pradesh and present collection of the species forms a new distributional record for India.

**Amorphophallus longistylus** Kurz (Rep. Andaman 50.1866, *nomen*) *ex* Hook. f., Fl. Brit. India 6: 515 (1893); Engl., Pflanzenr. IV. 23 (48): 83 (1911); Sivad. & Jaleel, Rheedea 8(1): 103 (1998).

**(Fig. 3)**.

*Types*: India, South Andaman, (no date), *Kurz* s.n. (*Holotype:* CAL; *Isotype:* K).

Tubers sub-globose, 5.0-6.5 cm diam. and 4-5 cm thick in vegetative phase, c. 6.2 cm diam. and 4.2 cm thick in reproductive phase. Petiole 43-71 cm long, surface uneven with white elongated or round thickened swellings or projections; pale greenish with greenish black and white mottles. Leaflets sessile, ovate to oblong, apex acuminate, 6.5-12.0 × 2.8-4.6 cm; base of leaflets on secondary rachises decurrent, leaflets on primary rachises usually not decurrent-based; greenish above, paler below with pinkish tinge along the veins, margins and tip; margin undulate. Peduncle 30.5-42.5 cm long, 1.0-1.2 cm diam. Spathe erect, ovate-lanceolate, 18.5-23.5 × 9.5-10.5 cm, tip acute, basal convolute tube 7.5-14.0 cm high and 3.5-4.0 cm diam.; upper expanded limb portion with longitudinal shallow folds on either side of mid-portion; light purplish with greenish black blotches and small dark spots on outside, more on basal tube; smooth, dark purplish and verrucose at base, pale purplish and smooth above with few light greenish-black blotches above within. Spadix longer than spathe, exserted, 24-26 cm long, stipitate; stipe short, 0.2-0.4 cm long; female zone 2.4-3.2 cm; male zone 3.2-3.7 cm long; appendix 17-20 cm long. Female flowers 60-80, loosely arranged, each 5-8 mm high, ovary 1.7-2.0 mm high, shallowly 3-5-lobed, greenish yellow, 3-5-locular, each locule with a single basal anatropous ovule; style 3-7 mm long, rarely with longitudinal ridges at upper portion, dark purplish; stigma 3-5-lobed, rarely 2-lobed, creamy. Male flowers dense, each c. 2 mm high, sessile, inconspicuously 2-lobed, yellowish-creamy with purplish tinge on connective at top. Spadix-appendix cylindric, gradually tapering to tip.

*Phenology:* Flowering: May - June; fruiting: Fruits could not be collected.

*Representative specimens examined:* **Andaman and Nicobar Islands**: South Andaman, date nil, *S. Kurz* s.n. (CAL, K); Middle Andaman, Panchawati, 7.12.1997, *Jaleel & Bobby Thomas* RIA 224 (leaf) (CALI); North Andaman, Ray Hill, 26.5.1999, *Jaleel* RIA 357 (leaf) (CALI). **Kerala**: Calicut University Botanic Garden, 2.5.1998, *Jaleel & Sivadasan* RIA 275 (infl.) (CALI); *Ibid.*, 21.5.1998, *Jaleel & Sivadasan* RIA 227 (leaf) (CALI) (Originally collected by Jaleel and Bobby Thomas (RIA 224) on 7.12.1997 from Panchawati, Midddle Andaman and introduced in the Calicut University Botanic Garden).

*Notes:* This is a very rare species and was left unknown until it was re-collected in 1997 from Middle Andaman about 131 years after it was first collected by Kurz (Sivadasan and Jaleel, 1998). The species is very distinct in having uneven surface of petioles and peduncles with bumps or swellings.

*Distribution*: Known to occur only in Andaman Islands.

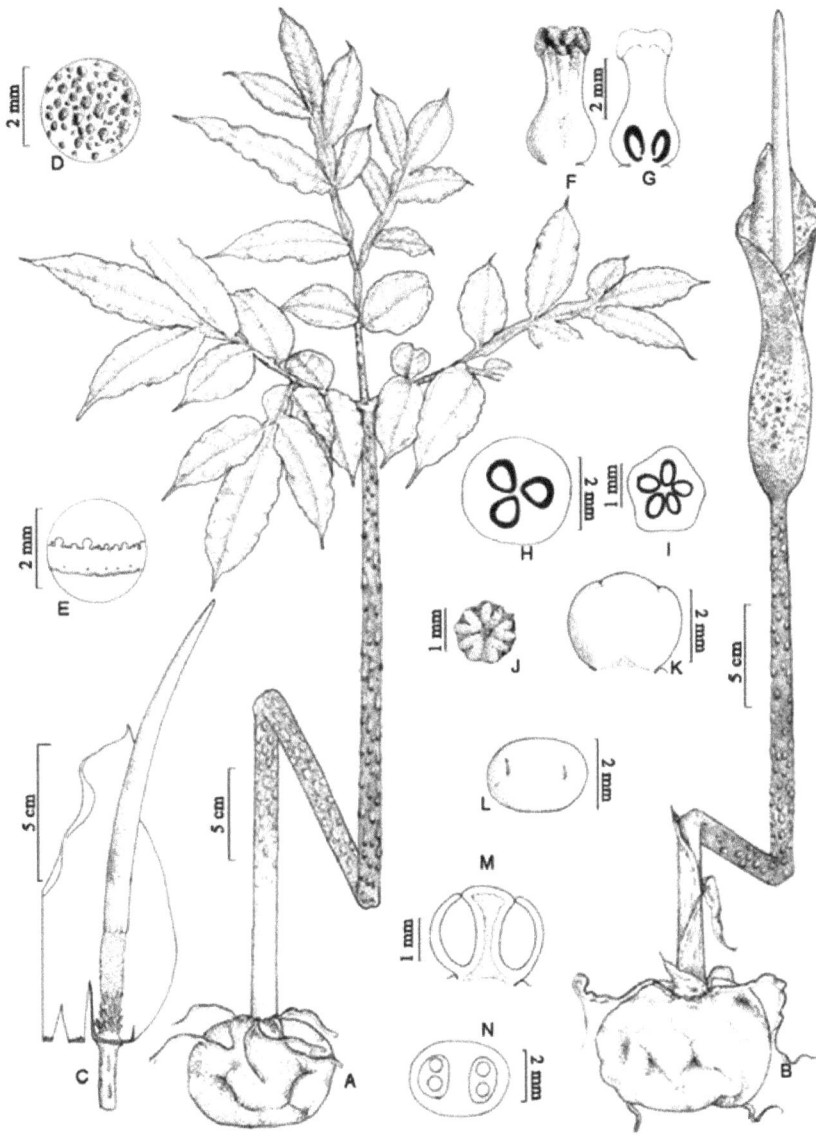

Fig. 3. *Amorphophallus longistylus* Kurz. A. Tuber with leaf; B. Tuber with inflorescence; C. Inflorescence - spathe cut-opened showing spadix; D. Small basal inside portion of spathe; E. A small basal portion of spathe - c.s.; F. Female flower; G. Female flower - l.s.; H. Ovary with three locules - c.s.; I. Ovary with five locules - c.s.; J. Stigma of five-loculed flower - view from top; K. Male flower - view from broad side; L. Male flower - view from top; M. Male flower - l.s.; N. Male flower - c.s.

**Amorphophallus napalensis** (Wall.) Bogner *et* Mayo in Bogner *et al*., Aroideana 8(1): 19 (1985); Hett. & Ittenbach, Aroideana 19: 103 (1996). *Thomsonia napalensis* Wall., Pl. Asiat. Rar.1: 83, t. 99 (1830); Hook. f., Fl. Brit. India 6: 518 (1893); Engl., Pflanzenr. IV. 23C (48): 56 (1911). *Pythonium wallichianum* Schott in Schott & Endl., Melet. Bot. 17 (1832).          (**Fig. 4**).

*Type*: Nepal, in mountain forests, flowering in June, t. 99 (Wallich, *Pl. Asiat. Rar.,* 1830).

Tubers subglobose or depressed-globose, 5.0-8.5 cm diam. and 4.5-6.5 cm thick in vegetative phase, 7.5-10.0 cm diam. and 5-8 cm thick in reproductive phase, offsets produced from tuber

(observed only in tubers of vegetative phase), each offset 3.5-7.5 cm long. Petiole smooth, 41.5-
78.0 cm long, pale green with more or less irregular or oval elongate, brown or dark brown
patches. Leaflets sessile, ovate-lanceolate, large leaflets 12.2-19.2 × 5.3-6.0 cm, small leaflets 5.5-
9.0 × 2.5-4.5 cm, marginal portion undulate, tip long-acuminate; upper surface green, lower
surface pale green. Peduncle smooth, 67-76 cm long. Spathe elongate-obovate with acute tip, 24-
27 cm long, differentiated into a basal convolute tube and open limb; tube 4.5-6.0 cm long and
5.5-7.0 cm diam., pale green with greenish brown at base outside, pale green within, changes to

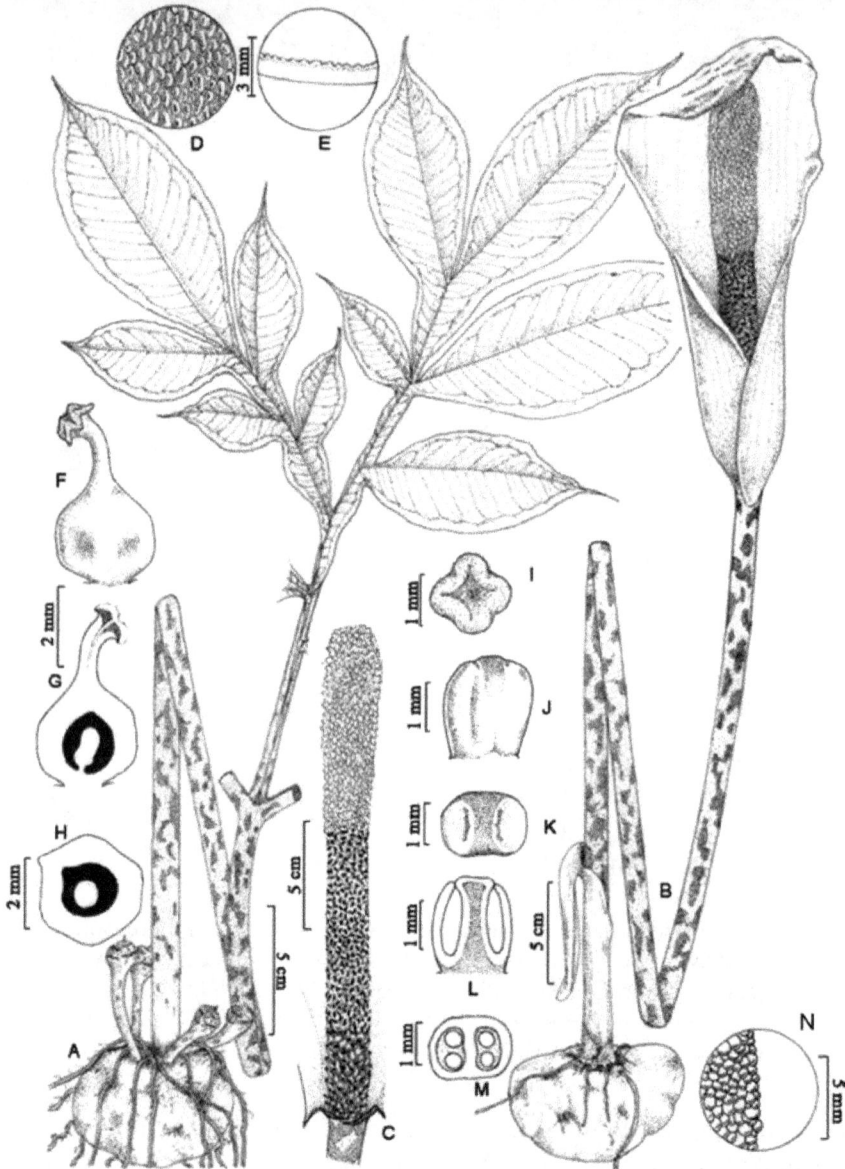

Fig. 4. *Amorphophallus napalensis* (Wall.) Bogner *et* Mayo. A. Tuber with leaf; B. Tuber with inflorescence;
C. Inflorescence - spathe partially removed showing spadix; D. Small basal inside portion of spathe; E.
Small basal portion of spathe - c.s.; F. Female flower; G. Female flower - l.s.; H. Ovary - c.s.; I. Stigma -
view from top; J. Male flower - view from broad side; K. Male flower - view from top; L. Male flower -
l.s.; M. Male flower - c.s.; N. A small lateral portion of spadix-appendix.

yellowish after male anthesis, mouth of tube wide, limb expanded, become apically cucullate. Spadix sessile, shorter than the spathe, 18-20 cm long; female zone 3-4 cm long; male zone 6-7 cm long; appendix 10-12 cm long. Female flowers dense, each 4-5 mm high, ovary sub-spherical, greenish, 2.0-2.5 mm high, unilocular with single basal anatropous ovule; style c. 2 mm long, apically bent towards spadix-axis; stigma 4-lobed. Male flowers dense, each 2.0-2.5 mm high, inconspicuously 2-lobed, pale yellowish with reddish brown tinge on connective at top. Spadix-appendix cylindric with obtuse tip, rough with prominent, short cylindric or obovoid warts, greenish when young, brownish yellow when mature, yellowish after anthesis. Fruits ovoid-elliptic, 1.4-1.5 cm long. Seeds ellipsoid, 1.2-1.3 cm long.

*Phenology:* Flowering: May - June; fruiting: July - September.

*Representative specimens examined*: **Sikkim**: Gangtok, Deurali, 23.9.1997, *Abdul Jaleel & Bobby Thomas* RIA 168 (leaf) (CALI); *Ibid.*, 24.9.1997, *Abdul Jaleel & Bobby Thomas* RIA 171 (infr.) (CALI). **Assam**: Assam, 29.5.1896, *Prain,* Acc. No. 496857 (leaf) (CAL); Chirapunji, 3.6.1956, *Rolla Seshagiri Rao* 2697 (infl.) (ASSAM). **Meghalaya**: Shillong, Woodlands, BSI Campus (Introduced; exact locality of original collection not known), 15.7.1967, *Verma* 35658 (infl.) (ASSAM); Shillong, Oakland, 24.6.1998, *Abdul Jaleel* RIA 289 (infl.) (CALI); **Arunachal Pradesh**: Sessa Orchid Sanctuary, 27.6.1998, *Abdul Jaleel* RIA 314 (leaf) (CALI).

*Notes*: *Amorphophallus napalensis* differs from other Indian species in having a verrucate appendix with short, prominent warts, style apically bent towards spadix-axis, and stigma 4-lobed.

*Distribution*: Bhutan, Nepal and India (Sikkim, Assam, Meghalaya and Arunachal Pradesh).

**Amorphophallus paeoniifolius** (Dennst.) Nicolson, Taxon 26: 338 (1977); Nicolson in Saldanha & Nicolson, Fl. Hassan Dist. App. II : 7 (1978) ("1976"). *Dracontium paeoniifolium* Dennst., Schlüssel Hort. Malab. : 13, 38 (1818); Manitz, Taxon 17: 449 (1968). *Arum campanulatum* Roxb., [Hort. Beng. : 66 (1814), *nom. nud.*] Pl. Corom. 3: 68 (1819), *nom. illegit.* (incl. type of *D. paeoniifolium* Dennst., 1818); Wight, Icon. Pl. Ind. Or. 3: 5 (1844). *Amorphophallus campanulatus* Decne., Nouv. Ann. Mus. Hist. Nat. Paris 3: 336 (1834).

*Lectotype*: Rheede's illustration of *Mulenschena* in Hort. Malab. 11: t. 19 (1692), vide Nicolson, Taxon 26: 338 (1977).

Tuber depressed-globose. Petiole up to 1.2 m long, smooth or muricate, mottled. Inflorescence short-peduncled, elongating after anthesis. Spathe broadly campanulate, c. 25 × 28 cm, convolute below and spreading above. Spadix sessile, differentiated into basal female portion, a subturbinate or subcylindric male portion and an apical naked sterile, sessile subglobose or elongate-conoid appendix, wrinkled at maturity, spongy within. Ovary light purplish or pale yellowish, 2-3-loculed, each locule with single anatropous ovule; style elongate; stigma yellowish, reniform or 2-3-lobed. Male flowers creamy-yellow.

*Notes:* The close morphological similarities of *Schena* (Hort. Malab. 11: 35, t. 18. 1692) and *Mulenschena* (Hort. Malab. 11: 37, t. 19. 1692) of Rheede and unawareness of the existence of an earlier legitimate epithet for *Mulenschena*, the name *Amorphophallus campanulatus* Decne. had been used for both the cultivated and wild elements represented by *Schena* and *Mulenschena* respectively. Realizing the existence of an earliest epithet for *Amorphophallus campanulatus,* Nicolson (1977) made a combination of name, viz. *Amorphophallus paeoniifolius* (Dennst.) Nicolson, as applying to *Amorphophallus campanulatus* (*sensu lato*). The wild and cultivated elements differ in many respects even though they resemble in general appearance and many other characteristics; and hence they are treated as two distinct varieties of *A. paeoniifolius*. Backer (1920) recognized the wild and cultivated elements as belonging to two distinct subspecific taxa and assigned the rank 'hoofdgroep' which is not valid. Backer and Bakhuizen van den Brink

(1968) replaced the rank 'hoofdgroep' with 'forma'. Detailed accounts on the identity and nomenclature of Rheede's *Schena* and *Mulenschena* have been provided by Suresh *et al.* (1983).

*Distribution*: India, Sri Lanka and Pacific islands. In India, found in almost all States.

## Key to the varieties of *Amorphophallus paeoniifolius*

1.  Petiole usually purplish brown with light pinkish blotches, strongly muricate especially at basal half; leaflet-bases strongly decurrent on primary rachises to the  main junction; spadix-appendix elongate-conoid, height more than breadth; style length about double the ovary height; stigma usually 2-lobed.                                                    var. *paeoniifolius*

-   Petiole usually greenish with white blotches, smooth, rarely slightly rough at basal half; leaflet-bases not decurrent to the junction of the primary rachises; spadix-appendix subglobose, breadth more than height; style length 3-4 times the ovary height; stigma usually 3-lobed.            var. *campanulatus*

**Amorphophallus paeoniifolius** (Dennst.) Nicolson var. **paeoniifolius.** Sivad. in Suresh, Sivad. & Manilal, Taxon 32: 128 (1983); Karth., Jain, Nayar & Sanjappa, Fl. Ind. Enum. Monocot. : 6 (1989); Sivad. in Manoharan, Biju, Nayar & Easa, Silent Valley-Whisp. Reas.: 230 (1999). [*Mulenschena* Rheede, Hort. Malab. 11: 37, t. 19 (1692)]. *Dracontium paeoniifolium* Dennst., Schlüssel Hort. Malab.: 13, 21, 38 (1818) ('*paeoniaefolium*'); Manitz, Taxon 17: 499 (1968). *Arum campanulatum* Roxb. [Hort. Beng. 65 (1814)], Pl. Corom. 3: 68 (1820), *nom. illegit.*; Wight, Ic. Pl. Ind. Or. 3: 5, t. 785 (as to leaf) (1844). *Candarum hookeri* Schott in Schott & Endl., Melet. Bot.: 17 (1832), *nom. illegit. Kunda verrucosa* Raf., Fl. Tellur. 2: 82 (1837), *nom. illegit. Amorphophallus rex* Prain, J. Asiat. Soc. Bengal 62: 79 (Aug. 1893); Hook. f., Fl. Brit. India 6: 514 (Sept. 1893); Engl., Pflanzenr. IV. 23C (48): 75 (1911). *Amorphophallus campanulatus* hoofdgroep *sylvestris* Backer, Determ.-tab. Jav. *Amorphophallus*: 2 (1920), (rankless name). *Amorphophallus campanulatus* f. *sylvestris* Backer *ex* Backer & Bakh., Fl. Java 3: 112 (1968).

**(Fig. 5)**.

*Type*:   Same as of the species. [*Lectotype*: Rheede's illustration of *Mulenschena* in Hort. Malab. 11: t. 19 (1692), vide Nicolson, Taxon 26: 338 (1977)].

Tubers depressed-globose, 9-14 cm diam. and 8-11 cm thickness in vegetative phase; 13-20 cm diam. and 7-9 cm thickness in reproductive phase; skin pale brown to dark brown with prominent root scars. Petiole rough, 55-105 cm long, dark brownish green with round to ovoid green blotches, extreme base pinkish with minute pale greenish dots above. Leaflets sessile, large leaflets 17-20 × 5-6 cm, small leaflets 8-10 × 2.5-3.5 cm, obovate, tip acute, base unequal and decurrent on rachis, greenish above, paler below. Peduncle short, rough, 3-6 cm long. Spathe margin undulate, tip acute, 10-21 × 12-22 cm; outside greenish yellow with green blotches at base, pale brownish with pale yellow blotches above; brownish green with prominent murications at base within, pale green above. Spadix sessile, 15.5-16.0 cm long; female zone 4-5 cm long; male zone 3.5-4.5 cm long; appendix 9.5-10.5 cm long, 7-8 cm diam. at base. Female flowers dense, each  with ovary c. 2 mm high, sub-globose, style 4-5 mm long, pinkish brown; stigma 2-lobed, rarely inconspicuously 3-lobed. Male flowers dense, each 4.0-4.5 cm high, cream-coloured. Spadix-appendix dark purplish red. Fruits green, reddish at maturity, each 1.6-1.8 cm long. Seeds 1-2, each 1.1-1.3 cm long.

*Phenology:* Flowering:  May - June; fruiting:  July - November.

*Representative specimens examined*: **Kerala**: Pathanamthitta Dist.: Sabarimala, 18.5.1997, *Sivadasan* CU 19152A (infl.) (CALI). Thrissur Dist.: Kuthiran, 20.5.1997, *Abdul Jaleel* RIA 55 (infl.) (CALI). Palakkad Dist.: Walayar, 28.7.1929, *Raju & Ratnavelu* 18644 (leaf & infr.) (MH); Walayar forest, 15.4.1979, *Sivadasan* CU 21442 (infl.) (CALI). Malappuram Dist.: Calicut University Campus, 26.6.1997, *Abdul Jaleel* RIA 96 (infl.) (CALI). Kozhikode Dist.: Mukkam, Pannicode, 30.3.1997, *Abdul Jaleel* RIA 65 (infl.) (CALI). Wayanad Dist.: Waithiri, 16.5.1977,

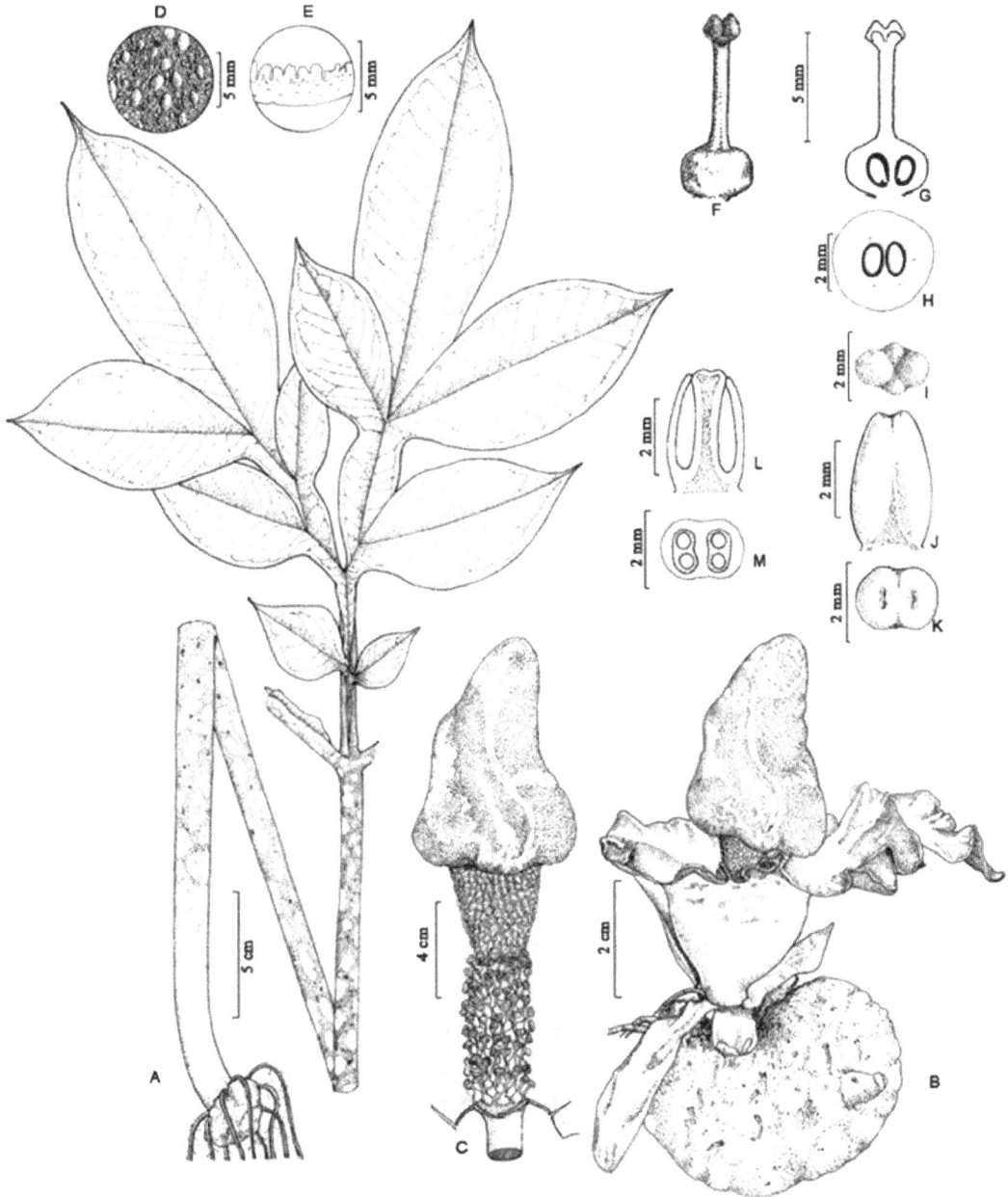

Fig. 5. *Amorphophallus paeoniifolius* (Dennst.) Nicolson var. *paeoniifolius*. A. Tuber with leaf; B. Tuber with inflorescence; C. Spadix; D. Small basal inside portion of spathe; E. A small basal portion of spathe - c.s.; F. Female flower; G. Female flower - l.s.; H. Ovary - c.s.; I. Stigma; J. Male flower - view from broad side; K. Male flower - view from top; L. Male flower - l.s.; M. Male flower - c.s.

*Sivadasan* CU 19175(CALI). Kannur Dist.: Taliparamba, 16.5.1982, *Nair* 73916 (infl.) (MH). **Karnataka**: Uduppi, Padigara, 25.5.1997, *Abdul Jaleel* RIA 58 (infl.) (CALI). **Andhra Pradesh**: Sreekakulam Dist.: 17.5.1979, *Subha Rao* 62460 (infl.) (MH). **Maharashtra**: Rajpurla, 11.9.1957, *Jain* 24248 (leaf) (CAL). **Madhya Pradesh**: Indore, Mandu, 20.9.1964, Acc. No. 5980 (BSA). **Orissa**: Dandarkarmya, Bichacakotta coffee plantation area, 27.5.1959, *Rao* 18585 (infl.) (ASSAM). **Bihar**: Borin HB, 13.12.1957, *Panigrahi* 11707 (leaf) (ASSAM); **Rajasthan**: Rajputana, Chitargo, 29.4.1896, *Prain* Acc. No. 496617 (infl.) (CAL). **West Bengal**: Botanical Garden, Calcutta (cultivated, originally collected from Bhitorgarh, Rajputana), 3.5.1895, *S. coll.*, s.n. (spathe) (CAL); Howrah Dist., Santra gatchi, 24.4.1896, *Prain*, Acc. No. 496599 (infl.) (CAL). **Assam**: Moshmai falls, 11.11.1938, *Biswas* 3928 (infr.) (CAL). **Tripura**: Agarthala, 10.5.1915, *Debbarmar*, Acc. No. 496616 (leaf) (CAL). **Andaman and Nicobar Islands**: South Andaman, Ograbraj, 30.11.1997, *Abdul Jaleel & Bobby Thomas* RIA 219 (leaf) (CALI); North Andaman, Kalipur, 13.12.1997, *Abdul Jaleel & Bobby Thomas* RIA 234 (infr.) (CALI); Nicobar islands: Car Nicobar, Lapathi – on the way to Tip Top, 12.5.1999, *Abdul Jaleel* RIA 339 (leaf) (CALI); Kamorta, Nancowry island, 13.5.1999, *Abdul Jaleel* RIA 342 (leaf) (CALI).

*Note*: A detailed account on identity of Rheede's (1692) *Mulenschena* was provided by Suresh *et al.* (1983) wherein nomenclatural history of *Amorphophallus paeoniifolius* var. *paeoniifolius* was elaborated.

**Amorphophallus paeoniifolius** (Dennst.) Nicolson var. **campanulatus** (Decne.) Sivad. in Suresh, Sivad. & Manilal, Taxon 32: 130 (1983); Nicolson in Dassan. & Fosb., Rev. Handb. Fl. Ceylon 6: 40 (1987); Nicolson, Suresh & Manilal, An Interpr. Hort. Malab. : 274 (1988); Karth., Jain, Nayar & Sanjappa, Fl. Ind. Enum. Monocot. : 6 (1989). [*Schena* Rheede, Hort. Malab. 11: 35, t. 18 (1692)]. *Dracontium polyphyllum* sensu Dennst., Schlüssel Hort. Malab.: 13, 38 (1818), non L. (1753). *Arum campanulatum* sensu auct. in part, not as to type of Roxb. (1820), *nom. illegit.*; Roxb., Pl. Corom. 3: t. 272 (1820); Hook., Bot. Mag. 55: t. 2812 (1828); Wight, Ic. Pl. Ind. Or. 3: 5, t. 782 (as to infl.) (1844). *Amorphophallus campanulatus* Decne., Nouv. Ann. Mus. Hist. Nat. Paris 3: 366 (1834); Blume, Rumphia 1: 139 (1837); Hook. f., Fl. Brit. India 6: 513 (1893); Engl., Pflanzenr. IV. 23C (48): 76 (1911). *Amorphophallus dubius* Blume, Rumphia 1: 142 (1837); Hook. f., Fl. Brit. India 6: 514 (1893); Engl., Pflanzenr. IV. 23C (48): 74 (1911). *Amorphophallus sativus* Blume, Rumphia 1: 145 (1837); Engl., Pflanzenr. IV. 23C (48): 109 (1911). *Amorphophallus campanulatus* var. *blumei* Prain, Bengal Pl.: 1109 (1903). *Amorphophallus campanulatus* hoofdgroep *hortensis* Backer, Determ.-tab. Jav. *Amorphophallus*: 2 (1920). *Amorphophallus campanulatus* f. *hortensis* Backer *ex* Backer & Bakh., Fl. Java 3: 112 (1968).

(**Fig. 6**).

*Type*: Timor, no date, *Gaudichaud* s.n. (P).

Tubers depressed-globose, 8-12 cm diam. and 7-10 cm thickness in vegetative phase; 16-23 cm diam. and 10-12 cm thickness in reproductive phase (huge-sized tubers are also produced); skin pale brownish, with prominent root scars. Petiole 75-150 cm long, green with ovoid to elongate ovoid pale green blotches and minute pale green spots in between, extreme base white. Leaflets sessile, long leaflets 18-22 × 5.0-6.5 cm; small leaflets 7.5-10.5 × 3.0-4.5 cm, obovate, tip acute, base unequal and decurrent on rachises, greenish above, paler below. Peduncle smooth, 6-8 cm long. Spathe tip acute, margin undulate, 19.0-23.5 × 16-19 cm, convolute portion greenish yellow, expanded portion purplish green with white blotches; dark purplish at base within with prominent murications, middle pale green, pale purplish above, margin pale green. Spadix sessile, 20.0-21.5 cm long, female zone 4.0-4.5 cm long; male zone 5.5-6.0 cm long; appendix 8.5-9.5 cm long. Female flowers dense, each with ovary sub-spherical, pale yellowish, 2.0-2.5 mm high;

style 7.0-7.5 mm long; stigma 3-lobed, rarely 2-lobed. Male flowers dense, each 2.5-3.5 mm high, yellowish-cream coloured. Spadix-appendix hemispherical.

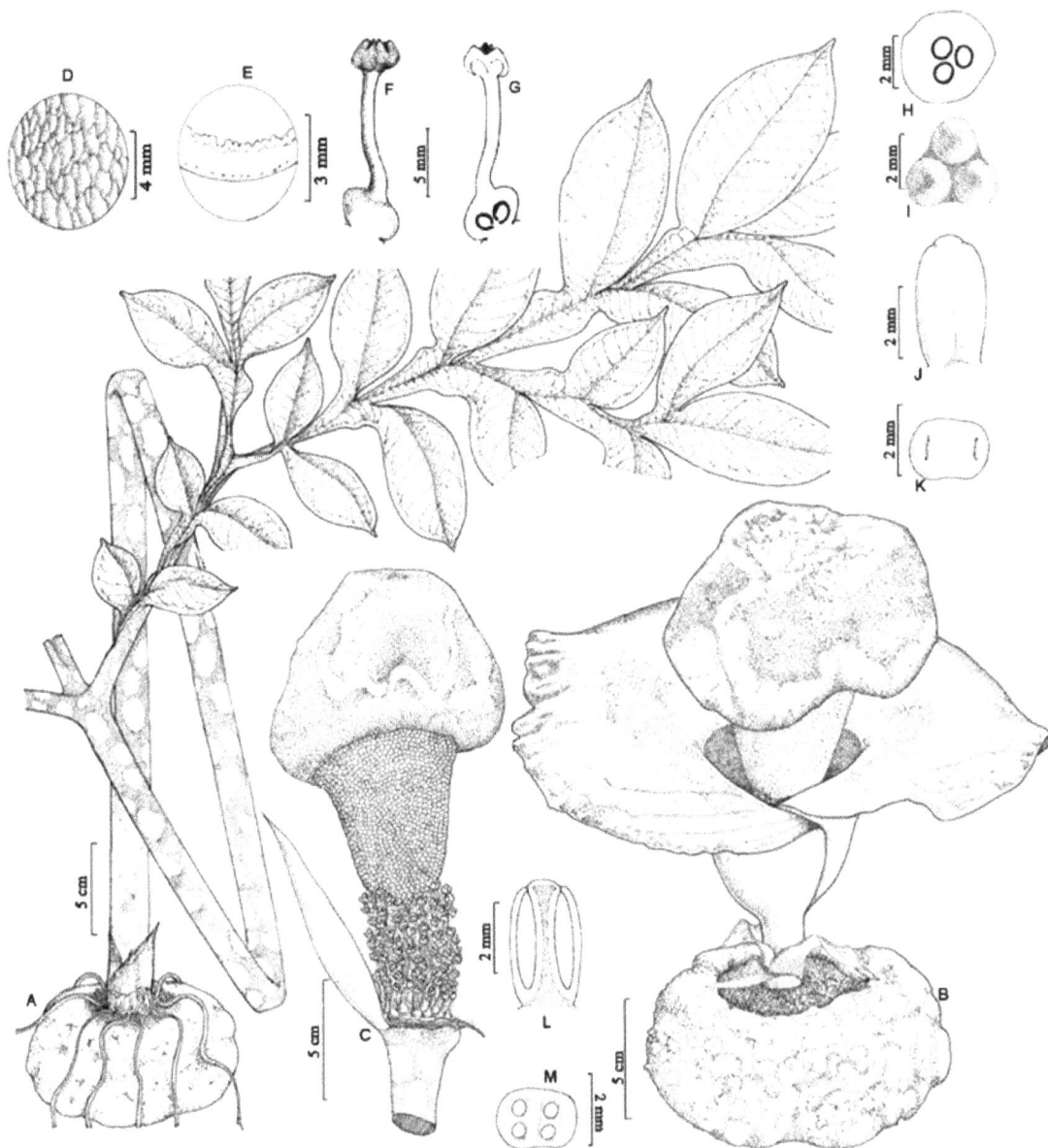

Fig. 6. *Amorphophallus paeoniifolius* (Dennst.) Nicolson var. *campanulatus* (Decne.) Sivad. A. Tuber with leaf; B. Tuber with inflorescence; C. Inflorescence - spathe partially removed showing spadix; D. Small basal inside portion of spathe; E. Small basal portion of spathe - c.s.; F. Female flower; G. Female flower - l.s.; H. Ovary - c.s.; I. Stigma; J. Male flower - view from broad side; K. Male flower - view from top; L. Male flower - l.s.; M. Male flower - c.s.

*Phenology:* Flowering: May - June; Fruiting: No fruit-setting.

*Representative specimens examined*: **Kerala**: Kozhikode Dist.: Areecode, 16.5.1997, *Abdul Jaleel* RIA 46 (infl.) (CALI); Vadakara, Azhiyur, 9.8.1997, *Abdul Jaleel* RIA 119 (leaf) (CALI).

Malappuram Dist.: Calicut University Campus, 10.8.2000, *Abdul Jaleel* RIA 386 (leaf) (CALI). Palakkad Dist.: Athicode, Chittur, 10.4.1976, *Sivadasan* CU 13132 (infl.) (CALI); Kumbalakode, Elavancherry, 24.4.1997, *Sivadasan* CU 19170 (infl.) (CALI).

*Notes*: The confusion in nomenclature of the wild and cultivated varieties of *Amorphophallus paeoniifolius* has been sorted out by Suresh *et al.* (1983) while providing correct identity of aroids described by Rheede (1692).

*Amorphophallus paeoniifolius* var. *campanulatus* representing the cultivated variety differs from var. *paeoniifolius* in texture of petiole, characters of female flowers and spadix-appendix. The former is having smooth or slightly rough greenish petiole with white blotches, leaflet-bases not decurrent to the junction of the petiolules, style of pistil more than thrice the height of the ovary, spadix-appendix round-obtuse to broadly hemispherical with height less than its breadth, and no fruit-setting. The latter is having strongly muricate purplish petiole with light pinkish blotches, leaflet-bases strongly ducurrent usually to the junction of the three main rachises, style more or less double  the height of the ovary, spadix-appendix long-conoidal, height more than its breadth, and fruit-setting.

*Distribution*: India, Sri Lanka, Java and Myanmar. In India, common in all States. Widely cultivated for the edible tubers.

**Taxonomic Analysis**

In India the genus *Amorphophallus* comprises three sections, viz. *Amorphophallus* sect. *Amorphophallus*, sect. *Rhaphiophallus*, and sect. *Conophallus*. Jaleel *et al.* (2011) stated the genus to have three sections namely *A.* sect. *Candarum*, sect. *Conophallus* and sect. *Rhaphiophallus*. But Engler's (1911) inclusion of *A. campanulatus* (=*A. paeoniifolius*), the type of the genus under sect. *Candarum* ("*Cundarum*") rendered the sectional name illegitimate. The sectional name which included the type of the genus has been correctly recognized here as *A.* sect. *Amorphophallus*.

*Amorphophallus* sect. *Rhaphiophallus* is the largest of the three sections of the genus and is represented in India by eight species (Jaleel *et al.*, 2011), viz. *A. bonaccordensis* Sivad. & N. Mohanan, *A. hohenackeri* (Schott) Engl. & Gehrm., *A. konkanensis* Hett. *et al.*, *A. longiconnectivus* Bogner, *A. margaritifer* (Roxb.) Kunth, *A. mysorensis* E. Barnes & C. E. C. Fisch., *A. smithsonianus* Sivad., and *A. sylvaticus* (Roxb.) Kunth. *Amorphophallus mysorensis* is with two varieties, viz. var. *mysorensis* and var. *bhandarensis* (S. R. Yadav, Kahalkar & Bhuskute) Sivad. & Jaleel. *Amorphophallus* sect. *Conophallus* is the second largest of the genus in India and comprises six species, viz. *A. bognerianus* Sivad. & Jaleel, *A. bulbifer* (Sims) Blume, *A. carnosus* Engl., *A. commutatus* (Schott) Engl., *A. nicolsonianus* Sivad. and *A. oncophyllus* Prain *ex* Hook. f. The species *A. commutatus* forms a complex with three varieties, viz. var. *commutatus*, var. *anmodensis* Sivad. & Jaleel, and var. *wayanadensis* Sivad. & Jaleel. The third and the smallest section is *A.* sect. *Amorphophallus* comprising five species, viz. *A. hirsutus* Teysm. & Binn., *A. kachinensis* Engl. & Gehrm., *A. longistylus* Kurz, *A. napalensis* (Wall.) Bogner & Mayo and *A. paeoniifolius* (Dennst.) Nicolson. The species *A. paeoniifolius* is with two varieties, viz. *A. paeoniifolius* var. *paeoniifolius*, and var. *campanulatus* (Decne.) Sivad.

Among the species of *A.* sect. *Amorphophallus*, the two species, viz. *A. hirsutus* and *A. kachinensis* are new addition to the flora of India and they have been collected from Great Nicobar Island of Andaman and Nicobar Islands, and Arunachal Pradesh respectively, and their discoveries formed first reports of their distribution in India. *Amorphophallus longistylus* was rediscovered and collected from Andaman Islands after 131 years from the time of its first collection.

*Amorphophallus paeoniifolius* is the widely distributed species in India represented in all states. During the present study it was found that maximum number of species of the section occur in Arunachal Pradesh and Andaman and Nicobar Islands. Arunachal Pradesh has 3 species, viz. *A. kachinensis*, *A. napalensis* and *A. paeoniifolius,* and Andaman and Nicobar Islands have 3 species, viz. *A. hirsutus*, *A. longistylus*, and *A. paeoniifolius*. *Amorphophallus longistylus* is strictly endemic to Andaman Islands with restricted distribution and according to the criteria D of section V of IUCN (2012), it is considered as Critically Endangered (CR).

The fast and indiscriminate deforestation and destruction of natural habitats of the species render most of the endemic species endangered and lead to extinction. Proper monitoring of habitats and distribution are essential for conservation of species of the genus *Amorphophallus* which comprised wild relatives of widely used and economically important cultivated species.

## Acknowledgements

The authors are thankful to Dr. Wilbert L. A. Hetterscheid, former Director of Botanical Gardens, Wageningen University, Netherlands for thorough review of a manuscript on revision of Indian *Amorphophallus* of which the present paper forms a part. The second author is extremely grateful towards Dr. Dan H. Nicolson, Smithsonian Institution, Washington, D.C., USA for valuable and critical discussions on nomenclatural problems during his work there. The facilities provided by authorities of various national and international herbaria mentioned under materials and methods for study of herbarium specimens are thankfully acknowledged. The authors appreciate with gratitude the valuable services rendered by Mr. V. B. Sajeev, Ernakulam, Kerala and Mr. Jayesh P. Joseph, Wayanad for the illustrations. The last four authors gratefully acknowledge the encouragements and support extended by the Deanship of Scientific Research, King Saud University, through the research group project No. RGP-VPP-135.

## References

Backer, C.A. 1920. Determinatietabel voor de Javaansche Soorten van *Amorphophallus* Bl. N.V. Boekhandel Visser & Co., Weltevreden, i+14 pp.

Backer, C.A. and Bakhuizen van den Brink, R.C. 1968. *Amorphophallus*. *In:* Backer, C.A. and R.C. Bakhuizen van den Brink, Flora of Java, **3**. Wolters – Noordhoff N.V., The Netherlands, pp. 111-113.

Blume, C.L. 1837. Rumphia, **1**. C.G. Sulpke, Leiden, Amsterdam, pp. viii+1-204.

Bogner, J., Mayo, S.J. and Sivadasan, M. 1985. New species and changing concepts in *Amorphophallus*. Aroideana **8**(1): 14-25.

Engler, A. 1911. Araceae-Lasioideae. *In*: Engler, A. (Ed.), Das Pflanzenreich IV. 23C**(48)**.Wilhelm Engelmann, Leipzig, pp. 1-130.

Hetterscheid, W.L.A. and Ittenbach, S. 1996. Everything you always wanted to know about *Amorphophallus*, but were afraid to stick your nose into !!!!!. Aroideana **19**: 7-131.

Hooker, J.D. 1894. *Amorphophallus*. *In*: Hooker, J.D., Flora of British India, Vol. **6**. L. Reeve & Co. Ltd., London, pp. 513-519.

IUCN. 2012. IUCN Red List Categories and Criteria: Version 3.1. Second edition. IUCN, Gland, Switzerland and Cambridge, UK, pp. iv+32.

Jaleel, V.A., Sivadasan, M., Alfarhan, A.H., Thomas, J. and Alatar, A.A. 2011. Revision of *Amorphophallus* Blume *ex* Decne. sect. *Rhaphiophallus* (Schott) Engl. (Araceae) in India. Bangladesh J. Plant Taxon. **18**(1): 1-26.

Jaleel, V.A., Sivadasan, M., Alfarhan, A.H., Thomas, J. and Alatar, A.A. 2012. Revision of *Amorphophallus* Blume *ex* Decne. sect. *Conophallus* (Schott) Engl. (Araceae) in India. Bangladesh J. Plant Taxon. **19**(2): 135-153.

McNeill, J., Barrie, F.R., Buck, W.R., Demoulin, V., Greuter, W., Hawksworth, D.L., Herendeen, P.S., Knapp, S., Marhold, K., Prado, J., Prud'homme van Reine, W.F., Smith, G.F., Wiersema, J.H. and Turland, N.J. 2012. International Code of Nomenclature for algae, fungi and plants (Melbourne Code) adopted by the Eighteenth International Botanical Congress Melbourne, Australia, July 2011. Regnum Vegetabile **154**. Koeltz Scientific Books, Germany, pp. xxx+1-240.

Nicolson, D.H. 1977. Nomina conservanda proposita: Proposal to change the typification of 723 *Amorphophallus*, nom. cons. (Araceae). Taxon **26**: 337-338.

Rheede, H.A. van, tot Draakestein. 1692. Hortus Indicus Malabricus, Vol. **11**. Johannis van Someren *et* Joannis van Dyck, Amsterdam, pp. v+134 pp. + 65 tabs.

Sivadasan, M. and Jaleel, V.A. 1998. Rediscovery of *Amorphophallus longistylus* (Araceae), a little known rare endemic species from Middle Andaman, India. Rheedea **8**(1): 103-106.

Sivadasan, M. and Jaleel, V.A. 2000. *Amorphophallus hirsutus* Teysm. *et* Binn. (Araceae): A new report from India. Rheedea **10**(2): 143-147.

Suresh, C.R., Sivadasan, M. and Mainilal, K.S. 1983. A commentary on Rheede's Aroids. Taxon **32**(1): 126-132.

# 16

# FOLIAR TRICHOMES OF *CROTON* L. (EUPHORBIACEAE: CROTONOIDEAE) FROM CHINA AND ITS TAXONOMIC IMPLICATIONS

HUAN-FANG LIU[1], YUN-FEI DENG AND JING-PING LIAO

*Key Laboratory of Plant Resources Conservation and Sustainable Utilization, Chinese Academy of Sciences, CN-510650 Guangzhou, PR China.*

*Keywords*: *Croton*; Foliar trichome; Taxonomy; Infrageneric classification.

## Abstract

Foliar trichomes of 21 species of the genus *Croton* L. from China have been examined using stereomicroscopy and scanning electron microscopy. Five trichome types characterized by their morphology are identified, *viz.*, stellate, lepidote, simple, dendritic and appressed-rosulate. Only stellate trichome is observed in most species, with only six species that are found to maintain two or three trichome types. Trichome types and density are useful for species identification and sectional classification for Chinese species. Based on the trichome types and other morphological characters, 21 Chinese species are proposed to be placed in five sections. *Croton crassifolius* belongs to sect. *Andrichnia*; *C. cascarilloides* belongs to sect. *Monguia*; *C. mangelong*, *C. kongensis*, *C. laevigatus* and *C. laniflorus* belong to sect. *Argyrocroton*; *C. lauioides*, *C. howii* and *C. damayeshu* belong to sect. *Adenophylli*. The remaining Chinese *Croton* species might be placed into sect. *Croton*. A key for Chinese *Croton* species based on trichome morphology is provided.

## Introduction

*Croton* L. (Euphorbiaceae *s.s.*) is one of the largest genera of flowering plants, with about 1300 species of herbs, shrubs, trees and occasionally lianas that are ecologically prominent and important elements of secondary vegetation in the tropical and subtropical regions worldwide (Webster, 1993; Radcliffe-Smith, 2001). *Croton* belongs to subfamily Crotonoideae (APG, 2009; Wurdack and Davis, 2009), is characterized by mostly lactiferous taxa having pollen with an unusual (crotonoid) exine pattern of triangular supratectal elements attached to a network of muri with short columellae (Nowicke, 1994). The synapomorphy that characterizes *Croton* is the inflexed conformation of the tips of the staminal filaments in bud, which causes the anthers to be introrsely inverted until anthesis (Berry *et al.*, 2005).

Because of the large number of species and extensive morphological variation, it has been proved difficult to define and delimit sections and subsections within the genus *Croton* (Webster *et al.*, 1996) despite the efforts of many taxonomists (Pax and Hoffmann, 1931; Webster, 1993). Webster (1993) established the most recent infrageneric classification, recognizing 40 sections in the genus mainly based on the New World taxa. Among them, three were reassigned generic status by Radcliffe-Smith (2001). Webster (1993) pointed out that his treatment of Old World taxa was much more cursory than that of New World taxa because of lack of familiarity with the living plants in Africa, Madagascar and Asia. Berry *et al.* (2005) presented a molecular systematic analysis of the genus *Croton* and tribe Crotoneae using nrITS and *trnL-trnF* DNA sequences data to test the validity of Webster's classification. van Ee *et al.* (2011) revised the infrageneric classification and proposed a new system for New World *Croton* dividing into four subgenera and 31 sections including some species described as new ones.

---

[1]Corresponding author. Email: hfliu@scbg.ac.cn

Trichomes have played an important role in plant taxonomy at generic, infragenetic and specific levels (Hardin, 1979; Theobald *et al.*, 1979) in groups of wide taxonomic range, such as *Cuphea* P. Browne (Lythraceae: Amarasinghe *et al.*, 1991), *Stachys* L. (Lamiaceae: Salmaki *et al.*, 2009) and *Chelonopsis* Miquel (Lamiaceae: Xiang *et al.*, 2010).

One of the most significant characters for infrageneric classification of *Croton* is the trichome morphology. Previous studies showed that trichome types have great variation within *Croton* (Webster *et al.* 1996; de Sá-Haiad *et al.*, 2009; Senakun and Chantaranothai, 2010). Müller (1866) characterized the taxa as having stellate and lepidote hairs. Solereder (1908) and Metcalfe and Chalk (1950) noted that stellate and lepidote hairs also occur in other genera of subfamily Crotonoideae. Webster *et al.* (1996) identified foliar trichome characteristics for 120 species from 36 sections in *Croton* and established the possible evolutionary relationships among the different sections based on trichome characters. Senakun and Chantaranothai (2010) observed 23 Thai species and recognized seven trichome types.

In China, 23 species of *Croton* are recorded, including 15 endemic species (Li and Esser, 2008). Among them, only five species are placed in the 40 sections of Webster (1993). Moreover, foliar trichomes in *Croton* from China are not well-studied. Chang (1996) and Li and Esser (2008) described two main trichome types in Chinese *Croton* species, peltate scales [same as lepidote of Webster *et al.* (1996)] and stellate. Previously, foliar trichomes of only seven Chinese *Croton* species have been observed (Webster, 1993; Senakun and Chantaranothai, 2010). In the present work, we characterize the foliar trichomes of young to mature leaves of 21 species from China, including 15 endemic species. The objectives of this study are to provide descriptions, illustrations, and a survey of the trichomes in these 21 species using stereomicroscopy and scanning electron microscopy (SEM) and to propose the infrageneric classification for Chinese species.

## Materials and Methods

Leaf samples of 21 species of *Croton* were obtained from dry specimens deposited at the herbarium of South China Botanical Garden, Chinese Academy of Sciences (IBSC). Both young and mature leaves were observed for each species. Two to four samples were examined for each species. A list of investigated materials is given in Table 1. Density of foliar trichomes was observed under Zeiss Stemi SV 11 stereomicroscopy, and photographed with an AxioCam MRC digital camera.

Both young and mature leaves were washed in 95% ethanol. Whole sections of young leaves and $0.5 \times 0.5$ cm of mature leaf fragments were bisected and mounted on copper stubs so that both adaxial and abaxial surfaces faced upwards. The mounts were air-dried, and coated with gold in a JFC-1600 sputter coater (JEOL Ltd, Tokyo, Japan). Observations and digital images were collected with a JEOL JSM-6360LV SEM (JEOL Ltd, Tokyo, Japan). The terminology follows Webster *et al.* (1996) and Senakun and Chantaranothai (2010).

## Results

The main types of the trichomes and their density among the *Croton* species studied are summarized in Table 2. Selected SEM micrographs of trichome types are presented in Figure 1. Foliar trichomes of 21 Chinese *Croton* can be separated into five types; stellate, lepidote, simple, dendritic and appressed-rosulate. Glandular trichomes are not observed.

**Type I.** Stellate trichome. This trichome type is characterized by its star-shaped form in one plane that is usually flattened onto the lamina with 0-30% webbing (Webster *et al.*, 1996; Senakun and Chantaranothai, 2010). Two subtypes are observed. Subtype Ia, appressed-stellate (radii webbed 0-15%), with porrect radius, occurs in *C. dinghuensis* H.S. Kiu, *C. euryphyllus* W.W. Sm.,

*C. lachnocarpus* Benth. (Fig. 1A), *C. merrillianus* Croizat, *C. tiglium* L., *C. yunnanensis* W.W. Sm., *C. chunianus* Croizat (Fig. 1B), *C. cnidophyllus* Radcl.-Sm. & Govaerts (Fig. 1C), *C. yanhuii* Y.T. Chang (Fig. 1D) and *C. crassifolius* Geiseler (Fig. 1E). Subtype Ib, stellate or stellate-rotate (radii webbed 15-30%) occurs in *C. lauioides* Radcl.-Sm. & Govaerts (Fig. 1F), stellate with porrect radius trichome which occurs in *C. howii* Merr. & Chun *ex* Y.T. Chang (Fig. 1G), and stellate trichome sometimes with porrect radius occurs in *C. damayeshu* Y.T. Chang (Fig. 1H).

**Table 1. List of *Croton* species used in the present study.**

| Species | Section | Locality | Voucher |
|---|---|---|---|
| 1. *Croton cascarilloides* Raeusch. | *Monguia* | Guangxi | X. W. Gao 55292 |
| 2. *\*C. chunianus* Croizat | *Croton* | Hainan | K. Z. Hou 71927 |
| 3. *\*C. cnidophyllus* Radcl.-Sm. & Govaerts | *Croton* | Yunnan | Menglian Expedition 9214 |
| 4. *C. crassifolius* Geiseler | *Andrichnia* | Guangdong | H. G. Ye 7742 |
| 5. *\*C. damayeshu* Y. T. Chang | *Adenophylli* | Yunnan | H. T. Tsai 5003 |
| 6. *\*C. dinghuensis* H. S. Kiu | *Croton* | Guangdong | S. T. Lin 30475 |
| 7. *\*C. euryphyllus* W.W.Sm. | *Croton* | Guangdong | H. S. Kiu 571 |
| 8. *\*C. hancei* Benth. | *Croton* | Guangxi | C. L. Tso 23429 |
| 9. *\*C. howii* Merr. & Chun ex Y. T. Chang | *Adenophylli* | Hainan | L. Tang 3303 |
| 10. *C. kongensis* Gagnep. | *Argyrocroton* | Hainan | Z. X. Li & F. W. Xing 1036 |
| 11. *C. lachnocarpus* Benth. | *Croton* | Guangdong | B. Y. Chen 2036 |
| 12. *\*C. laevigatus* Vahl | *Argyrocroton* | Hainan | K. Z. Hou 73784 |
| 13. *C. laniflorus* Geiseler | *Argyrocroton* | Hainan | S. H. Chun 11196 |
| 14. *\*C. laui* Merr. & F. P. Metcalf | *Croton* | Hainan | Z. X. Li 2540 |
| 15. *\*C. lauioides* Radcl.-Sm. & Govaerts | *Adenophylli* | Hainan | C. Wang 34386 |
| 16. *\*C. mangelong* Y.T. Chang | *Argyrocroton* | Yunnan | K. M. Feng 20258 |
| 17. *\*C. merrillianus* Croizat | *Croton* | Hainan | C. L. Tso 43813 |
| 18. *\*C. purpurascens* Y.T. Chang | *Croton* | Guangdong | H. S. Kiu 562 |
| 19. *C. tiglium* L. | *Croton* | Guangxi | Z. S. Chung 808393 |
| 20. *\*C. yanhuii* Y.T. Chang | *Croton* | Yunnan | W. Z. Li 85725 |
| 21. *\*C. yunnanensis* W.W. Sm. | *Croton* | Yunnan | Department of Biology, Yunnan University 757 |

*Species endemic to China.

**Type II.** Lepidote trichome. The individual lepidote hair resembles an appressed-stellate hair but has radii that are connected by webbing so that the trichome forms a more or less shield-like scale (Webster *et al.*, 1996). In our study, this type includes three subtypes. Subtype IIa, stellate-lepidote (radii webbed 30-50%), sometimes with porrect radius, occurs in *C. laevigatus* Vahl (Fig. 1I) and *C. laniflorus* Geiseler. Subtype IIb, dentate-lepidote (radii webbed 50-80%), sometimes with porrect radius, occurs on the adaxial surface of *C. cascarilloides* Raeusch. (Fig. 1J). Subtype IIc, lepidote-subentire (radii webbed 80-100%) occurs in *C. kongensis* Gagnep., *C. mangelong* Y.T. Chang and the abaxial surface of *C. cascarilloides* (Fig. 1K).

**Type III.** Simple trichome. This type is stiffly erect, directed upward from an inclined base (Payne, 1978). This type is only found in *C. crassifolius* (Fig. 1E).

**Table 2. Some characteristic features of trichomes in the examined *Croton* species.**

| Species | | Adaxial surface | | | | | | | | | | Abaxial surface | | | | | | | | | |
|---|---|---|---|---|---|---|---|---|---|---|---|---|---|---|---|---|---|---|---|---|---|
| | | I | | II | | | III | IV | V | p | density | I | | II | | | III | IV | V | p | density |
| | | a | b | a | b | c | | | | | | a | b | a | b | c | | | | | |
| 1. *C. cascarilloides* | YL | - | - | - | + | - | - | - | - | St + | dense | - | - | - | - | + | - | - | - | - | dense |
| | ML | - | - | - | - | - | - | - | - | - | 0 | - | - | - | - | + | - | - | - | - | dense |
| 2. *C. chunianus* | YL | - | - | - | - | - | - | - | - | - | 0 | + | - | - | - | - | - | - | - | + | sparse |
| | ML | - | - | - | - | - | - | - | - | - | 0 | - | - | - | - | - | - | - | - | - | 0 |
| 3. *C. cnidophyllus* | YL | + | - | - | - | - | - | + | - | + | dense | - | - | - | - | - | - | + | - | - | dense |
| | ML | - | - | - | - | - | - | - | - | - | 0 | - | - | - | - | - | - | + | - | - | dense |
| 4. *C. crassifolius* | YL | + | - | - | - | - | + | + | - | + | dense | - | - | - | - | - | - | + | - | - | dense |
| | ML | + | - | - | - | - | + | + | - | + | dense | - | - | - | - | - | - | + | - | - | dense |
| 5. *C. damayeshu* | YL | - | + | - | - | - | - | - | - | + | im | + | - | - | - | - | - | - | - | St + | dense |
| | ML | - | - | - | - | - | - | - | - | - | 0 | - | - | - | - | - | - | - | - | - | 0 |
| 6. *C. dinghuensis* | YL | + | - | - | - | - | - | - | - | + | sparse | - | - | - | - | - | - | - | + | - | sparse |
| | ML | - | - | - | - | - | - | - | - | - | 0 | - | - | - | - | - | - | - | - | - | 0 |
| 7. *C. euryphyllus* | YL | + | - | - | - | - | - | - | - | + | im | + | - | - | - | - | - | - | - | + | dense |
| | ML | - | - | - | - | - | - | - | - | - | 0 | - | - | - | - | - | - | - | - | - | 0 |
| 8. *C. hancei* | YL | - | - | - | - | - | - | - | + | - | dense | - | - | - | - | - | - | - | + | + | dense |
| | ML | - | - | - | - | - | - | - | - | - | 0 | - | - | - | - | - | - | - | - | - | sparse |
| 9. *C. howii* | YL | - | - | - | - | - | - | - | - | - | 0 | - | + | - | - | - | - | - | + | + | sparse |
| | ML | - | - | - | - | - | - | - | - | - | 0 | - | - | - | - | - | - | - | - | - | 0 |
| 10. *C. kongensis* | YL | - | - | - | - | + | - | - | - | - | im | - | - | - | - | + | - | - | - | - | dense |
| | ML | - | - | - | - | - | - | - | - | - | 0 | - | - | - | - | - | - | - | - | - | sparse |
| 11. *C. lachnocarpus* | YL | + | - | - | - | - | - | - | - | + | dense | - | - | - | - | - | - | + | - | - | dense |
| | ML | + | - | - | - | - | - | - | - | + | im | - | - | - | - | - | - | + | - | - | dense |

**Table 2. contd.**

| Species | | Adaxial surface | | | | | | | | | | Abaxial surface | | | | | | | | | |
|---|---|---|---|---|---|---|---|---|---|---|---|---|---|---|---|---|---|---|---|---|---|
| | | I | | II | | | III | IV | V | p | density | I | | II | | | III | IV | V | p | density |
| | | a | b | a | b | c | | | | | | a | b | a | b | c | | | | | |
| 12. *C. laevigatus | YL | - | - | - | - | - | - | - | - | - | 0 | - | - | + | - | - | - | - | - | St+ | dense |
| | ML | - | - | - | - | - | - | - | - | - | 0 | - | - | + | - | - | - | - | - | St+ | sparse |
| 13. C. laniflorus | YL | - | - | - | - | - | - | - | - | - | 0 | - | - | + | - | - | - | - | - | St+ | im |
| | ML | - | - | - | - | - | - | - | - | - | 0 | - | - | + | - | - | - | - | - | St+ | im |
| 14. *C. laui | YL | - | - | - | - | - | - | + | - | - | dense | - | - | - | - | - | - | + | - | - | dense |
| | ML | - | - | - | - | - | - | - | - | - | 0 | + | - | - | - | - | - | + | - | + | sparse |
| 15. *C. lauioides | YL | - | + | - | - | - | - | - | - | - | im | - | + | - | - | - | - | - | - | - | dense |
| | ML | - | + | - | - | - | - | - | - | - | sparse | - | + | - | - | - | - | - | - | - | im |
| 16. *C. mangelong | YL | - | - | - | - | - | - | - | - | - | 0 | - | - | - | - | + | - | - | - | - | dense |
| | ML | - | - | - | - | - | - | - | - | - | 0 | - | - | - | - | + | - | - | - | - | sparse |
| 17. *C. merrillianus | YL | - | - | - | - | - | - | - | - | - | 0 | + | - | - | - | - | - | - | - | + | dense |
| | ML | - | - | - | - | - | - | - | - | - | 0 | + | - | - | - | - | - | - | - | + | dense |
| 18. *C. purpurascens | YL | - | - | - | - | - | - | - | + | + | im | - | - | - | - | - | - | - | + | + | im |
| | ML | - | - | - | - | - | - | - | - | - | 0 | - | - | - | - | - | - | - | - | - | 0 |
| 19. C. tiglium | YL | + | - | - | - | - | - | - | - | - | im | + | - | - | - | - | - | - | - | + | dense |
| | ML | + | - | - | - | - | - | - | - | + | sparse | + | - | - | - | - | - | - | - | + | sparse |
| 20. *C. yanhuii | YL | + | - | - | - | - | - | + | - | + | dense | + | - | - | - | - | - | + | - | + | dense |
| | ML | + | - | - | - | - | - | + | - | + | im | + | - | - | - | - | - | + | - | + | dense |
| 21. *C. yunnanensis | YL | + | - | - | - | - | - | - | - | + | dense | + | - | - | - | - | - | - | - | + | dense |
| | ML | + | - | - | - | - | - | - | - | + | sparse | + | - | - | - | - | - | - | - | + | sparse |

- = absent, + = present, YL = young leaf, ML = mature leaf, p = porrect, St⁺ = sometimes with porrect radius, im = intermediate, I = stellate trichome, Ia = appressed-stellate, Ib = stellate, II = lepidote trichome, IIa = stellate-lepidote, IIb = dentate-lepidote, IIc = lepidote- subentire, III = simple trichome, IV = dendritic trichome, V = appressed-rosulate trichome,* species endemic to China.

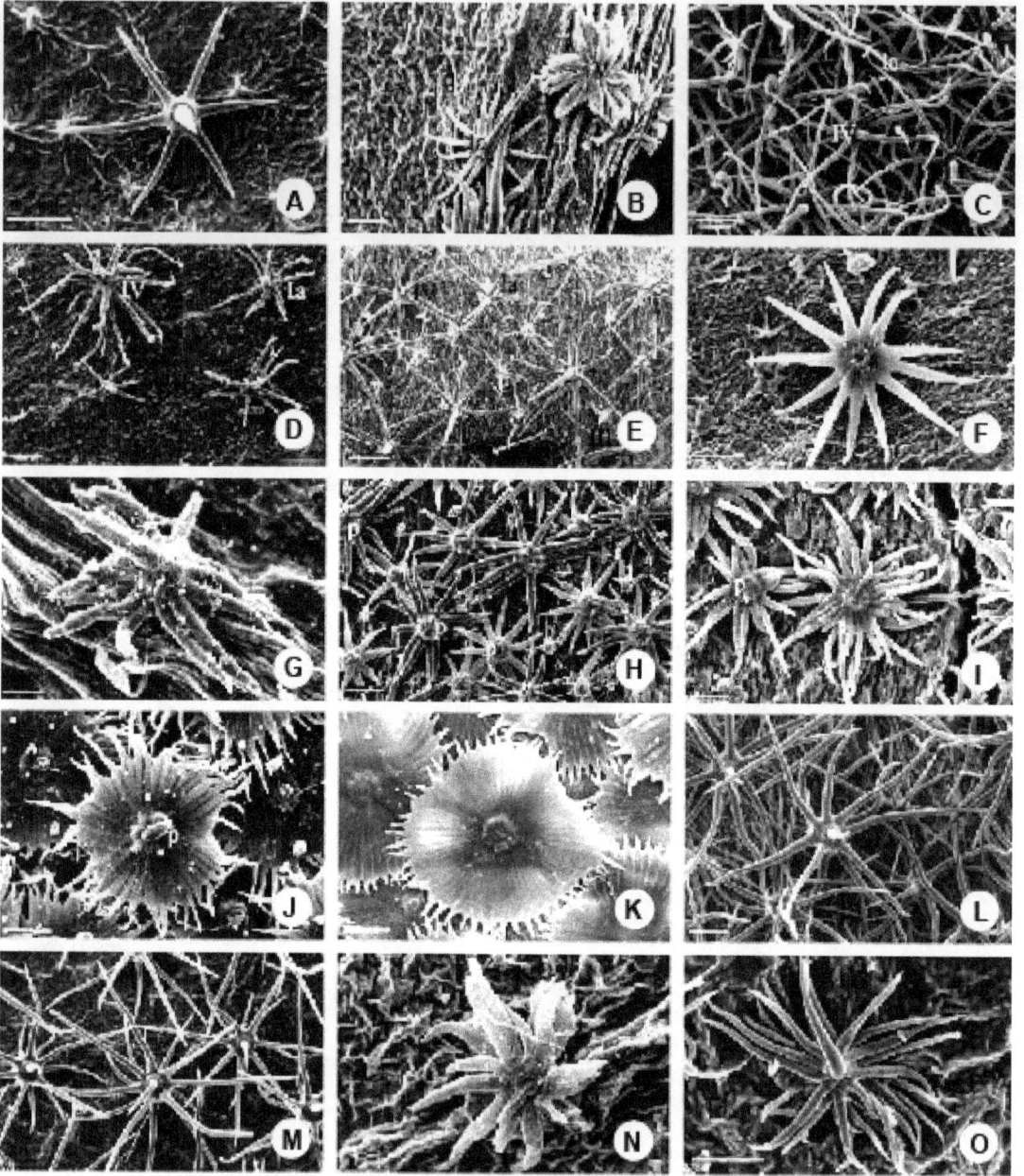

Fig. 1. SEM micrographs of trichomes in *Croton*: (A) Appressed-stellate with porrect radius trichome in *C. lachnocarpus* on Adm; (B) Appressed-stellate with porrect radius (Ia) and appressed-rosulate (V) trichomes in *C. chunianus* on Aby; (C) Appressed-stellate with porrect radius (Ia) and dendritic (IV) trichomes in *C. cnidophyllus* on Adm; (D) Appressed-stellate with porrect radius (Ia) and dendritic (IV) trichomes in *C. yanhuii* on Adm; (E) Appressed-stellate with porrect radius (Ia), simple (III) and dendritic (IV) trichomes in *C. crassifolius* on Abm; (F) Stellate trichome in *C. lauioides* on Ady; (G) Stellate with porrect trichome in *C. howii* on Aby; (H) Stellate with sometimes porrect radius (p) trichome in *C. damayeshu* on Aby; (I) Stellate-lepidote with sometimes porrect radius (p) trichome in *C. laevigatus* on Aby; (J) Dentate-lepidote with sometimes porrect radius (p) trichome in *C. cascarilloides* on Ady; (K) Lepidote-subentire trichome in *C. cascarilloides* on Abm; (L) Dendritic trichome in *C. crassifolius* on Abm; (M) Dendritic trichome in *C. lachnocarpus* on Abm; (N) Appressed-rosulate trichome in *C. dinghuensis* on Aby; (O) Appressed-rosulate with porrect trichome in *C. purpurascens* on Ady. Abm = the abaxial surface of mature leaf; Aby = the abaxial surface of young leaf; Adm = the adaxial surface of mature leaf; Ady = the adaxial surface of young leaf. Scale bars = 200 μm (E), 100 μm (A-D, H, J-K, L-M, O), 50 μm (F-G, I, N).

**Type IV.** Dendritic trichome. This type has the radii inserted at different levels on an axis (Webster *et al.*, 1996). Dendritic trichome with porrect radius occurs in *C. cnidophyllus, C. crassifolius* (Figs 1E, L), *C. lachnocarpus* (Fig. 1M), *C. laui* Merr. & F.P. Metcalf and *C. yanhuii* (Fig. 1D).

**Type V.** Appressed-rosulate trichome. This type resembles stellate ones, but differs in the larger number of radii that are not all in a single whorl (Webster *et al.*, 1996). This type includes appressed-rosulate and appressed-rosulate with porrect radius. Appressed-rosulate trichome occurs in *C. dinghuensis* (Fig. 1N) and *C. hancei* Benth. Appressed-rosulate trichome with porrect radius occurs in *C. chunianus* (Fig. 1B), *C. hancei* and *C. purpurascens* Y.T. Chang (Fig. 1O).

Density of trichome distribution is variable on different surfaces even within the same species. In general, trichomes are much denser on the abaxial surface than on the adaxial surface. Among 21 observed species, six species are glabrous on the adaxial surface even when they are at very young stage: *C. chunianus, C. howii, C. laevigatus, C. laniflorus, C. mangelong* and *C. merrillianus*. The density of trichomes decreases drastically with leaf development on both surfaces of *C. yunnanensis* and on the abaxial surface of *C. hancei, C. kongensis, C. laevigatus, C. laui, C. mangelong* and *C. tiglium*.

In some species, trichomes are observed on both surfaces when young, but fall off completely on either both surfaces or on a single surface when mature. For example, trichomes fall off completely on both surfaces in *C. damayeshu, C. dinghuensis, C. euryphyllus* and *C. purpurascens*, or on the adaxial surface in *C. cascarilloides, C. cnidophyllus, C. hancei, C. kongensis* and *C. laui*, or on the adaxial surface in *C. chunianus* and *C. howii*.

Variation in trichome type can be used to differentiate the *Croton* species examined in this study. In most species, only stellate trichome is observed. Only six species are found to have two or three trichome types. In *C. chunianus*, a few appressed-rosulate and few appressed-stellate with porrect radius trichomes are observed on the abaxial surface (Fig. 1B) and its adaxial surface is glabrous even very young. In *C. cnidophyllus*, a few dendritic and few appressed-stellate trichomes occur on the adaxial surface (Fig. 1C), and only dendritic trichome occurs on the abaxial surface. In *C. crassifolius*, three types of trichomes (appressed-stellate with porrect radius, simple and dendritic) are observed on the adaxial leaf surface (Fig. 1E), and only dendritic trichome occurs on the abaxial surface (Fig. 1L). This species can also be easily distinguished from other Chinese *Croton* species by possessing simple trichome which is not found in any other species. In *C. dinghuensis*, appressed-stellate with porrect radius trichome is found on the adaxial surface and appressed-rosulate trichome is observed on the abaxial surface (Fig. 1N). In *C. lachnocarpus*, it is observed that appressed-stellate with porrect radius trichome occurs on the adaxial surface (Fig. 1A) and dendritic trichome occurs on the abaxial surface (Fig. 1M). The dendritic with porrect radius trichome type and few appressed-stellate with porrect radius trichomes are found in *C. yanhuii* (Fig. 1D). In addition, two subtypes of trichomes are found in *C. cascarilloides*; dentate-lepidote trichome sometimes with porrect radius occurs on the adaxial surface (Fig. 1J), and leptidote-subentire trichome occurs on the abaxial surface (Fig. 1K).

## Discussion

Among the 21 species we observed, seven species were also observed earlier by Webster *et al.* (1996) and Senakun and Chantaranothai (2010). Compared to their works, trichomes of *C. lachnocarpus, C. laevigatus* and *C. tiglium* are characterized as identical to their observation. Webster (1993) reported that *C. kongensis* had stellate trichome and was accordingly placed into Sect. *Cascarilla*, but we observed the lepidote-subentire trichome type as observed by Senakun and Chantaranothai (2010) based on Thai material. We cannot discuss more about the differences of the observation between our studies and Webster (1993), because we did not see the material observed by them. Only lepidote-subentire trichome was observed in *C. cascarilloides* by Senakun and Chantaranothai (2010), but we find that dentate-lepidote trichome sometimes with porrect radius occurs on the adaxial surface and lepidote-subentire trichome occurs on the abaxial surface. Senakun and Chantaranothai (2010) observed three types of trichomes (fasciculate, dendritic and glandular) in *C. crassifolius*, but we find that appressed-stellate with porrect radius, simple, and dendritic trichomes occur on the adaxial surface, and dendritic trichome occurs on the abaxial surface. Our observation accords with the previous studies (Webster *et al.*, 1996; Chayamarit and van Welzen, 2005; Li and Esser, 2008). We could not check their voucher specimen of *C. crassifolius* observed by Senakun and Chantaranothai (2010) and therefore presume that their material was misidentified. Webster *et al.* (1996) indicated that the number and length of radii of trichomes could vary considerably on different leaves of a single specimen. It is supported by our observation. It also showed that the number and length of radii vary considerably even on same leaf of a single specimen. For example, it has 6-17 radii, 0.18-1.1 mm in diam. in *C. yanhuii* (Fig. 1D).

The foliar trichome is one of the most important characters to define sections in the genus *Croton* (Webster, 1993). According to the trichome types and other morphological characters, Chinese *Croton* species can be divided into five sections. Two species, *C. cascarilloides* and *C. crassifolius*, have multifid styles. They can be easily distinguished from each other by foliar trichome type. Webster (1993) placed *C. crassifolius* into sect. *Croton*, which had lepidote trichome. However, *C. crassifolius* has stellate trichome and it might be a member of sect. *Andrichnia*. Webster (1993) uncertainly listed *C. cascarilloides* under both sect. *Anisophyllum*, having appressed-stellate trichome, and sect. *Monguia*, having lepidote trichome. This is the same as Senakun and Chantaranothai (2010) observed, *C. cascarilloides* has lepidote trichome and it is suggested to be placed in sect. *Monguia*. Among species with bifid styles, *C. mangelong*, *C. kongensis*, *C. laevigatus* and *C. laniflorus* are different from other Chinese species in having lepidote trichome and belong to sect. *Argyrocroton* which is characterized by the bifid style and lepidote trichomes. *Croton laevigatus* and *C. laniflorus* were placed by Müller (1866) in sect. *Decapetalon*, however, they are not related to the species of sect. *Decapetalon* because they have glands on the leaf blades while eglandular in sect. *Decapetalon*. *Croton lauioides*, *C. howii* and *C. damayeshu* have stellate trichome and belong to sect. *Adenophylli*. *Croton hancei* and *C. purpurascens* with appressed-rosulate trichomes and the remaining Chinese *Croton* species having appressed-stellate trichomes might be placed into sect. *Tiglium* according to the classification of Webster (1993). However, the correct name for sect. *Tiglium* is sect. *Croton* because the section including *C. tiglium*, the type of the genus. Although Webster *et al.* (1996) superseded Small's (1913) choice of *C. tiglium* as the lectotype of *Croton* and designated *C. aromaticus* L., the valid lectotype of the genus is *C. tiglium* (Britton, 1918; van Ee and Berry, 2010).

A key to species of *Croton* from China is provided as follows.

1. With simple trichome ........................................................................................ *C. crassifolius*
1. Without simple trichome
  2. With dentritic trichome
    3. Only dentritic trichome ............................................................................... *C. laui*
    3. With dentritic and appressed-stellate trichome
      4. Dentritic trichome only on abaxial surface ........................................ *C. lachnocarpus*
      4. Dentritic trichome on both surfaces
        5. Without appressed-stellate trichome on abaxial surface ............. *C. cnidophyllus*
        5. With appressed-stellate trichome on abaxial surface ................... *C. yanhuii*

2. Without dentritic trichome
   6. With lepidote trichome
      7. With dentate-lepidote trichome                                       *C. cascarilloides*
      7. Without dentate-lepidote trichome
        8. With lepidote-subentire trichome
           9. With lepidote-subentire trichome on the adaxial surface            *C. kongensis*
           9. Without lepidote-subentire trichome on the adaxial surface      *C. mangelong*
        8. Without lepidote-subentire trichome
          10. Stellate-lepidote trichome is intermediate when mature          *C. laniflorus*
          10. Stellate-lepidote trichome is sparse when mature                *C. laevigatus*
   6. Without lepidote trichome
          11. Glabrous on adaxial surface
            12. With appressed-rosulate trichome                            *C. chunianus*
            12. Without appressed-rosulate trichome
              13. With stellate trichome                                    *C. howii*
              13. Without stellate-trichome                               *C. merrillianus*
          11. With trichome on adaxial surface
            14. With appressed-rosulate trichome
              15. Only appressed-rosulate trichome
                16. All trichome fallen down when mature              *C. purpurascens*
                16. Trichome not fallen down when mature                   *C. hancei*
              15. With appressed-rosulate and appressed-stellate trichome     *C. dinghuensis*
            14. Without appressed-rosulate
              17. With stellate trichome
                18. All trichome fallen down when mature              *C. damayeshu*
                18. Trichome not fallen down when mature                  *C. lauioides*
              17. Without stellate trichome
                19. All trichome fallen down when mature              *C. euryphyllus*
                19. Trichome not fallen down when mature
                  20. Trichome is dense on adaxial surface when young      *C. yunnanensis*
                  20. Trichome is intermediate on adaxial surface when young      *C. tiglium*

## Acknowledgements

This work was supported by the National Natural Science Foundation of China (31200246, 31100240, 31200176), South China Botanical Garden-Shanghai Institute of Plant Physiology & Ecology Joint Fund and the Foundation of Key Laboratory of Plant Resources Conservation and Sustainable Utilization, South China Botanical Garden, Chinese Academy of Sciences. We thank the curator of the herbarium, South China Botanical Garden, Chinese Academy of Sciences (IBSC) for his permission to access their collections. We also thank Dr. Chelsea Specht and Dr. Yuping Lin working in University of California, Berkeley for their English revision.

## References

Amarasinghe, V., Graham, S.A. and Graham, A. 1991. Trichome morphology in the genus *Cuphea* (Lythraceae). Bot. Gaz. **152**: 77-90.

APG (Angiosperm Phylogeny Group). 2009. An update of the Angiosperm Phylogeny Group classification for the orders and families of flowering plants: APG III. Bot. J. Linn. Soc. **161**: 105-121.

Berry, P.E., Hipp, A.L., Wurdack, K.J., Ee, B.V. and Riina, R. 2005. Molecular phylogenetics of the giant genus *Croton* and tribe Crotoneae (Euphorbiaceae *sensu stricto*) using ITS and trnL-trnF DNA sequence data. Am. J. Bot. **92**: 1520-1534.

Britton, N.L. 1918. Flora of Bermuda. Charles Scribner's Sons, New York.

Chang, Y.T. 1996. *Croton. In*: Kiu, H.S. (Ed.), Flora Reipublicae Popularis Sinicae, Vol. **44**, No. 2. Science Press, Beijing.

Chayamarit, K. and van Welzen, P.C. 2005. *Croton. In:* Santisuk, T. and Larsen, K. (Eds), Flora of Thailand. Vol. **8**, No. 1. The Forest Herbarium, National Park, Wildlife and Plant Conservation Department, Bangkok, pp. 189-226.

de Sá-Haiad, B., Serpa-Ribeiro, A.C.C., Barbosa, C.N., Pizzini, D., de O. Leal D., de Senna-Valle L., de Santiago-Fernandes, L.D.R. 2009. Leaf structure of species from three closely related genera from tribe Crotoneae Dumort. (Euphorbiaceae s.s., Malpighiales). Plant Syst. Evol. **283**: 179-202.

Hardin, J.W. 1979. Patterns of variation in foliar trichomes of eastern North American *Quercus*. Am. J. Bot. **6**(5): 576-585.

Li, B.T. and Esser, H. 2008. *Croton. In:* Wu, Z.Y., Raven, P.H. and Hong, D.Y. (Eds), Flora of China, Vol. **11**. Science Press, Beijing & Missouri Botanical Garden Press, St. Louis, pp. 258-264.

Metcalfe, C.R. and Chalk, L. 1950. Anatomy of the Dicotyledons, Vol. **2**. Clarendon Press, Oxford, pp. 1207-1235.

Müller, J. 1866. *Croton. In*: de Candolle, A.P. (Ed.), Prodromus Systematis Naturalis Regni Vegetabilis **15**(2). Paris, Strasbourg & London.

Nowicke, J.W. 1994. A palynological study of Crotonoideae (Euphorbiaceae). Ann. Mo. Bot. Gard. **81**: 245-269.

Pax, F. and Hoffmann, K. 1931. Euphorbiaceae, Crotoneae. *In*: Engler, A. and Prantl, K. (Eds), *Die Natürlichen Pflazmfamiliem,* ed. 2, 19c: Leipzig: Wilhelm Engelmann, pp. 83-88.

Payne, W.W. 1978. A glossary of plant hair terminology. Brittonia **30**: 239-255.

Radcliffe-Smith, A. 2001. Genera Euphorbiacearum. Royal Botanic Gardens, Kew.

Salmaki, Y., Zarre, S., Jamzad, Z. and Bräuchler, C. 2009. Trichome micromorphology of Iranian *Stachys* (Lamiaceae) with emphasis on its systematic implication. Flora **204**: 371-381.

Senakun, C. and Chantaranothai, P. 2010. A morphological survey of foliar trichomes of *Croton* L. (Euphorbiaceae) in Thailand. Thai. For. Bull. (Bot.) **38**: 167-172.

Small, J.K. 1913. *Croton. In:* Britton, N.L. and Brown, A. (Eds), An illustrated flora of the Northern United States, ed. 2, Vol. **2**. New York: Charles Scribner's Sons, pp. 454-455.

Solereder, H. 1908. Euphorbiaceae. *In:* Systematic Anatomy of the Dicotyledons, Vol. **2**. Oxford University Press, Oxford. pp. 739-763.

Theobald, W.L., Krahulik, J.L. and Rollins, R.C. 1979. Trichome description and classification. *In:* Metcalfe, C.R. and Chalk, L. (Eds), Anatomy of the Dicotyledons, 2nd edition. Clarendon Press, Oxford, pp. 40-53.

van Ee, B.W. and Berry, P.E. 2010. Typification notes for *Croton* (Euphorbiaceae). Harvard Papers in Botany **15**: 73-84.

van Ee, B.W., Riina, R and Berry, P.E. 2011. A revised infrageneric classification and molecular phylogeny of New World *Croton* (Euphorbiaceae). Taxon **60**: 791-823.

Webster, G.L. 1993. A provisional synopsis of the sections of the genus *Croton* (Euphorbiaceae). Taxon **42**: 793-823.

Webster, G.L., Del-arco-aguilar, M.J. and Smith, B.A. 1996. Systematic distribution of foliar trichome types in *Croton* (Euphorbiaceae). Bot. J. Linn. Soc. **121**: 41-57.

Wurdack, K.J. and Davis, C.C. 2009. Malpighiales phylogenetics: Gaining ground on one of the most recalcitrant clades in the angiosperm tree of life. Am. J. Bot. **96**(8): 1551-1570.

Xiang, C.L., Dong, Z.H., Peng, H. and Liu, Z.W. 2010. Trichome micromorphology of the East Asiatic genus *Chelonopsis* (Lamiaceae) and its systematic implications. Flora **205**: 434-441.

# 17

# AN ENUMERATION TO THE ORCHIDS AND THEIR CONSERVATION STATUS IN GREATER SYLHET, BANGLADESH

M.M. Islam, M.K. Huda[1] and M. Halim[2]

*Department of Botany, University of Chittagong, Chittagong 4331, Bangladesh*

*Keywords*: Conservation; Diversity; Orchidaceae; Sylhet.

## Abstract

The present investigation deals with enumeration including diversity, ecology and conservation of the family Orchidaceae of greater Sylhet region of Bangladesh. Extensive field trips were made at 11 different sites of this region during early monsoon, late monsoon and winter seasons. Relevant literature and different herbaria were consulted to gather information about the orchids of this region. Orchidaceae is represented in greater Sylhet by 75 species under 49 genera. Out of these, 25 species are terrestrial, 48 are epiphytic, one is saprohytic and one is hemiepiphytic. Presence of 37 monotypic genera indicates a narrow diversity in Orchidaceae of this area. The present investigation revealed that 26 orchid species are restricted and distributed only in Sylhet region in Bangladesh. The currently accepted taxonomic nomenclature, synonyms, habit, flowering time, present conservation status and geographical distribution are provided under each taxon.

## Introduction

Orchidaceae is one of the largest flowering plant families, represented by about 1000 genera and 20,000 species with cosmopolitan distribution, primarily in the tropics and rarely in arctic regions (Chowdhery, 1998). A preliminary checklist of family Orchidaceae for Bangladesh was made by Huda *et al.* (1999) with an enumeration of 160 species and 2 varieties under 63 genera for Bangladesh. Of them, 106 taxa were epiphytic and remaining 56 were terrestrial. Huda *et al.* (2001) added some new records for the family of Orchidaceae from Bangladesh. The distribution of terrestrial orchids in Bangladesh was compiled as a check list mainly on the basis of previous records, literature survey and herbarium collections (Khanam *et al.,* 2001).

Diversity and ecology of the orchids in the south-eastern part of Bangladesh have also been studied by Huda (2000). Earlier reports indicate that Sylhet region was rich in orchid diversity (Hooker 1890a, b; Prain, 1903). Some research works on orchids from Bangladesh were done sporadically, *viz.* Huda *et al.* (1999), Huda (2000, 2008, 2008a), Ahmed and Pasha (1993, 1993a, 1994, 1998, 1998a, 1998b, 1999) as part of their floral exploration but focus has not been given to the diversity and ecology of orchids of Sylhet region. Valuable herbarium specimens of orchids from the greater Sylhet regions, collected by many taxonomists of Bangladesh are housed at the different herbaria of Bangladesh, *viz.* Bangladesh National Herbarium (DACB), Dhaka University Salar Khan Herbarium (DUSH), Herbarium of Chittagong University (HCU), Herbarium of Bangladesh Forest Research Institute (HBFRI) and Herbarium of BCSIR laboratory (HBCSIR). The present study was, therefore, undertaken with a view to examine the previous specimens and relevant literatures, and also to conduct field investigation for collecting specimens of the family occurring in the area, particularly in the greater Sylhet region mainly in the forest areas of Sylhet,

[1]Corresponding author. Email: mkhuda70@hotmail.com
[2]Chittagong Education Board, Chittagong, Bangladesh.

Moulavi Bazar, Sunamganj and Habiganj of Bangladesh for a taxonomic treatment of the family Orchidaceae.

## Materials and Methods

*Study area*

The floristic study of the family Orchidaceae was conducted from July 2006 to January 2010 in Lawachara National Park, Madhabkunda Eco-Park, Bangladesh Tea Research Institute (BTRI) Campus in Moulvi Bazar; Rema-Kalenga Wildlife Sanctuary and Satchari National Park in Habiganj; Jaflong, Tamabil, Sripur, Jaintapur, Tilagarh forest beat in Sylhet district and Sadar of Sunamganj district. Eleven different sites in four districts of greater Sylhet region were visited to study the diversity and ecology of orchids.

*Collection of specimens*

Orchid specimens with detailed information were collected both in the flowering and non-flowering stage from the study area through eight field trips each consisting of 4 to 5 days in early monsoon (March to May), monsoon (June to July), late monsoon (August to October) and winter (November to February).

*Herbarium and literature survey*

Both living and herbarium specimens were examined and studied carefully at the Herbarium of Chittagong University (HCU). Herbarium specimens of orchids collected in the present study were studied and matched with herbarium specimens available at DACB, DUSH, HBFRI, HBCSIR, and HCU. Local orchid experts were consulted to identify some specimens and to confirm some critical specimens. In order to compare the description, nomenclature and geographical distribution and uses, Roxburgh (1814, 1832) Hooker (1890a, b), Prain (1903), Heinig (1925), Bruhl (1926), Sinclair (1956), Abraham and Vatsala (1981), Joseph (1987) and Huda (2000) were consulted.

Abundance status was measured based on observation, availability of the species in the field, herbarium specimen preserved at DACB, DUSH, HCU and HBFRI and literature survey following Misra (2000) and Rao (1998). Categories for Abundance status based on their availability are mentioned as Specimen deficient (no collection or herbarium specimen is available in any herbarium of Bangladesh), Rare (only one herbarium specimen found but no further collection made after record), Scarce (one or two herbarium specimens available and collected once or twice after record), Occasional (few herbarium specimens available and collected from one or few localities from other parts of Bangladesh also), and Common (usually occur in the different areas of greater Sylhet and other parts of Bangladesh).

## Taxonomic enumeration to the species

The present study identified 75 species belonging to 49 genera of Orchidaceae in the greater Sylhet region. Enumeration is presented below alphabetically. An asterisk (*) at the beginning of the species name indicates its occurrence from the greater Sylhet region only, on the other hand, another asterisk (*) used at the end of synonym in the enumeration to indicate that it is the first recorded name, if applicable. Flowering time (Fl.) of the species is presented numerically from 1 to 12 for January to December, respectively. Categories of Abundance status follows the flowering time.

1.  **Acampe papillosa** (Lindl.) Lindl., Fol. Orchid. 2 (1853). *Saccolabium papillosum* Lindl. (1841); *Gastrochilus papillosus* (Lindl.) O. Kuntze (1891). Reported by its synonym from Chittagong and the Sundarbans by Prain (1903). Epiphytic. *Fl.*: 8–9. Common. *Distribution*: India

and Bangladesh (Chittagong, Cox's Bazar, Bandarban, Rangamati, Khagrachari and greater Sylhet). *Specimen examined*: Habiganj: Rema-Kalenga; 25.02.2007, M. Islam 01 (HCU).

2.  ***Acanthephippium sylhetense*** Lindl., Gen. Sp. Orchid. Pl.: 177 (1833). Reported from Sylhet by Hooker (1890a). Terrestrial. *Fl.*: 4–5. Specimen deficient. *Distribution*: China, Fiji Island, India and Bangladesh (Sylhet).

3.  **Aerides crispa** Lindl., Gen. Sp. Orchid. : 239 (1833). *Aerides lindleyana* Wight, *Ic.* t. 1677. (1851). Reported by Ahmed *et al.* (1989). Epiphytic. *Fl.*: 3–5. Rare. *Distribution*: Southern India to Myanmar, Bangladesh (Chittagong and Sylhet). *Specimen examined*: Sylhet: Lama Bazar; 21.3.1988, M. Ahmed 122 (HCU).

4.  **Aerides multiflorum** Roxb., Pl. Cor. 3: 68, t. 271 (1820). *Aerides affine* Lindl. (1833); *Aerides multiflora* var. *dactyloides* Mokter *et al.* (1989). Roxburgh (1832) reported from Sylhet. Epiphytic. *Fl.*: 5–6. Occasional. *Distribution*: India, Malaysia, Philippines, Thailand and Bangladesh (Cox's Bazar, Rangamati and Sylhet). *Specimen examined*: Sylhet: Lama Bazar (Near college road); 21.03. 1986, M. Ahmed, 97 (HCU).

5.  **Aerides odoratum** Lour., Fl. Cochinch. 2: 525 (1790). *Aerides cornutum* Roxb. (1832). Reported by its synonym from Dhaka by Roxburgh (1832). Epiphytic. *Fl.*: 5–6. Common. *Distribution*: China, India, Malaysia, Myanmar, Nepal and Bangladesh (distributed in most of the areas of south-east part of Bangladesh and greater part of Sylhet). *Specimens examined*: Sylhet: Jainta bazar; 10.10.81; Moyeen 70 (HCU); Lawachara; 27. 02. 07, M. Islam and M. K. Huda 11 (HCU).

6.  ***Aerides suavissima*** Lindl. N. Journ. Hort. Soc. Iv. : 263 (1858). Hooker, f. (1890b) reported from Sylhet. Epiphytic. *Fl.*: 5–6. Rare. *Distribution*: Malaysia, Myanmar and Bangladesh (Sylhet). *Specimen examined*: Sylhet: Tamabil; 23.03.86, M. Ahmed 130 (HCU).

7.  **Agrostophyllum khasianum** Griff. Calcutta J. Nat. Hist. 4: 376, t. 19. (1844). *Appendicula hasseltii* Wight. (1851). Reported from Sylhet by Ahmed *et al.* (1989a). Epiphytic. *Fl.*: 4-5. Scarce. *Distribution*: India and Bangladesh (Cox' Bazar and Sylhet). *Specimen examined*: Sunamganj: Bagan bari, Sadar; 25.4.1986, M. Ahmed 116 (HCU).

8.  ***Anaectochilus roxburghii*** (Wall.) Lindl., Gen. Sp. Orchid. Pl.: 499 (1840). *Chrysobaphus roxburghii* Wall. (1826); *Anaectochilus yungianus* Hu (1971); *Zeuxine roxburghii* (Lindl.) Hiroe (1971). Lindley (1830–40) reported it from Sylhet. Terrestrial. *Fl.*: 4–6. Specimen deficient. *Distribution*: Bhutan, China, India, Laos, Thailand, Vietnam and Bangladesh (Sylhet).

9.  **Arundina graminifolia** (D. Don) Hochr. in Bull. New York Bot. Gard. 6: 270 (1910). *Blettia graminifolia* D. Don (1825); *Limodorum graminifolia* Buch.-Ham. *ex* D. Don (1825); *Arundina bambusifolia* Lindl. (1830); *Cymbidium bambusifolium* Roxb. (1832). Reported from Chittagong by Roxburgh (1814, 1832). Terrestrial. *Fl.*: 12–3. Occasional. *Distribution*: China, India, Malaysia, Myanmar, Philippines, Sri Lanka and Bangladesh (Bandarban, Chittagong, Cox's Bazar, Khagrachari, Rangamati and Sylhet). *Specimen examined*: Dhaka: 07.09. 46, S. K. Sen (DUSH).

10. **Brachycorythis helferi** (Rchb. f.) Summerh. in Kew Bull. 1955: 235 (1955). *Gymnadenia helferi* Rchb. f. (1872); *Habenaria helferi* (Rchb. f.) Hook. f. (1890b). Terrestrial. Fl.: 8–9. Scarce. Distribution: India, Myanmar and Bangladesh (Bandarban and Sylhet). Uddin *et al.* (2000) reported it from Habiganj. *Specimen examined:* Bandarban: Chimbuk hills, 08. 09. 99, M.A. Rahman *et al.* 5744c (HCU).

11. **Bulbophyllum lilacinum** Ridl. in J. Linn. Soc. 32: 276 (1896). Epiphytic. *Fl.*: 10–11. Common. *Distribution*: India, Malaya Peninsula and Bangladesh throughout Cox's Bazar district (Khan and Halim, 1987) and Sylhet. *Specimen examined*: Sylhet: Tamabil, 12. 06. 07, M. Islam 02 (HCU).

12. ***Calanthe densiflora** Lindl., Gen. Sp. Orchid. Pl.: 250 (1833). *Alimorchis densiflora* (Lindl.) Kuntze (1891); *Calanthe kazuoi* Yamamoto (1930). Reported from Sylhet by Lindley (1830–40). Terrestrial. *Fl.*: 10–12. Specimen deficient. *Distribution*: Bhutan, China, India, Japan, Nepal, Vietnam and Bangladesh (Sylhet).

13. ***Calanthe puberula** Lindl., Gen. Sp. Orchid. Pl.: 252 (1833). *Alismorchis puberula* (Lindl.) Kuntze (1891); *Calanthe amoena* Smith (1921); *C. lepida* Smith (1921). Lindley (1830–40) reported from Sylhet. Terrestrial. *Fl.*: 10–12. Specimen deficient. *Distribution*: Bhutan, China, India, Myanmar, Taiwan, Vietnam and Bangladesh (Sylhet).

14. **Cephalantheropsis gracilis** (Lindl.) S. Y. Hu. in Quart. J. Taiwan Mus. 25 (3-4): 213 (1972). *Calanthe gracilis** Lindl. (1833). Terrestrial. *Fl.*: 10. Rare. *Distribution*: India and Bangladesh (Bandarban and Sylhet). Lindley (1830–40) reported from Sylhet by synonym. *Specimen examined*: Bandarban: Teracha mukh; 09. 09. 99, M. A. Rahman *et al.* 5787a (HCU).

15. **Cleisostoma subulatum** Blume, Bijdr. : 363 (1825). *Sarcanthus secundus** Griff. (1851); *Sarcanthus subulatus* (Blume) Rchb. f. (1857); *Saccolabium secundum* (Griff.) Ridl. (1907). Reported by its synonym from Sylhet by Hooker (1890b). Epiphytic. *Fl.*: 8. Rare. *Distribution*: Bhutan, Cambodia, India, Malaysia, Myanmar, Philippines, Thailand and Bangladesh (Chittagong and Sylhet) No specimen examined.

16. **Coelogyne cristata** Lindl., Coll. Bot. : t. 33 (1821). *Cymbidium speciosissimum* Don (1825). Hooker (1890b) reported it from Sylhet. Epiphytic. *Fl.*: 3–4. Rare. *Distribution*: Bhutan, India, Nepal and Bangladesh (Cox's Bazar and Sylhet).

17. ***Coelogyne punctulata** Lindl., Coll. Bot.: sub t. 33 (1821). *Cymbidium nitidum** sensu Roxb. (1814). *Coelogyne ocellata* Lindl. (1830); *C. goweri* Rchb. f. (1869); *C. nitida sensu* (Roxb.) Hook. f. (1890b). Reported by its synonym from Sylhet by Roxburgh (1832). Epiphytic. *Fl.*: 3–4. Specimen deficient. *Distribution*: Bhutan, India, Myanmar, China, Nepal and Bangladesh (Sylhet).

18. **Cymbidium aloifolium** (L.) Sw. in Nov. Act. Soc. Upsal. 6: 73 (1799). *Epidendrum aloifolium* L. (1753); *Epidendrum pendulum* Roxb. (1795); *Cymbidium bicolor* Lindl. (1833); *Cymbidium erectum* Wight (1851). Reported from Chittagong by Heinig (1925). Epiphytic. *Fl.*: 4–6. Common. *Distribution*: India, Myanmar to Java, Sri Lanka and Bangladesh (Commonly distributed in Chittagong, Chittagong Hill Tracts, Sylhet and Cox's Bazar). *Specimen examined*: Habiganj: Rema-Kalenga National Forest, 25. 02. 07, M. Islam 03 (HCU).

19. **Dendrobium amoenum** Wall in Lindl., Gen. Sp. Orchid. Pl.: 78 (1830). *Dendrobium egertoniae* Lindl. (1847); *D. mesochlorum* Lindl. (1847); *D. amoena* (Wall. *ex* Lindl.) Kuntze (1891). Hooker (1890b) reported it from Sylhet. Epiphytic. *Fl.*: 6. Rare. *Distribution*: India, Myanmar and Bangladesh (Sylhet). *Specimen examined*: Sylhet: Haripur, 08.07.1981, Moyeen 57 (HCU).

20. **Dendrobium aphyllum** (Roxb.) C.E.C. Fischer in Gamble, Fl. Pres. Madras 3: 1416 (1928). *Limodorum aphyllum* Roxb. (1795); *Dendrobium pierardi** Roxb. *ex* Hook. (1822); *D. aphyllum* var. *cucullatum* (Hook. f.) Sarkar (1984). Roxburgh (1832) reported it from Chittagong by synonym. Epiphytic. *Fl.*: 4–5. Common. *Distribution*: India, Myanmar and Bangladesh

(Chittagong, Chittagong Hill Tracts, Cox's Bazar and Sylhet). *Specimen examined*: Sylhet: Jaintapur; 12. 06. 07, M. Islam and M. K. Huda 12 (HCU).

21. **\*Dendrobium chryseum** Rolfe in Gard. Chron. Ser. 3, 3: 233 (1888). *Dendrobium clavatum* Wall. *ex* Lindl. (1852); *Callista clavata* (Wall. *ex* Lindl.) Kuntze (1891); *Dendrobium tibeticum* Schltr. (1921); *D. clavatum* var. *aurantiacum* (Rchb. f.) Tang & Wang (1951). Reported from Sylhet by Hooker (1890b). Epiphytic. *Fl.*: 6. Specimen deficient. *Distribution*: Bhutan, China, India, Myanmar, Nepal, Thailand, Vietnam and Bangladesh (Sylhet).

22. **Dendrobium fimbriatum** Hook., Exot. Fl. : t. 71 (1823). *Dendrobium paxtonii* Paxt. (1839); *D. fimbriatum* var. *oculatum* Hook. f. (1890b); *Callista oculata* (Hook.) Kuntze (1891). Reported from Chittagong by Prain (1903). Epiphytic. *Fl.*: 3–5. Rare. *Distribution*: India and Bangladesh (Chittagong Hill tracts, Cox's Bazar and Sylhet). *Specimen examined*: Sylhet: Jaintapur, 23.03.86, M. Ahmed 1000 (HCU).

23. **Dendrobium formosum** Roxb. *ex* Lindl. in Wall., Pl. Asiat. Rar. 1: 34, t. 29 (1830). *Dendrobium formosum* Roxb. (1814); *D. infundibulum sensu* Rchb. f. (1887); *Callista formosa* (Roxb. *ex* Lindl.) Kuntze (1891). Reported from Sylhet by Roxburgh (1832). Epiphytic. *Fl.*: 5. Rare. *Distribution*: Bhutan, India, Nepal and Bangladesh (Chittagong, Cox's Bazar and Sylhet). *Specimens examined*: Cox's Bazar: Ukhia, 07.05.84, D.K. Das and M.K. Alam 5015 (HBFRI); Ukia, 08.08.81; Moyeen 63 (HCU).Sylhet: Tamabil, 19.05.1983, Mia 909 (DACB).

24. **Dendrobium lindleyi** Steud., Nomencl. Bot. ed. 2: 490 (1840). *Dendrobium aggregatum\** Roxb. (1814); *Callista aggregata* (Roxb.) Kuntze (1891). Reported by its synonym from Cox's Bazar by Sinclair (1956) and from Sylhet by Uddin *et al.* (2002). Epiphytic. *Fl.*: 3–5. Occasional. *Distribution*: Bhutan, China, India, Laos, Myanmar, Thailand, Vietnam and Bangladesh (Chittagong Hill Tracts, Cox's Bazar and Sylhet). *Specimen examined*: Rangamati: Sitapahar, Kaptai, 08.01.95, Mezanul Hoque 7344 (HBFRI).

25. **\*Dendrobium macrostachyum** Lindl., Gen. Sp. Orchid. Pl. : 78 (1830). *Dendrobium gamblei* King & Pantl. (1897). Reported from Sylhet by Hossain (2002). Epiphytic. *Fl.*: 5–6. Rare. *Distribution*: India, Sri Lanka and Bangladesh (Sylhet). *Specimen examined*: Sylhet: Near Forest School. 22.03.96, Ahmed 96 (HCU).

26. **Dendrobium moschatum** (Buch-Ham) Sw. in Schltr. Neim. J. Bot. 1: 943 (1805). *Dendrobium calceolaria* Carey *ex* Hook. f. (1825-26); *Epidendrum moschatum* Buch-Ham. (1800). Reported from Chittagong by Prain (1903). Epiphytic. *Fl.*: 5–7. Rare. *Distribution*: India and Bangladesh (Chittagong, Chittagong Hill Tracts, Cox's Bazar and Sylhet). *Specimen examined*: Sylhet: New Forest School; 13.03.56, M. S. Khan Collection number: not available (DUSH).

27. **Dendrobium parishii** Rchb. f. in Bot. Zeit. 21(31): 237 (1863). *Callista parishii* (Rchb. f.) Kuntze (1891). Epiphytic. *Fl.*: 3–5. Scarce. *Distribution*: India, Malaysia, Myanmar, Thailand and Bangladesh (Bandarban, Rangamati and Sylhet). *Specimens examined*: Rangamati: Naniarchar, 26.09.1998, M. K. Huda & S. B. Uddin 469 (HCU); Sylhet: Jaflong; 23.02.1986, M. Ahmed *sn* (HCU).

28. **\*Dendrobium pulchellum** Roxb *ex* Lindl., Gen. Sp. Orchid. Pl.: 82 (1830). *Dendrobium pulchellum* Roxb. (1814); *D. dalhausieanum* Wall. (1844); *Callista pulchella* (Roxb. *ex* Lindl.) Kuntze (1861). Reported from Sylhet by Roxburgh (1814, 1832). Epiphytic. *Fl.*: 2–4. Specimen deficient. *Distribution*: Bhutan, India and Bangladesh (Sylhet).

29. **\*Dendrobium ruckeri** Lindl. in Bot. Reg. 29: t. 60, misc. 25, no. 38 (1843). *Dendrobium ramosum\** sensu Lindl., Gen. & Sp. Orchid Pl : 82 (1830). Reported from Sylhet by Hooker (1890b). Epiphytic. *Fl.*: not known. Specimen deficient. *Distribution*: India and Bangladesh (Sylhet).

30. **Didymoplexis pallens** Griff., Calcutta J. Nat. Hist. 4: 383, t. 17 (1844). *Leucorchis sylvatica* Bl. (1849); *Arethusa ecristata* Griff. (1851); *Apetelon minutum* Wight (1852); *Gastrodia pallens* (Griff.) F. Mueller (1870); *Didymoplexis brevipes* Ohwi. (1937). Reported from Bengal by Hooker (1890b). Saprophytic. *Fl.*: 4–5. Scarce. *Distribution*: Afghanistan, Australia, Bhutan, India, Japan, Malaysia to Philippines, New Guinea, Thailand, and Bangladesh (Comilla, Dhaka, Gazipur, Panchagarh and Sylhet by Khanam *et al.*, 2001).

31. **Eria pubescens** (Hook. f.) Lindl. in Edw., Bot. Reg. 11: t. 904 (1825). *Dendrobium pubescens* Hook. f. (1890b); *Eria flava\** Lindl. (1830). Reported by its synonym from Cox's Bazar by Ahmed *et al.* (1989b). Epiphytic. *Fl.*: 2–4. Scarce. *Distribution*: India and Bangladesh (Chittagong and Sylhet). *Specimens examined*: Sylhet: Jaflong, 22.03.1989, M. Ahmed and Pasha 135 (HCU); Tamabil, 12. 06. 07, M. Islam and M. K. Huda, 04 (HCU).

32. **\*Erythrodes humilis** (Bl.) J. J.Smith, Bull, Dep. Agric. Indes Neerl. 13: 11 (1907). *Physurus humilis* Bl., Orch. Archip. Ind. 96: t. 27, f. 2 (1859); *Physurus blumei\** Lindl. (1840). Reported by its synonym from Sylhet by Lindley (1830 - 40). Terrestrial. *Fl.*: 1–3. Specimen deficient. *Distribution*: Borneo, India, Java, Sri Lanka and Bangladesh (Sylhet).

33. **Gastrochilus calceolaris** (Buch-Ham. *ex* J. E. Smith) D. Don, Prodr. Fl. Nepal.: 32 (1852). *Aerides calceolaris* Buch.-Ham. *ex* J. E. Smith (1819); *Epidendrum calceolare* Buch.-Ham. (1825); *Sarcochilus nepalensis* Spreng. (1826); *Saccolabium calceolare\** (Buch.-Ham. *ex* J. E. Smith) Lindl. (1833); *Aerides leopardium* Wall. *ex* Lindl. (1838); *A. leopardorum* Wall. (1890). Reported by its synonym from Sylhet by Hooker (1890b). Epiphytic. *Fl.*: 3–5. Specimen deficient. *Distribution*: Bhutan, China, India, Myanmar, Nepal, Thailand, Vietnam to Malaysia and Bangladesh (Cox's Bazar and Sylhet).

34. **\*Gastrochilus inconspicuous** (Hook. f.) Kuntze, Revis. Gen. Pl. 2: 661 (1891). *Saccolabium inconspicuum* Hook. f. (1890b); *Cymbidium incospicuum* Wall. *ex* Hook. f. (1895); *Luisia inconspicua* Hook. f. (1898). Reported from Sylhet by Ahmed and Pasha (1998a). Epiphytic. *Fl.*: 6–7. Rare. *Distribution*: Bhutan, India, Nepal and Bangladesh (Sylhet). *Specimen examined*: Sylhet, 05.06.81, M. Ahmed *sn* (HCU).

35. **Geodorum densiflorum** (Lam.) Schltr. in Fedde, Repert. 4: 259 (1929). *Limodorum densiflorum* Lam. (1792); *Geodorum dilatatum\** R. Br. (1813); *Geodorum purpureum* R. Br. (1813); *Limodorum candidum* Roxb. (1814). Reported by its synonym from Sylhet by Roxburgh (1814, 1832). Terrestrial. *Fl.*: 4. Occasional. *Distribution*: Australia, Bhutan, China, Fiji, India, Malaysia, Myanmar, New Guinea, Samoa, Solomon Island, Sri Lanka, Tonga and Bangladesh (Chittagong, Chittagong Hill tracts, Gazipur, Mymensigh, Tangail and Sylhet). *Specimens examined*: Rangamati: Kaptai, 30.08.99, M.A. Rahman *et al.* 5365 (HCU).

36. **Goodyera procera** (Wall. *ex* Ker-Gawl.) Hook. f., Exot. Fl. 1(3): t. 39 (1823). *Neottia procera* Wall. *ex* Ker-Gawl. (1822); *Goodyera carnea* A. Rich. (1841); *Epipactis procera* (Ker-Gawl.) Eaton (1908). Reported from Sylhet by Lindley (1830) Terrestrial. *Fl.*: 3–5. Occasional. *Distribution*: Bhutan, China, India, Japan, Malaysia, Myanmar, Sri Lanka, Taiwan, Philippines and Bangladesh (Chittagong, Cox's Bazar and Sylhet). *Specimen examined*: Bandarban: Ali Kadam; 30.05.98, M.A. Rahman *et al.* 2882b (HCU).

37. **Habenaria digitata** Lindl. Gen. Sp. Orchid. Pl.: 307 (1835). *Habenaria trinervia* Wight (1851); *H. graveolens* Duthie (1906). Hooker (1890b) reported it from Sylhet. Terrestrial. *Fl.*: 8–11. Specimen deficient. *Distribution*: India, Myanmar and Bangladesh (Chittagong and Sylhet).

38. **Hetaeria affinis** (Griff.) Seidenf. in Oasis, Suppl. 2: 9 (2001). *Goodyera affinis* Griff. (1851); *Cerochilus rubens* Lindl. (1854); *Rhamphidia rubens* (Lindl.) Lindl. (1857); *Hetaeria rubens*\* (Lindl.) Bentham *ex* Hook. f. (1890). Reported by its synonym from Chittagong by Bruhl (1926). Terrestrial. *Fl.*: 3–4. Scarce. *Distribution*: Bhutan, China, India, Myanmar, Thailand, Vietnam and Bangladesh (Chittagong, Mymensingh and Sylhet). *Specimen examined*: Moulvi Bazar: Lawachara, 27. 02. 07, M. Islam 05 (HCU).

39. **Luisia filiformis** Hook. f., Fl. Brit. India 6(1): 23 (1890). *Luisia grovesi* Hook. f. (1890); *L. volurcris* sensu King & Pantl. (1898); *L. gamblei* Durand (1906). Reported from Sylhet by Hooker (1890b). Epiphytic. *Fl.*: 3–4. Scarce. *Distribution*: Bhutan, India, Laos, Thailand, Vietnam and Bangladesh (Cox's Bazar and Sylhet). *Specimen examined*: Sylhet: Jaintapur, 12. 06. 07, M. Islam 13 (HCU).

40. **Luisia trichorhiza** (Hook. f.) Bl. Rumphia 4: 50 (1849). *Vanda trichorhiza* Hook. f. (1825); *Cymbidium triste sensu* Lindl. (1833). Reported from Sylhet by Ahmed and Pasha (1998b). Epiphytic. *Fl.*: 3–5. Occasional. *Distribution*: Bhutan, India, Myanmar, Thailand and Bangladesh (Chittagong, Cox's Bazar and Sylhet). *Specimens examined*: Sylhet: Lama Bazar; 24.03.86, M. Ahmed 102 (HCU); Cox's Bazar: Whykong Reserve Forest; 10.09.99, M.K. Huda *et al.* 5834 (HCU).

41. **Luisia volucris** Lindl., Fol. Orchid. 1 (1852). Reported from Sylhet by Hooker (1890b). Epiphytic. *Fl.*: 3–4. Rare. *Distribution*: India, Sikkim and Bangladesh (Chittagong and Sylhet).

42. **Malaxis acuminata** D. Don, Prodr. Fl. Nepal. : 29 (1825). *Microstylis wallichii* Lindl. (1830); *M. biloba* Lindl.(1829); *Malaxis biloba* (Lindl.) Ames (1908); *Malaxis wallichii* (Lindl.) Deb (1962). Reported from Sylhet by Lindley (1830–40). Terrestrial. *Fl.*: 7–9. Rare. *Distribution*: Bhutan, Cambodia, India, Java, Malaysia, Myanmar, Nepal, Philippines, Sumatra, Thailand, Vietnam and Bangladesh (Sylhet and Rangamati). *Specimen examined*: Rangamati: Kaptai,27.06.98, M.A. Rahman *et al.* 3229 (HCU).

43. \***Malaxis biaurita** Lindl., Gen. Sp. Orchid. : 20 (1830). Report from Sylhet by Lindley (1830–40). Terrestrial. *Fl.*: 7–10. Specimen deficient. *Distribution*: India and Bangladesh (Sylhet).

44. **Micropera rostrata** (Roxb.) Balakr. in J. Bombay Nat. Hist. Soc. 67 (1): 66 (1970). *Aerides rostrata*\* Roxb. (1814); *Camarotis purpurea* Lindl. (1832); *Micropera pallida sensu* Lindl. (1833); *Camarotis pallida* (Lindl.) Lindl. (1859); *C. rostrata* (Roxb.) Roxb. (1864); *Sarcochilus purpureus* (Lindl.) Benth. *ex* Hook. f. (1890); *Micropera purpurea* (Lindl.) Pradhan (1979). Reported by synonym by Roxburgh (1814, 1832) from Chittagong and Sylhet. Epiphytic. *Fl.*: 5–6. Occasional. *Distribution*: India, Malaysia, Myanmar, Thailand and Bangladesh (Chittagong, Chittagong Hill Tracts and Sylhet).

45. \***Nervilia juliana** (Roxb.) Schltr. In Bot. Jahrb. Syst. 45: 402 (1911). *Arethusa juliana* Roxb. (1814); *Epipactis juliana* Roxb. (1832); *Pogonia juliana* (Roxb.) Lindl. (1832). Reported from Sylhet by Jayaweera (1981). Terrestrial. *Fl.*: Not known. Specimen deficient. *Distribution*: India, Sri Lanka and Bangladesh (Sylhet).

46. \***Oberonia mannii** Hook. f. Ic. Pl. : t. 2003 (1890). Reported from Sylhet by Hooker (1890a). Epiphytic. Fl.: Not known. Specimen deficient. *Distribution*: India and Bangladesh (Sylhet).

47. **Oberonia mucronata** (D. Don.) Ormerod & Seidenfaden in Seidenfaden, Contrib. Orch. Flora Thailand XIII: 20 (1997). *Stelis mucronata* D. Don. (1825); *Cymbidium iridifolium*\* Roxb. (1832); *Oberonia iridifolia* Lindl., (1830); *Malaxis iridifolia* (Roxb.) Rchb. f. (1861); *Oberonia denticulata* var. *iridifolia* (Roxb.) S. Misra (1989). Reported by its synonym from Sylhet by Roxburgh (1814, 1832). Epiphytic. *Fl.*: 8–9. Occasional. *Distribution*: Bhutan, China, India, Indonesia, Malaysia, Nepal, Philippines and Bangladesh (Chittagong and Sylhet). *Specimens examined*: Sylhet: Lathitila Rain Forest, 29.11.83, M. K. Alam 4742 (HBFRI); Lama Bazar, 21.03.86, M. Ahmed 103 (HCU),: Madhabkunda Eco Park,26. 02. 07, M. Islam 06 (HCU).

48. **Oberonia rufilabris** Lindl., Sert. Orch. : t. 8 A (1838). *Malaxis rufilabris* (Lindl.) Rchb. f. (1861). Reported from Sylhet by Hooker (1890a) and from Cox's Bazar by Huda (2000). Epiphytic. *Fl.*: 8–9. Scarce. *Distribution*: Bhutan, Cambodia, India, Malaysia, Myanmar, Nepal, Thailand, Vietnam and Bangladesh (Cox's Bazar and Sylhet). *Specimen examined*: Cox's Bazar: Panerchara Tulabagan; 30.01.99, M.K. Huda *et al.* 5315 (HCU).

49. **Oberonia wallichii** Hook. f.  Fl. Brit. India. 6: 681 (1890). *Oberonia iridifolia*  Wall. Cat. 1948/2 in part. Reported from Sylhet by Hooker (1890a). Epiphytic. *Fl.*: Not known. Specimen deficient. *Distribution*: India and Bangladesh (Cox's Bazar and Sylhet).

50. ***Paphiopedilum insigne** (Wall. *ex* Lindl.) Pfitz., in Engler, Bot. Jahrb. 19: 41 (1894). *Cypripedium insigne* Wall. *ex* Lindl. (1840). Reported from Sylhet by Lindley (1830-1840). Terrestrial. *Fl.*: 10–3. Specimen deficient. *Distribution*: Bhutan, India and Bangladesh (Sylhet).

51. ***Paphiopedilum venustum** (Wall.) Pfitzer *ex* Stein, Orchid.- Buch: 489 (1892). *Cypripedium venustum* Wall. (1820); *Cypripedium pardinum* Rchb. f. (1869). Reported from Sylhet by Lindley (1830-1840). Terrestrial. *Fl.*: 3–5. Specimen deficient. *Distribution*: Bhutan, India and Bangladesh (Sylhet).

52. **Papilionanthe teres** (Roxb.) Schltr. in Orchis 9: 78 (1915). *Dendrobium teres* Roxb. (1832); *Vanda teres*\* (Roxb.) Lindl. (1833). Reported by its synonym from Chittagong by Roxburgh (1814, 1832). Epiphytic. *Fl.*: 3–4. Occasional. *Distribution*: Bhutan, China, India, Myanmar, Nepal, Thailand, Vietnam and Bangladesh (Chittagong, Chittagong Hill Tracts, Cox's Bazar and Sylhet). *Specimens examined*: Rangamati: Kaptai, 23.05.83, M. N. Islam *sn* (HBFRI); Cox's Bazar: Sylhet: Madhabkundu, 16.06.07; M. Islam, 10 (HCU).

53. **Pelatantheria insectifera** (Rchb. f.) Ridl. in J. Linn. Soc. 32: 373 (1896). *Sarcanthus insectifer* Rchb. f. (1857). Reported from Chittagong by Hooker (1890b). Epiphytic. *Fl.*: 9–12. Scarce. *Distribution*: Bhutan, India, Myanmar, Nepal, Thailand and Bangladesh (Cox's Bazar and Sylhet). *Specimen examined*: Habiganj: Rema-Kalenga, 25.02.07, M. Islam 07(HCU).

54. **Peristylus goodyeroides** (D. Don.) Lindl., Gen. Sp. Orchid. : 299 (1835). *Habenaria goodyeroides*\* D. Don (1825). Reported by its synonym from Sylhet by Hooker (1890b). Terrestrial. *Fl.*: 5–7. Occasional. *Distribution*: China, India, Indonesia, Malaysia, Philippines and Bangladesh (Chittagong, Rangamati and Sylhet). *Specimen examined*: Rangamati: Khajachara, 26. 07. 97, M.A. Rahman *et al.* 1579 (HCU).

55. **Phaius tancarvilleae** (Banks *ex* L "Herit) Blume, Mus. Bot. 2: 177 (1856). *Limodorum tancanvilleae* Banks in L "Herit. (1789); *Blettia tanearvilleae* Ait (1813); *Phajus veratrifolius* Wall *ex* Lindl. (1831); *Phaius wallichii*\* Lindl (1831); *Phaius blumei* var. *assamica* Rchb. f. (1882). Reported by its synonym from Sylhet by Hooker (1890b). Terrestrial. *Fl.*: 2–3. Scarce. *Distribution*: Australia, Bhutan, China, India, Indonesia, Malaysia, Myanmar, New Guinea, Pacific Islands, Sri Lanka, and Bangladesh (Runctia Forest,  Gazni and Sylhet).

56. **\*Phalaenopsis taenialis** (Lindl.) E.A. Christenson & Pradhan in Selbyana 9: 168 (1986). *Aerides taeniale* Lindl. (1833); *Doritis taenialis* (Lindl.) Hook. f. (1890); *Kingiella taenialis* (Lindl.) Rolfe (1917); *Kingidium taenialis\** (Lindl.) P.F. Hunt (1970). Reported by its synonym from Sylhet by Ahmed *et al.* (1992). Epiphytic. *Fl.*: 5. Rare. *Distribution*: Bhutan, China, India, Nepal and Bangladesh (Sylhet). *Specimen examined*: Sylhet: Jainta Bazar, 24.03.86, M. Ahmed *sn* (HCU).

57. **Pholidota imbricata** Hook. f., exot. Fl. 2: t. 138 (1825). *Cymbidium imbricatum\** Roxb. (1832); *Coelogyne imbricata* (Roxb.) Rchb. f. (1861); *Pholidota asamica* Regel. (1890); *Pholidota pallida sensu* Holtum (1964). Reported by its synonym from Chittagong and Sylhet by Roxburgh (1814, 1832). Epiphytic. *Fl.*: 6–7. Occasional. *Distribution*: Australia, China, India, Laos, Malaysia, Myanmar, Nepal, Nicobar Island, New Guinea, Pacific Islands, Sri Lanka, Thailand and Bangladesh (Chittagong, Chittagong Hill Tracts, Cox's Bazar and Sylhet). *Specimen examined*: Cox's Bazar: Whykong Reserve Forest, 11.09.99, M.K. Huda *et al.* 5848 (HCU).

58. **\*Podochilus khasianus** Hook. f., Fl. Brit. India 6: 81 (1890). *Podochilus chinensis* Schltr. (1924). Reported from Sylhet by Hooker (1890b). Epiphytic. *Fl.*: 3–5. Specimen deficient. *Distribution*: Bhutan, China, India and Bangladesh (Sylhet).

59. **Pomatocalpa decipiens** (Lindl.) J. J. Smith, Natuurk. Tijdscr. Ned. Indie 72: 33 (1912). *Cleisostoma decipiens* Lindl. (1844); *Saccolabium decipiens* (Lindl.) Alston (1931). Reported from Habiganj by Uddin *et al.* (1999). Epiphytic. *Fl.*: 3. Scarce. *Distribution*: Sri Lanka and Bangladesh (Sylhet). *Specimen examined*: Habiganj: Chunarughat, Rema Kalenga Wild life Sanctuary, Kalega beat, Habiganj: 18.03.99 (DACB and DUSH).

60. **Rhynchostylis retusa** (L.) Blume, Bijdr. : 286, t. 49 (1825). *Epidendrum retusum* L. (1753); *Aerides guttatum\** Roxb. (1832); *Saccolabium rheedii* Wight (1851); *Saccolabium guttatum* Lindl. (1833); *Saccolabium berkeleyi* Rchb. f. (1883). Reported by its synonym from Dhaka by Roxburgh (1814, 1832). Epiphytic. *Fl.*: 5–7. Common. *Distribution*: Bhutan, India, Malaysia, Myanmar, Nepal, Philippines, Sri Lanka and Bangladesh (throughout Bangladesh). *Specimen examined*: Sylhet: Shreepur, 12. 06. 07, M. Islam 08 (HCU).

61. **Robiquetia spathulata** (Bl.) J. J. Sm. in Nat. Tijdschr. Ned. Ind. 72: 114 (1912). *Cleisostoma spathulatum* Bl. (1825); *Saccolabium densiflorum* Lindl. (1832); *Cleisostoma spicatum* Lindl. (1847). Reported from Sylhet by Seidenfaden (1988). Epiphytic. *Fl.*: 5–7. Scarce. *Distribution*: China, India, Indo-china, Indonesia, Malaysia, Myanmar, Singapore, Thailand and Bangladesh (Cox's Bazar and Sylhet).

62. **Robiquetia succisa** (Lindl.) Seid. & Garay in Bot. Tidsskr. 67: 119 ( 1972). *Sarcanthus succisus* Lindl. (1826); *Oecoclades paniculata* Lindl.(1833); *Saccolabium parvulum* Lindl. (1859); *S. buccosum\** Rchb. f. (1871); *Robiquetia paniculata* (Lindl.) J. J. Smith (1912); *Sarcanthus henryi* Schltr. (1919). Reported from Sylhet by Hooker (1890b). Epiphytic. *Fl.*: 6–7. Scarce. *Distribution*: Bhutan, Cambodia, China, India, Laos, Myanmar, Thailand, Vietnam and Bangladesh (Chittagong, Chittagong Hill Tracts and Sylhet).

63. **Saccolabiopsis pusilla** (Lindl.) Seidenf. & Garay in Bot. Tidsskr. 67: 118, f. 33 (1972). *Saccolabium pusillum* Lindl. (1858); *Saccolabium pumilio\** Rchb. f. (1890). Reported by its synonym from Sylhet by Hooker (1890b). Epiphytic. *Fl.*: 4–6. Scarce. *Distribution*: Bhutan, India, Myanmar and Bangladesh (Rangamati and Sylhet). *Specimen examined*: Chittagong: Gondamara, Dhoplachari, Chandanaish, 24.07.99, M.K. Huda *et al.* 5152 (HCU).

64. **\*Saccolabium cephalotes** Hook. f., Fl. Brit. India 5: 63 (1890). *Acampe cephalotes* Lindl. Reported from Sylhet by Hooker (1890b). Epiphytic. *Fl.*: Not known. Specimen deficient. *Distribution*: India and Bangladesh (Sylhet).

65. **\*Schoenorchis gemmata** (Lindl.) J. J. Smith in Natuurk. Tijdschr. Ned.-Indie. 72: 100 (1912). *Saccolabium gemmata* Lindl. (1838); *S. geminatum\** (Lindl.) Hook. f.(1890); *Cleisostoma gemmatum* (Lindl.) King & Pantl. (1898). Reported by its synonym from Jyantia (Sylhet) by Hooker (1890b). Epiphytic. *Fl.*: 5–6. Specimen deficient. *Distribution*: Bhutan, Cambodia, China, India, Laos, Myanmar, Thailand, Vietnam and Bangladesh (Sylhet).

66. **Smitinandia micrantha** (Lindl.) Holttum in Gard. Bull. Singapore, 25: 106 (1969). *Saccolabium micranthum\** Lindl. (1833); *Cleisostoma micranthum* (Lindl.) King & Pantl. (1898). Reported by its synonym from Sylhet by Hooker (1890b). Epiphytic. *Fl.*: 7–9. Scarce. *Distribution*: Bhutan, India, Laos, Malaysia, Myanmar, Nepal, Thailand, Vietnam and Bangladesh (Cox's Bazar and Sylhet).

67. **\*Spathoglottis pubescens** Lindl., Gen. Sp. Orchid. : 120 (1831). *Spathoglottis pubescens* var. *parviflora* (Lindl.) Hook. f. (1890). Collected from East Bengal by Griffith, CAL- 5194 (Huda *et al.*, 1999) and reported from Sylhet by Lindley (1830 - 40). Terrestrial. *Fl.*: 6–9. Specimen deficient. *Distribution*: Bhutan, China, India, Myanmar and Bangladesh (Sylhet).

68. **Staurochilus ramosus** (Lindl.) Seidenf., in Opera Bot. 95: 95 (1988). *Saccolabium ramosum* Lindl. (1833); *Aerides ramosum* Wall. (1833); *Cleisostoma ramosum\** (Lindl.) Hook. f. (1890); *Gastrochilus ramosus* (Lindl.) Kuntze (1891); *Sarcanthus ramosus* (Lindl.) J. J. Smith (1912); *Pomatocalpa ramosum* (Lindl.) Summerh. (1948). Reported by its synonym from Sundarbans by Hooker (1888–90) and from Cox's Bazar by Seidenfaden (1988). Epiphytic. *Fl.*: 5. Common. *Distribution*: Bhutan, India, Myanmar, Thailand and Bangladesh (Chittagong, Chittagong Hill Tracts, Cox's Bazar and Sylhet). *Specimen examined*: Habiganj: Rema-Kalenga, 25. 02. 07, M. Islam 09 (HCU).

69. **\*Tainia latifolia** (Lindl.) Rchb. f. in Bonplandia 5: 54 (1857). *Ania latifolia* Lindl. (1831); *Mitopetalum latifolium* (Lindl.) Bl. (1856); *Eulophia hastate* Lindl. (1859); *Tainia hastata* (Lindl.) Hooker (1890); *T. khasiana* Hook. f.(1890). Reported from Sylhet by Hook. f. (1890a). Terrestrial. *Fl.*: 9–11. Specimen deficient. *Distribution*: Bhutan, China, India, Laos, Myanmar, Thailand, Vietnam and Bangladesh (Sylhet).

70. **\*Tropidia angulosa** (Lindl.) Bl., Coll. Orchid. : 122 (1859). *Decasinea angulosa* Wall. (1832); *Cnemidia angulosa* Lindl. (1840); *Govindova nervosa* Wight (1853); *Tropidia govindovii* Bl. (1858). Reported from Sylhet by Hooker (1890b). Terrestrial. *Fl.*: 8–11. Rare. *Distribution*: Bhutan, China, India, Malaysia, Myanmar, Sumatra, Thailand, and Bangladesh (Sylhet).

71. **Tropidia curculigoides** Lindl., Gen. Sp. Orchid. Pl. : 497 (1840). *Tropidia squamata* Bl. (1859); *T. assamica* Bl. (1858); *T. graminea* Bl. (1859); *T. formosana* Rolfe (1895). Report from Sylhet by Hooker (1890b). Terrestrial. *Fl.*: 9–12 . Occasional. *Distribution*: Bhutan, China, India, Malaysia, Myanmar and Bangladesh (Cox's Bazar and Sylhet). *Specimen examined*: Rangamati: Kaptai, Sitapahar, 03.09.99, M.A. Rahman *et al.* 5591 (HCU).

72. **\*Vanda crisata** Lindl., Gen. Sp. Orchid. Pl. : 216 (1833). *Aerides cristatum* Wall. (1832); *Trudelia cristata* (Lindl.) Senghas (1988). Reported from Sylhet by Hooker (1890b). Epiphytic. *Fl.*: 5. Rare. *Distribution*: Bhutan, India, Nepal and Bangladesh (Sylhet).

73. **Vanilla parishii** Rchb. f., Otia Bot. Hamb. : 39 (1878). Reported from Rangamati by Prain (1903). Hemi-epiphytic. *Fl.*: Not known. Scarce. *Distribution*: India, Myanmar and Bangladesh (Chittagong, Cox's Bazar and Sylhet).

74. ***Zeuxine flava** (Wall. *ex* Lindl.) Trimen, Syst. Cat. Fl. Pl. : 90 (1885). *Etaeria flava* Lindl. (1832); *Monochilus flavum* Wall. *ex* Lindl. (1840). Reported from Sylhet by Seidenfaden (1978). Terrestrial. *Fl.*: 5. Rare. *Distribution*: Bhutan, India, Nepal, Thailand and Bangladesh (Sylhet). *Specimen examined*: Sylhet: Tamabil; 11.03.56, M. S. Khan (DUSH).

75. **Zeuxine nervosa** (Wall. *ex* Lindl.) Bentham *ex* C.B. Clarke in J. Linn. Soc., Bot. 25: 73 (1889). *Monochilus nervosum* Wall. *ex* Lindl. (1840); *Haplochilus nervosum* (Wall. *ex* Lindl.) D. Dietrich (1852). Reported from Comilla and Sylhet by Hooker (1890b). Terrestrial. *Fl.*: 3–4. Scarce. *Distribution*: Bhutan, China, India, Philippines, Taiwan, and Bangladesh (Comilla, Mymensingh and Sylhet). *Specimen examined*: Mymensingh; 04.03.77, M. Rahman (DACB).

## Discussion

A total of 75 orchid species were recorded from greater Sylhet, out of which 48 species are epiphytic, 25 species are terrestrial, one is saprophytic (*Didymoplexis pallens* Griff.) and another one is hemiepiphytic (*Vanilla parishii* Rchb. f.). Based on literature and the present field work, 26 species were found to occur in Sylhet region only. Thirty-seven out of 49 genera are monotypic indicating a narrower diversity of the family in the studied region. Herbarium specimens of 53 orchid species are available in different herbaria of Bangladesh, namely DACB, DUSH, HCU, HBFRI and HBCSIR. Only 17 orchid species were found in the present survey and are housed at HCU. 22 orchid species were not found in the last 50 years from greater Sylhet region or other parts of Bangladesh.

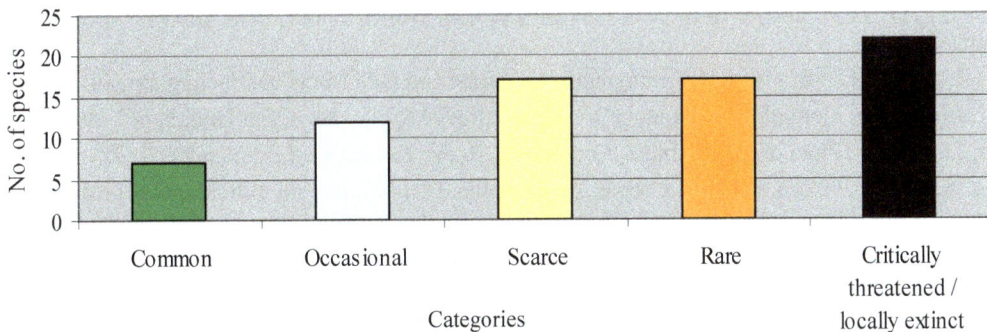

Fig. 1. Abundance status of the orchid species distributed in Sylhet.

Orchids are under great threats to their existence in the natural habitats due to biotic pressures, like illegal felling of large host trees, clearing off forest floor, forest fire and collection of horticultural valuable orchid species. Abundance status based on the availability of different orchid species in the natural habitat of the studied area and previous reports is presented in Fig. 1. It indicates that 30% orchid species in the Sylhet region have no data of their present occurrence and probably locally extinct from this area or critically threatened. So, both *in-situ* and *ex-situ* conservation measures along with public awareness programmes need to be undertaken.

## References

Abraham, A and Vatsala, P. 1981. Introduction to Orchids. Tropical Botanic Garden and Research Institute, Trivandram, India, 533 pp.

Ahmed, M. and Pasha, M.K. 1993. A taxonomic account of *Sarcanthus* Lindl. (Orchidaceae) from Bangladesh. J. Econ. Tax. Bot. **17**(2): 487–491.

Ahmed, M. and Pasha, M.K. 1993a. A taxonomic account of *Thrixspermum* Lour. (Orchidaceae) from Bangladesh. J. Asiat. Soc. Bang. Sci. **19**(1): 35–42.

Ahmed, M. and Pasha, M.K. 1994. A taxonomic account of *Hetaeria* Bl. (Orchidaceae) from Bangladesh. Chittagong Univ. Stud. Part II: Sci. **18**(2): 179–182.

Ahmed, M. and Pasha, M.K. 1998. A taxonomic account of *Eulophia* R. Br. (Orchidaceae) from Bangladesh. J. Asiat. Soc. Bang. Sci. **24**(1): 11–22.

Ahmed, M. and Pasha, M.K. 1998a. *Gastrochilus inconspicuum* (Hooker, f.) Seidenf. (Orchidaceae) - A new angiospermic record for Bangladesh. Bangladesh J. Bot. **27**(2): 143–146.

Ahmed, M. and Pasha, M.K. 1998b. A taxonomic account of *Luisia* Gaud. (Orchidaceae) from Bangladesh. J. Bom. Nat. Hist. Soc. **95**: 301–306.

Ahmed, M. and Pasha, M.K. 1999. A taxonomic account of *Robiquetia* Gaud. (Orchidaceae) from Bangladesh. J. Bom. Nat. Hist. Soc. **96**(3): 499–502.

Ahmed, M., Pasha, M.K. and Khan, M.A.A. 1989. A taxonomic account of *Aerides* Lour. (Orchidaceae) from Bangladesh. Bangladesh J. Bot. **18**(2): 147–155.

Ahmed, M., Pasha, M.K. and Khan, M.A.A. 1989a. *Agrostophyllum khasianum* Griff. - A new record for Bangladesh. Bangladesh J. Bot. **18**(1): 101–104.

Ahmed, M., Pasha, M.K. and Khan, M.A.A. 1989b. *Eria flava* Lindl. (Orchidaceae) - A new record for Bangladesh. Bangladesh J. Bot. **18**(2): 223–226.

Ahmed, M., Pasha, M.K. and Khan, M.A.A. 1992. *Kingidium taenialis* (Lindl.) P. F. Hunt (Orchidaceae) - A new record for Bangladesh. Bangladesh J. Bot. **21**(2): 283–285.

Bruhl, P. 1926. A guide to the orchids of Sikkim. Calcutta and Simla, Thacker, Spink & Co. 208 pp.

Chowdhery, H.J. 1998. Orchid flora of Arunachal Pradesh. Bishen Singh Mahendra Pal Singh, Dehra Dun, India, 824 pp.

Heinig, R.L. 1925. List of plants of Chittagong collectorate and Hill Tracts. Darjeeling, pp. 68–70.

Hooker, J.D. 1890a. Flora of British India (Orchideae), Reeve & Co., Kent, England. **5**: 667–864.

Hooker, J.D. 1890b. Flora of British India (Orchideae), Reeve & Co., Kent, England. **6**: 1–224.

Hossain, A.B.M.E. 2002. A taxonomic report on the genus *Dendrobium* Sw. (Orchidaceae) from Bangladesh. Bangladesh J. Plant Taxon. **9**(2): 47–55.

Huda, M. K., Rahman, M.A. and Wilcock, C.C. 1999. A preliminary checklist of orchid taxa occurring in Bangladesh. Bangladesh J. Plant Taxon. **6**: 69–85.

Huda, M. K. 2000. Diversity, Ecology, Reproductive Biology and Conservation of orchids of South East Bangladesh. Doctoral thesis. University of Aberdeen. UK. 266 pp. (Unpublished).

Huda, M.K. Rahman, M.A. and Wilcock, C.C. 2001. Notes on the Orchidaceae of Bangladesh-1 : Some new records. Bangladesh J. Plant Taxon. **8**(2): 9–17.

Huda, M.K. 2008. An up to date enumeration of the family Orchidaceae from Bangladesh. J. Orchid Soc. India **21**(1-2): 35–49.

Huda, M.K. 2008a. Orchidaceae *In*: Ahmed, Z.U., Hassan, M.A., Begum, Z.N.T., Khondker, M., Kabir, S.M.H., Ahmad, M., Ahmed, A.T.A., Rahman, A.K.A. and Haque, E.U. (Eds), Encyclopedia of Flora and Fauna of Bangladesh. Vol. **12**. Angiosperms: Monocotyledons (Orchidaceae- Zingiberaceae). Asiatic Society of Bangladesh, Dhaka, pp. 1–148.

Jayaweera, D.M.A. 1981. Orchidaceae. *In*: Dassanayakake, M.D. and Fosberg, F.R. (Eds), Flora of Ceylon, Vol. **2**. Amerind Publishing Co. Ltd. New Delhi, pp. 1-386.

Joseph, J. 1987. Orchids of Nilgiris. Botanical Survey of India, Govt. Press of India, 186 pp.

Khanam, M., Uddin, M.Z., Khan, M.S. and Hassan, M.A. 2001. Our present knowledge on the terrestrial orchidaceous taxa from Bangladesh. Bangladesh J. Plant Taxon. **8**(2): 35–49.

Khan, M.S. and Halim, M. 1987. *Bulbophyllum lilacinum* Ridley - A new angiospermic record for Bangladesh. Bangladesh J. Bot. **16**(2): 203–205.

Lindley, J. 1830-40. The genera and species of orchidaceous plants. Indian reprint 1983. Bishen Singh Mahendra Pal Singh, Dehra Dun, India.

Misra, S. 2000. Untamed Orissa- Wild Orissa, Bhubanswar, Orissa in Anonymous (Ed.) pp-25–30.

Pasha, M.K. 1984. Taxonomic studies of orchids of Chittagong division, Project report. Sponsored by the University Grant Commission, Dhaka, Bangladesh.

Prain, D. 1903. Bengal Plants, Vol. **2**. Calcutta, pp. 750–777.

Rao, T.A. 1998. Conservation of wild orchids of Kodagu in the Western Ghats, India, pp. 192–230.

Roxburgh, W. 1814 (reprint 1980). Hortus Benghalensis. Boerhavve Press, Leiden.

Roxburgh, W. 1832. Flora Indica (Gynadria Monandria). ed.**2**,2: 609–622.

Seidenfaden, G. 1978. Orchid genera in Thailand 6. Neottoideae. Dansk Bot. Ark. **32**: 1-195.

Seidenfaden, G. 1988. Orchid genera in Thailand XIV. fifty nine Vandoid genera. Opera Bot. **95**: 1–398.

Sinclair, J. 1956. The Flora of Cox's Bazar, East Pakistan. Bull. Bot. Soc. Beng. **9**(2): 107–108.

Uddin, M.Z., Khan, M.S., Hassan, M.A. and Khanam, M. 1999. *Pomatocalpa decipiens* (Lindl.) J. J. Smith - A new orchid record for Bangladesh. Bangladesh J. Bot. **28**(2): 169–171.

Uddin, M.Z., Khan, M.S., Hassan, M.A. and Khanam, M. 2000. *Brachycorythis helferi* (Rchb. f.) Summerh. - A new orchid record for Bangladesh. Bang. J. Plant Taxon. **7**(1): 73–75.

Uddin, M.Z., Khan, M.S. and Hassan, M.A. 2002. An annotated checklist of angiospermic flora of Rema-Kalenga Wildlife Sanctuary (Habiganj) in Bangladesh –I. Liliopsida (Monocots). Bangladesh J. Plant Taxon. **9**(2): 57–66.

# 18

# FLORISTIC DIVERSITY (MAGNOLIIDS AND EUDICOTS) OF BARAIYADHALA NATIONAL PARK, CHITTAGONG, BANGLADESH

MOHAMMAD HARUN-UR-RASHID[1], SAIFUL ISLAM AND SADIA BINTE KASHEM

*Department of Botany, University of Chittagong, Chittagong 4331, Bangladesh*

*Keywords*: Plant diversity; Baraiyadhala National Park; Conservation management.

## Abstract

An intensive floristic investigation provides the first systematic and comprehensive account of the floral diversity of Baraiyadhala National Park of Bangladesh, and recognizes 528 wild taxa belonging to 337 genera and 73 families (Magnoliids and Eudicots) in the park. Habit analysis reveals that trees (179 species) and herbs (174 species) constitute the major categories of the plant community followed by shrubs (95 species), climbers (78 species), and two epiphytes. Status of occurrence has been assessed for proper conservation management and sustainable utilization of the taxa resulting in 165 (31.25%) to be rare, 23 (4.36%) as endangered, 12 (2.27%) as critically endangered and 4 species (0.76%) are found as vulnerable in the forest. Fabaceae is the dominant family represented by 75 taxa, followed by Rubiaceae (47 taxa), Malvaceae (28 species), Asteraceae (27 species) and Euphorbiaceae (24 species). Twenty-three families represent single species each in the area.

## Introduction

Baraiyadhala National Park as one of the important Protected Areas (PAs) of Bangladesh that lies between $22^040.489'$-$22^048'$N latitude and $90^040'$-$91^055.979'$E longitude and located in Sitakundu and Mirsharai Upazilas of Chittagong district. The forest is under the jurisdiction of Baraiyadhala Forest Range of Chittagong North Forest Division. The park encompasses 2,933.61 hectare (7,249 acres) area and is classified under Category II of the International IUCN classification of protected areas (Hossain, 2015). Formerly, it was a part of the Reserved Forest of Chittagong North Forest Division, and was declared as Baraiyadhala National Park on 6th April, 2010 by the Ministry of Environment and Forest through a Gazzette notification no. MoEF/For-Sec-02/02 National Park/10/2010/210 dated 06/04/2010 under the Provisions of Article 23(1) of the Bangladesh Wildlife (Preservation) (Amendment) Act, 1974; which has now been altered by Wildlife (Conservation and Security) Act, 2012 (Hossain, 2015). The park comprises three blocks: Baraiyadhala and Wahidpur Blocks under Baraiyadhala Forest Beat, and Kunderhat Block under Bartakia Forest Beat.

The landscape of the forest is characterized by hills, valleys, gullies and numerous water streams and covered mainly by secondary degraded forests and plantations. Floristically Baraiyadhala National Park is rich and diverse. Hence, it is very important to take proper steps to conserve this natural forest; and to explore, document and analyze the species diversity occurring in the Baraiyadhala National Park before disappearing from nature.

However, the forest has not been botanized to determine the plant species diversity, their status of occurrence and conservation measures. Therefore, the present investigation aims to explore, collect, and document the Angiospermic (Magnoliids and Eudicots) plant resources in the forest, their conservation management and sustainable utilization.

---

[1]Corresponding author. Email: haruncu@gmail.com

## Material and Methods

The survey of the flora has been made through repeated field visits during February 2016 to July 2017 in Baraiyadhala National Park. Random sampling and collections of fertile specimens have been made for identification and voucher specimens have been preserved at Herbarium of Chittagong University (HCU). Collected specimens have been critically examined, studied and identified. Identifications have been confirmed by consulting standard literature and specimens and taxonomists of HCU and Bangladesh National Herbarium (DACB).

Nomenclature has been updated following recent literature (Ahmed *et al.*, 2008, 2009a,b,c,d; Rashid and Rahman, 2011, 2012), and confirmed with consulting The Plant List (2013) (http://www.theplantlist.org). Families are arranged according to the classification of Angiosperm Phylogeny Group (APG IV, 2016). The taxa are listed alphabetically under each family along with their habit, Bangla name, and status of occurrence (Table 1). Local names of many plants have been noted from local people during field trips and/or consulting Prain (1903), Heinig (1926), Huq (1986), Das and Alam (2001), and Dey *et al.* (1999).

## Results and Discussion

The present inventory provides the first systematic and comprehensive account of the floral diversity of the forest and recognizes 528 wild taxa belonging to 337 genera and 73 families (Magnoliids and Eudicots) in the park (Table 1).

Baraiyadhala National Park presents diverse habitat including hills, valleys, gullies and numerous water streams and covered mainly by secondary degraded forests. Some patches of the forest are planted with *Acacia auriculiformis* (Akashmoni), *Artocarpus chama* (Chapalish), *Chukrasia tabularis* (Chikrassi), *Eucalyptus* sp. (Eucalyptus), *Gmelina arborea* (Gamar), *Dipterocarpus turbinatus* (Garjan), *Swietenia mahagoni* (Mehogini), *Azadirachta indica* (Neem), *Aphanamixis polystachya* (Pitraj), *Tectona grandis* (Segun), *Hopea odorata* (Telsur), *Toona ciliata* (Toon) etc. The palms, rattans and bamboos mostly occupy the valleys. Common shrubs, herbs, grasses and babanas are fragmented to degraded habitats. A few individuals of Boilam (*Anisoptera scaphula*), Civit (*Swintonia floribunda*) and Lohakat (*Xylia xylocarpa* var. *kerrii*) are still available as characteristic elements of the forest. Some epiphytic species of *Cymbidium, Dendrobium, Drynaria, Raphidophra,* and *Phothos* are distributed in the forest area, while luxuriant growth of Aroids, Begonias, Bryophytes and Pteridophytes is observed in the natural moist habitats of the park. One of the most characteristic features of this forest is the occurrence of three indigenous gymnospermic species, *Cycas pectinata* Buch.-Ham., *Podocarpus neriifolius* D. Don and *Gnetum montanum* Markgr. In Bangladesh, Baraiyadhala National Park is the only home of *C. pectinata*. The biodiversity of the area is highly imperiled due to anthropogenic activities, including habitat destruction, over-exploitation, unsustainable hunting and all of these three species are critically endangered and are on the verge of extinction in the forest. However, Nguyen (2010) categorizes *C. pectinata* as Vulnerable (VU).

The park is dominated by trees and herbs consisting of 179 (33.90%) and 174 (32.95%) species respectively, followed by 18% shrub (95 spp.), 14.77% climbers (78 spp.), and two epiphytes. Fabaceae appears as the largest family with 75 taxa, followed by Rubiaceae (47 taxa), Malvaceae (28 spp.), Asteraceae (27 spp.) and Euphorbiaceae (24 spp.). Twenty-three families are represented by single species each in the study area. Ten dominant families (Fig. 2) constitute 320 species amounting 60.60% of the total species reported from the park, while remaining 63 families comprise only 39.40% of total species.

Status of occurrence has been assessed for proper conservation management and sustainable utilization of the natural resources of the forest. A total of 165 species (31.25%) are found to be rarely distributed in the forest, while 23 (4.36%) are assessed as endangered, 12 (2.27%) as

**Table 1. Plant diversity in Baraiyadhala National Park.**

| Family | Taxa | Bangla name | Habit | Status of occurrence |
|--------|------|-------------|-------|----------------------|
| Piperaceae | *Peperomia pellucida* (L.) Kunth | Luchi pata | H | Common |
| | *Piper longum* L. | Pipul | H | Rare |
| | *P. rhytidocarpum* Hook. f. | Ban pipul | CS | Common |
| | *P. sylvaticum* Roxb. | Ban Pan | Cr | Common |
| Aristolochiaceae | *Aristolochia indica* L. | Ishwarmul | Cl | Rare |
| | *A. saccata* Wall. | Ishwarmul | Cl | Rare |
| | *A. tagala* Cham. | Ishwarmul | WT | Rare |
| Myristicaceae | *Knema erratica* (Hook. f. & Thom.) Sinclair | - | T | Critically endangered |
| Magnoliaceae | *Magnolia champaca* (L.) Baill. *ex* Pierre | Champa | T | Common |
| Annonaceae | *Artabotrys caudatus* Wall. *ex* Hook. f. & Thomson | - | WC | Endangered |
| | *Desmos chinensis* Lour. | Sotoyalang | S | Common |
| | *D. dumosus* (Roxb.) Saff. | - | WC | Endangered |
| | *Fissistigma rubiginosum* (A. DC.) Merr. | - | WC | Endangered |
| | *F. wallichii* (Hook.f. & Thom.) Merr. | - | WC | Endangered |
| | *Uvaria dioeca* Roxb. | Tasbi | T | Vulnerable |
| | *U. hamiltonii* Hook. f. & Thom. | Latkan | WC | Endangered |
| | *U. littoralis* (Blume) Blume | Bagh-runga | WC | Vulnerable |
| Lauraceae | *Actinodaphne gullavara* (Buch.-Ham. *ex* Nees) Almeida | Tejmatan | T | Common |
| | *Litsea glutinosa* (Lour.) C.B. Rob. | Menda | T | Common |
| | *L. monopetala* (Roxb.) Pers. | Menda | T | Rare |
| | *Machilus gamblei* King *ex* Hook. f. | Nala-amsi | T | Rare |
| Menispermaceae | *Cocculus hirsutus* (L.) Theob. | Jaljamani | | Rare |
| | *Parabaena sagittata* Miers. | Jaljamani | Cl | Common |
| | *Stephania glabra* (Roxb.) Miers | Musahanilata | WC | Rare |
| | *S. japonica* (Thunb.) Miers | Nimukha | WC | Common |
| | *S. reticulata* Forman | - | Cl | Common |
| | *Tinospora crispa* (L.) Hook. f. & Thom. | Gulancha | WC | Endangered |
| Sabiaceae | *Meliosma pinnata* (Roxb.) Maxim. | Attalia | T | Rare |
| Proteaceae | *Helicia excelsa* (Roxb.) Blume | - | T | Rare |
| Dilleniaceae | *Dillenia pentagyna* Roxb. | Banchalta | T | Critically Endangered |
| | *D. scabrella* (D. Don) Roxb. *ex* Wall. | Ajuli | T | Common |
| Vitaceae | *Ampelocissus barbata* (Wall.) Planch. | Jarila-lahari | CS | Common |
| | *Cayratia japonica* (Thunb.) Gagnep. | - | Cl | Common |
| | *C. trifolia* (L.) Domin | Amal lata | Cl | Common |
| | *Cissus assamica* (M.A. Lawson) Craib | Amasha-lata | WC | Common |
| | *C. elongata* Roxb. | Dhemna | Cl | Common |
| | *C. javana* DC. | Bichitra-lata | Cl | Common |
| | *C. pentagona* Roxb. | Sona-lota | Cl | Rare |
| | *Leea aequata* L. | Kakjangha | S | Rare |
| | *L. asiatica* (L.) Ridsdale | Mach | S | Rare |

**Table 1 Contd.**

| Family | Taxa | Bangla name | Habit | Status of occurrence |
|--------|------|-------------|-------|----------------------|
| Vitaceae | *L. guineensis* G. Don | Phupharia | S | Rare |
| | *L. indica* (Burm.f.) Merr. | Kurkurji | T | Common |
| | *L. macrophylla* Roxb. *ex* Hornem. | - | T | Rare |
| | *Tetrastigma angustifolia* (Roxb.) Deb | Nekung riubi | Cl | Rare |
| | *T. bracteolatum* (Wall.) Planch. | Golgoli lata | Cl | Rare |
| | *T. hookeri* (M.A. Lawson) Planch. | Horina-lata | WC | Common |
| | *T. leucostaphylum* (Dennst.) Alston | Nekung | WC | Rare |
| Fabaceae | *Abrus precatorius* L. | Kunch | WC | Common |
| | *A. pulchellus* Wall. *ex* Thwaites | Kaichagula | TS | Rare |
| | *Adenanthera pavonina* L. | Raktachandan | T | Common |
| | *Albizia odoratissima* (L. f.) Benth. | Tetuya Koroi | T | Common |
| | *A. richardiana* (Voigt) King & Prain | Gagan Siris | T | Common |
| | *A. saman* (Jacq.) Merr. | Rendi koroi | T | Common |
| | *Bauhinia acuminata* L. | Sada kancon | S | Common |
| | *B. purpurea* L. | Devakanchan | T | Common |
| | *B. scandens* L. | Gendi lata | WC | Common |
| | *B. variegata* L. | Raktakancan | T | Common |
| | *Butea monosperma (*Lam.) Taub. | Palash | T | Common |
| | *Caesalpinia bonduc* (L.) Roxb. | Natakaranga | SS | Rare |
| | *C. digyna* Rottler | Umulkuchi | S | Rare |
| | *Cajanus scarabaeoides* (L.) Thouars | - | Tw | Rare |
| | *Calliandra umbrosa* (Wall.) Benth. | Chotobetmara | Tree | Common |
| | *Cassia fistula* L. | Sonalu | Tree | Common |
| | *C. javanica* subsp. *nodosa* (Roxb.) K. Larsen & S.S. Larsen | Bon-sonalu | Tree | Rare |
| | *C. obtusifolia* L. | Chakunda | U | Common |
| | *Codariocalyx gyroides* (Link) Hassk. | - | S | Rare |
| | *C. motorius (Houtt.)* H.Ohashi | Gorachand | S | Rare |
| | *Crotalaria acicularis* Buch.-Ham. *ex* Benth. | - | H | Common |
| | *C. albida* Roth | - | H | Common |
| | *C. bracteata* DC | - | S | Common |
| | *C. calycina* Schrank | - | H | Common |
| | *C. dubia* Graham | - | H | Common |
| | *C. ferruginea* Benth. | - | H | Common |
| | *C. incana* L. | Chotojhunjhuna | H | Rare |
| | *C. pallida* Aiton | Jhun-jhuni | H | Common |
| | *C. tetragona* Roxb. *ex* Andrews | - | H | Rare |
| | *C. verrucosa* L. | Bansan | U | Common |
| | *Dalbergia sericea* G. Don | Sristi | T | Rare |
| | *D. lanceolaria* L. f. | Chakemdia | T | Rare |
| | *D. malabarica* Prain | - | T | Rare |
| | *D. spinosa* Roxb. | Ananta kantha | LS | Common |
| | *D. stipulacea* Roxb. | Dadbari | T | Rare |
| | *D. volubilis* Roxb. | Ankilata | WC | Rare |
| | *Dendrolobium triangulare* (Retz.) Schindl. | Bir Jarwar | S | Rare |
| | *Derris robusta* (DC.) Benth. | Jangaria | T | Rare |

**Table 1 Contd.**

| Family | Taxa | Bangla name | Habit | Status of occurrence |
|---|---|---|---|---|
| Fabaceae | *D. scandens* (Roxb.) Benth. | Amkurchi | WC | Common |
| | *Desmodium gangeticum* (L.) DC. | Salpani | S | Common |
| | *D. heterocarpon* (L.) DC. | - | S | Common |
| | *D. heterophyllum* (Willd.) DC. | Bonmotorshuty | H | Common |
| | *D. laxiflorum* DC. | - | S | Common |
| | *D. triflorum* (L.) DC. | Kulalia | H | Common |
| | *Entada gigas* (L.) Fawc. & Rendle | Gila | Cl | Rare |
| | *E. rheedii* Spreng. | Gilalata | WC | Rare |
| | *Erythrina variegata* L. | Mandar | T | Rare |
| | *Flemingia macrophylla* (Willd.) Merr. | Bara salphan | LS | Common |
| | *F. strobilifera* (L.) W.T.Aiton | Sim busak | S | Common |
| | *Gliricidia sepium* (Jacq.) Walp. | Bashantamanjuri | MT | Rare |
| | *Mimosa diplotricha* Sauvalle | Bara lajjabati | CS | Common |
| | *M. himalayana* Gamble | Jharua | CS | Rare |
| | *M. pudica* L. | Lajja bati | H | Common |
| | *Mucuna monosperma* Wight | Soash guri | WC | Rare |
| | *M. pruriens* (L.) DC. | Al-kushi | Cl | Common |
| | *Phyllodium pulchellum* (L.) Desv. | Jatsalpani | S | Common |
| | *Pithecellobium jiringa* (Jack) Merr. | Kuramara Gach | S | Common |
| | *Pueraria tuberosa* (Willd.) DC. | Shimia | H | Common |
| | *P. phaseoloides* var. *subspicata* (Benth.) Maesen | - | H | Common |
| | *Saraca asoca* (Roxb.) Willd. | Ashok | T | Rare |
| | *S. indica* L. | Ashok | T | Common |
| | *Senna alata* (L.) Roxb. | Dad mardon | S | Common |
| | *S. occidentalis* (L.) Link | Boro kalkasunda | H | Rare |
| | *S. siamea* (Lam.) H. S. Irwin & Barneby | Minjiri | T | Common |
| | *S. sophera* (L.) Roxb. | Kalkashunda | S | Common |
| | *S. tora* (L.) Roxb. | Chakunda | H | Common |
| | *Sesbania bispinosa* (Jacq.) Wight. | Dhaincha | H | Common |
| | *Spatholobus parviflorus* (DC.) Kuntze | Goalia lata | Cl | Rare |
| | *Tadehagi triquetrum* (L.) H. Ohashi | - | U | Common |
| | *Tephrosia candida* (Roxb.) DC. | Bilokhoni | S | Common |
| | *T. purpurea* (L.) Pers. | Bon neel | H | Common |
| | *Uraria crinita* (L.) DC. | Dieng-kha-riu | S | Common |
| | *U. rufescens* (DC.) Schindl. | Belai leza | S | Common |
| | *Vicia sativa* L. | Ankari | H | Common |
| | *Xylia xylocarpa* var. *kerrii* (Craib & Hutch.) I.C. Nielsen | Lohakat | T | Rare |
| Polygalaceae | *Salomonia ciliata* (L.) DC. | - | H | Common |
| | *Xanthophyllum flavescens* Roxb. | Ajensak | T | Common |
| Rhamnaceae | *Gouania napalensis* Wall. | - | CS | Common |
| | *G. tiliifolia* Lam. | - | SS | Rare |
| | *Ziziphus glabrata* Heyne *ex* Roth. | Pahari boroi | S | Rare |

**Table 1 Contd.**

| Family | Taxa | Bangla name | Habit | Status of occurrence |
|---|---|---|---|---|
| Rhamnaceae | *Z. oenopolia* (L.) Mill. | Banboroi | S | Common |
| Ulmaceae | *Trema orientalis* (L.) Blume | Jiban | T | Common |
| Moraceae | *Artocarpus lakoocha* Roxb. | Deua | T | Common |
| | *Ficus benjamina* L. | Pakur | T | Common |
| | *F. benghalensis* L. | Bot | T | Common |
| | *F. fistulosa* Reinwdt. *ex* Blume | - | ST | Common |
| | *F. tinctoria* subsp. *gibbosa* (Blume) Corner | Bot | T | Common |
| | *F. heterophylla* L. f. | Bhui dumur | CS | Common |
| | *F. hispida* L. f. | Kakdumur | ST | Common |
| | *F. punctata* Thunb. | - | Cl | Common |
| | *F. semicordata* Buch.-Ham. *ex* Sm. | Jagadumur | ST | Common |
| | *Streblus asper* Lour. | Sheora | T | Common |
| Urticaceae | *Boehmeria glomerulifera* Miq. | Borthurthuri | S | Common |
| | *Dendrocnide sinuata* (Blume) Chew | - | S | Common |
| | *Oreocnide integrifolia* (Gaudich.) Miq. | Horhuta | ST | Common |
| | *Pilea microphylla* Liebm. | Mariccha lata | H | Common |
| | *Pouzolzia sanguinea* (Blume) Merr. | - | S | Common |
| | *P. zeylanica* (L.) Benn. | Kullaruki | H | Common |
| | *Sarcochlamys pulcherrima* Gaudich | Maricha | LS | Common |
| Fagaceae | *Castanopsis indica* (Roxb. *ex* Lindl.) DC. | Batna | T | Endangered |
| Betulaceae | *Alnus nepalensis* G. Don | - | T | Endangered |
| Cucurbitaceae | *Actinostemma tenerum* Griff. | Golapata | H | Common |
| | *Citrullus colocynthis* (L.) Schrad. | Indrayan | Cl | Endangered |
| | *Coccinia grandis* (L.) Voigt | Telakucha | H | Common |
| | *Hodgsonia macrocarpa* (Bl.) Cogn. | Pathligular | Cl | Critically Endangered |
| | *Thladiantha cordifolia* (Bl.) Cogn. | - | Cl | Rare |
| | *Trichosanthes cordata* Roxb. | Bhuikakra | Cl | Rare |
| | *T. tricuspidata* Lour. | Makal | H | Rare |
| Datiscaceae | *Tetrameles nodiflora* R. Br. | Chundul | TT | Rare |
| Begoniaceae | *Begonia roxburghii* (Miq.) A. DC. | - | H | Endangered |
| Celastraceae | *Euonymus attenuatus* Wall. *ex* Laws. | - | S | Rare |
| | *E. glaber* Roxb. | - | ST | Rare |
| Connaraceae | *Connarus paniculatus* Roxb. | Katgular | WC | Rare |
| Oxalidaceae | *Biophytum sensitivum* (L.) DC. | Jhalali | H | Rare |
| | *Oxalis corniculata* L. | Amrul | H | Common |
| Elaeocarpaceae | *Elaeocarpus tectorius* (Lour.) Poir. | Jalpai | T | Rare |
| Clusiaceae | *Garcinia cowa* Roxb. *ex* Choisy | Kao-gola | MT | Rare |
| Hypericaceae | *Cratoxylum sumatranum* subsp. *neriifolium* (Kurz) Gogelein | Nerikath | LS | Common |
| Achariaceae | *Hydnocarpus kurzii* (King) Warb. | Chaulmoogra | T | Rare |
| Passifloraceae | *Adenia trilobata* (Roxb.) Engl. | Akandaphal | Cl | Common |
| | *Passiflora foetida* L. | Jhumka-lata | H | Rare |
| Salicaceae | *Flacourtia jangomas* (Lour.) Racusch. | Paniamala | MT | Endangered |
| Euphorbiaceae | *Acalypha indica* L. | Muktajhuri | H | Common |
| | *Astraea lobata* (L.) Klotzsch | - | H | Common |
| | *Balakata baccata* (Roxb.) Esser | Katagola | T | Rare |

**Table 1 Contd.**

| Family | Taxa | Bangla name | Habit | Status of occurrence |
|---|---|---|---|---|
| Euphorbiaceae | *Chaetocarpus castanocarpus* (Roxb.) Thwaites | Bul kakra | MT | Common |
| | *Chrozophora rottleri* (Geisel.) A. Juss. *ex* Spreng. | Khudi okra | H | Common |
| | *Cnesmone javanica* Blume | Paharibichuti | WC | Common |
| | *Croton bonplandianus* Baill. | Bondhone | H | Common |
| | *C. caudatus* Geisel. | Nanbhanti | S | Common |
| | *Euphorbia hirta* L. | Dudhiya | H | Common |
| | *E. thymifolia* L. | Swetkan | H | Rare |
| | *Falconeria insignis* Royle | Belua | T | Rare |
| | *Jatropha gossypifolia* L. | Lal Bherendha | S | Common |
| | *Macaranga denticulata* (Bl.) Müll.Arg. | Bura, Jagra | T | Common |
| | *M. peltata* (Roxb.) Müll.Arg. | - | T | Common |
| | *Mallotus nudiflorus* (L.) Kulju & Welzen | Pitali | T | Rare |
| | *M. philippensis* (Lam.) Müll.Arg. | Kamela | ST | Rare |
| | *M. repandus* (Willd.) Müll.Arg. | Gunti | ST | Rare |
| | *M. roxburghianus* Müll.Arg. | Nim puteli | S | Rare |
| | *M. tetracoccus* (Roxb.) Kurz | Kumari-bura | MT | Rare |
| | *Manihot esculenta* Crantz | Kasava | S | Rare |
| | *Ricinus communis* L. | Verenda | H | Common |
| | *Shirakiopsis indica* (Willd.) Esser | Hura | ST | Rare |
| | *Suregada multiflora* (A. Juss.) Baill. | Maricha | ST | Rare |
| | *Tragia involucrata* L. | Bichuti | H | Rare |
| Phyllanthaceae | *Actephila excelsa* (Dalzell) Müll.Arg. | - | ST | Common |
| | *Antidesma acidum* Retz. | Chutki | T | Common |
| | *A. bunius* (L.) Spreng. | Banshial buka | MT | Rare |
| | *A. ghaesembilla* Gaertn. | Khudijam | MT | Rare |
| | *A. velutinosum* Bl. | - | ST | Rare |
| | *Aporosa aurea* Hook. f. | Kechuan | T | Common |
| | *A. octandra* (Buch.-Ham. *ex* D. Don) Vickery | Pat khorulla | ST | Common |
| | *A. wallichii* Hook.f. | Karullah | T | Rare |
| | *Baccaurea ramiflora* Lour. | Latkon | MT | Rare |
| | *Bischofia javanica* Bl. | Kanjalbhadi | T | Common |
| | *Breynia retusa* (Dennst.) Alston | Silpati | S | Rare |
| | *Bridelia stipularis* (L.) Bl. | Pat khowi | ST | Common |
| | *B. tomentosa* Bl. | Khoi | ST | Rare |
| | *B. verrucosa* Haines | - | ST | Common |
| | *Glochidion assamicum* (Müll.Arg.) Hook.f. | - | ST | Rare |
| | *G. lanceolarium* (Roxb.) Voigt | Bhauri | ST | Common |
| | *G. multiloculare* (Rottler *ex* Willd.) Voigt | Keotomi | ST | Common |
| | *G. sphaerogynum* (Müll.Arg.) Kurz | - | T | Rare |
| | *Phyllanthus attenuatus* Miq. | Panjuli | S | Rare |
| | *P. niruri* L. | Bhuimala | H | Common |
| | *P. reticulatus* Poir. | Chitka | S | Common |
| | *P. sikkimensis* Müll.Arg. | Sikim-amla | LS | Common |

**Table 1 Contd.**

| Family | Taxa | Bangla name | Habit | Status of occurrence |
|---|---|---|---|---|
| Combretaceae | *Combretum album* Pers. | Kali gumuchi | SS | Rare |
| | *C. apetalum* Wall. *ex* Kurz | Dolhara | S | Rare |
| | *Getonia floribunda* Roxb. | Guicha lata | SS | Rare |
| | *Terminalia alata* Heyne *ex* Roth. | Asal | TT | Rare |
| | *T. bellirica* (Gaertn.) Roxb. | Bohera | T | Rare |
| Lythraceae | *Ammannia multiflora* Roxb. | - | H | Common |
| | *Lagerstroemia parviflora* Roxb. | Baturi | MT | Rare |
| | *Woodfordia fruticosa* (L.) Kurz | Dhatriphul | S | Common |
| | *Duabanga grandiflora* (Roxb. *ex* DC.) Walp. | Bandarhulla | T | Rare |
| Onagraceae | *Ludwigia adscendens* (L.) H. Hara | Kesardam | CH | Common |
| | *L. hyssopifolia* (G. Don) Exell | - | H | Common |
| Myrtaceae | *Syzygium nervosum* A. Cunn. *ex* DC. | Bhutijam | T | Rare |
| | *S. fruticosum* DC. | Khudijam | ST | Rare |
| Melastomataceae | *Melastoma malabathricum* L. | Bantejpzta | S | Common |
| | *Osbeckia aspericaulis* Hook. f. *ex* Tri. | - | S | Common |
| Crypteroniaceae | *Crypteronia paniculata* Blume | Goru-mara | T | Rare |
| Burseraceae | *Protium serratum* (Wall. ex Colebr.) Engl. | Gutgutya | T | Rare |
| Anacardiaceae | *Bouea oppositifolia* (Roxb.) Adelb. | Uri Aam | T | Rare |
| | *Buchanania lancifolia* Roxb. | Chikki | T | Rare |
| | *Drimycarpus racemosus* (Roxb.) Hook.f. *ex* Marchand. | Nala-amshi | T | Critically Endangered |
| | *Holigarna caustica* (Dennst.) Oken | Barola | T | Critically Endangered |
| | *Lannea coromandelica* (Houtt.) Merr. | Jiga | T | Rare |
| | *Mangifera laurina* Bl. | Jangali aam | T | Vulnrerable |
| | *M. sylvatica* Roxb. *ex* Wall. | Uri Aam | T | Critically Endangered |
| | *Semecarpus subpanduriformis* Wall. | Beula | T | Rare |
| | *Spondias pinnata* (L. f.) Kurz | Amra | T | Rare |
| | *Swintonia floribunda* Griff. | Chundul | T | Critically Endangered |
| Sapindaceae | *Allophylus cobbe* (L.) Raeusch. | Chita | S | Common |
| | *A. villosa* (Roxb.) Blume | - | ST | Common |
| Sapindaceae | *Cardiospermum halicacabum* L. | Phutka | H | Common |
| | *Lepisanthes rubiginosum* (Roxb.) Leenh. | Baraharina | ST | Rare |
| | *L. senegalensis* (Poir.) Leenh. | Gotaharina | S | Common |
| | *Xerospermum laevigatum* Radlk. | Muraillah lichu | T | Rare |
| Rutaceae | *Aegle marmelos* (L.) Corr. | Bel | T | Rare |
| | *Acronychia pedunculata* (L.) Miq. | Bon jamir | ST | Rare |
| | *Citrus aurantiifolia* (Christm.) Sw. | Lebu | S | Common |
| | *Clausena heptaphylla* (Roxb.) Wight & Arn. | Karanphul | S | Common |
| | *Glycosmis pentaphylla* (Retz.) DC. | Sheora | S | Common |
| | *Micromelum minutum* Wight & Arn. | Koroiphula | ST | Common |
| | *Paramignya scandens* (Griff.) Craib. | Bannebu | ST | Rare |

**Table 1 contd.**

| Family | Taxa | Bangla name | Habit | Status of occurrence |
|---|---|---|---|---|
| Meliaceae | *Aglaia edulis* (Roxb.) Wall. |  | T | Critically endangered |
|  | *Aphanamixis polystachya* (Wall.) R. Parker | Pitraj | T | Common |
|  | *Azadirachta indica* A. Juss. | Neem | T | Common |
|  | *Chisocheton paniculatus* Hiern | Rata | T | Rare |
|  | *Chukrasia tabularis* A. Juss. | Chikrassi | T | Common |
|  | *Melia azedarach* L. | Bokhain | T | Rare |
|  | *Swietenia mahagoni* (L.) Jacq. | Mehogini | T | Common |
|  | *Toona ciliata* M. Roem | Toon | T | Common |
| Malvaceae | *Abelmoschus moschatus* Medik. | Mushakdana | H | Common |
|  | *Abroma augusta* (L.) L. f. | Ulatkambal | ST | Common |
|  | *Abutilon indicum* (L.) Sweet | Petari | H | Common |
|  | *Bombax ceiba* L. | Shimul | T | Common |
|  | *B. insigne* Wall. | Ban-simul | T | Rare |
|  | *Brownlowia elata* Roxb. | Masjot | T | Critically endangered |
|  | *Byttneria pilosa* Roxb. | Harjora-lata | WC | Common |
|  | *Ceiba pantandra* (L.) Gaertn. | Pahari tula | T | Rare |
|  | *Corchorus aestuans* L. | Titapat | H | Rare |
|  | *Firmiana colorata* (Roxb.) R. Br. | Patagota | T | Critically endangered |
|  | *Grewia asiatica* L. | Pholsa | ST | Common |
|  | *G. laevigata* Vahl | Panisara | ST | Common |
|  | *G. nervosa* (Lour.) Panigrahi | Asar | ST | Common |
|  | *G. serrulata* DC. | Panisara | T | Rare |
|  | *Hibiscus vitifolius* L. | Bon-karpas | H | Common |
|  | *Malachra capitata* L. | Bon vindi | H | Rare |
|  | *Pterospermum acerifolium* (L.)Willd. | Musigondha | T | Endangered |
|  | *P. semisagittatum* Buch.-Ham. *ex* Roxb. | Bara assar | T | Endangered |
|  | *Sida acuta* Burm. f. | Kureta | H | Common |
|  | *S. cordata* (Burm. f.) Borss. Waalk. | Jhumka | H | Common |
|  | *S. cordifolia* L. | Berela | H | Common |
|  | *S. mysorensis* Wight & Arn. | Chatehata | H | Common |
|  | *Sterculia foetida* L. | Udal | MT | Endangered |
|  | *S. hamiltonii* (Kuntze) Adelb. | - | ST | Endangered |
|  | *S. villosa* Roxb. *ex* Smith | Udal | MT | Endangered |
|  | *Triumfetta pilosa* Roth | - | H | Common |
|  | *T. rhomboidea* Jacq. | Ban okra | H | Common |
|  | *Urena lobata* L. | Banokra | H | Common |
| Bixaceae | *Bixa orellana* L. | Latkon | ST | Rare |
| Dipterocarpaceae | *Anisoptera scaphula* (Roxb.) Pierre. | Boilsur | T | Critically endangered |
|  | *Dipterocarpus costatus* | Garjan | T | Rare |
|  | *D. turbinatus* Gaertn. | Teli garjan | T | Common |
|  | *Hopea odorata* Roxb. | Telsur | T | Common |
| Capparaceae | *Capparis spinosa* L. | Kabra | S | Rare |
|  | *Cleome rutidosperma* DC. | Begunehurhury | H | Common |
|  | *C. viscosa* L. | Holdehurhury | H | Common |

**Table 1 Contd.**

| Family | Taxa | Bangla name | Habit | Status of occurrence |
|---|---|---|---|---|
| Capparaceae | *Crateva religiosa* G. Forst. | Barun | ST | Common |
| | *Capsella bursa-pastoris* (L.) Medik. | Capsala | H | Common |
| Brassicaceae | *Rorippa indica* (L.) Hiern | Ban-saruisha | H | Common |
| | *R. palustris* (L.) Besser | Panisarisha | H | Common |
| Loranthaceae | *Dendrophthoe pentandra* (L.) Miq. | - | S | Rare |
| | *Macrosolen cochinchinensis* (Lour.) Van Tiegh. | Chota banda | S | Rare |
| | *Scurrula gracilifolia* (Schult.) Danser | Pargacha | S | Rare |
| | *S. parasitica* L. | Parula | S | Rare |
| Polygonaceae | *Persicaria chinensis* (L.) H.Gross | Mohicharan sak | H | Common |
| | *P. hydropiper* (L.) Delarbre | Bishkatali | H | Common |
| | *P. orientalis* (L.) Spach | Panimarich | H | Common |
| | *Polygonum viscosum* Buch.-Ham. *ex* D. Don | Athalo bishkatali | H | Rare |
| | *Rumex vesicarius* L. | Tokpalong | H | Rare |
| Amaranthaceae | *Achyranthes aspera* L. | Apang | H | Common |
| | *Alternanthera philoxeroides* (Mart.) Griseb. | Sachishak | H | Common |
| | *A. sessilis* (L.) R. Br. *ex* DC. | Chanchi | H | Common |
| | *Amaranthus spinosus* L. | Kantanotey | H | Common |
| | *A. viridis* L. | Notey shak | H | Common |
| | *Celosia argentea* L. | Shet morogha | H | Common |
| | *Chenopodium album* L. | Batua Shak | S | Common |
| | *Cyathula prostrata* (L.) Blume | Shyontula | H | Common |
| | *Gomphrena celosioides* Mart. | - | H | Common |
| Nyctaginaceae | *Boerhavia repens* L. | Punarnava | H | Common |
| Molluginaceae | *Glinus oppositifolius* (L.) A. DC. | Gimashakh | H | Common |
| | *Mollugo stricta* L. | Khetpapra | H | Rare |
| Portulacaceae | *Portulaca oleracea* L. | Nune shak | H | Common |
| Lecythidaceae | *Barringtonia acutangula* (L.) Gaertn. | Hijal | T | Common |
| Ebenaceae | *Diospyros malabarica* (Desr.) Kostl. | Deshi gab | T | Rare |
| | *D. racemosa* Roxb. | Gab | MT | Endangered |
| Primulaceae | *Ardisia humilis* Vahl | Chauldhoa | S | Common |
| | *A. paniculata* Roxb. | - | ST | Rare |
| | *A. sanguinolenta* Blume | - | ST | Endangered |
| Primulaceae | *A. solanacea* (Poir.) Roxb. | Bon -jam | S | Common |
| | *Maesa indica* (Roxb.) A. DC. | Sirkhi | S | Common |
| | *M. ramentacea* (Roxb.) A. DC. | Naricha | ST | Common |
| Actinidiaceae | *Saurauia roxburghii* Wall. | Bhola kadam | S | Rare |
| Rubiaceae | *Canthium angustifolium* Roxb. | Katamalli | S | Common |
| | *Ceriscoides campanulata* (Roxb.) Tirveng. | Tirveng | T | Endangered |
| | *Chassalia curviflora* var. *ophioxyloides* (Wall.) Deb & B.Krishna | - | S | Rare |
| | *Dentella repens* (L.) J.R. Forst. & G. Frost. | Bhuipat | H | Common |
| | *Gardenia coronaria* Buch.-Ham. | Torgular | MT | Rare |
| | *Haldina cordifolia* (Roxb.) Ridsd. | Kalakadam | T | Rare |
| | *Hedyotis scandens* Roxb. | Bish lata | H | Common |
| | *Hyptianthera stricta* (Roxb. *ex* Schult.) Wight & Arn. | - | S | Common |

**Table 1 Contd.**

| Family | Taxa | Bangla name | Habit | Status of occurrence |
|---|---|---|---|---|
| Rubiaceae | *Ixora athrorantha* Bremek. | - | ST | Common |
| | *I. balakrishnanii* Deb & Rout | Bhantjhara | ST | Common |
| | *I. cuneifolia* Roxb. | Beophul | S | Common |
| | *I. nigricans* R.Br. *ex* Wight & Arn. | Nikranga | ST | Common |
| | *I. pavetta* Andr. | Swetrangan | ST | Common |
| | *I. pubirama* Bremek. | Keamosi | S | Common |
| | *I. spectabilis* Wall. *ex* G. Don | - | ST | Common |
| | *I. subsessilis* Wall. *ex* G. Don | Rengchan | S | Common |
| | *Knoxia sumatrensis* (Retz.) DC. | - | H | Common |
| | *Lasianthus hirsutus* (Roxb.) Merr. | Kala long | S | Common |
| | *Meyna spinosa* Roxb. *ex* Link | Mainakata | ST | Rare |
| | *Mitracarpus hirtus* (L.) DC. | - | H | Common |
| | *Mitragyna diversifolia* (Wall. *ex* G. Don) Havil. | Phul-kadam | T | Common |
| | *M. parvifolia* var. *microphylla* (Kurz) Ridsdale | Putikadam | T | Near threatened |
| | *Morinda angustifolia* Roxb. | Banamali | ST | Common |
| | *M. citrifolia* L. | Ronch | ST | Common |
| | *M. persicifolia* Buch.-Ham. | - | S | Common |
| | *Mussaenda roxburghii* Hook. f. | Sildaura | S | Common |
| | *Neolamarckia cadamba* (Roxb.) Boss. | Kadom | T | Common |
| | *Neonauclea sessilifolia* (Roxb.) Merr. | Kom | T | Near threatened |
| | *Oldenlandia auricularia* (L.) K. Sch. | Mutia lata | H | Common |
| | *O. corymbosa* L. | Ketpapra | H | Common |
| | *O. diffusa* (Willd.) Roxb. | - | H | Common |
| | *Ophiorrhiza mungos* L. | Kalashana | H | Common |
| | *O. rugosa* var. *prostrata* (D. Don) Deb & Mondal | Jari | H | Common |
| | *Oxyceros kunstleri* (King & Gamble) Tirveng. | Maish kata | CS | Rare |
| | *Paederia foetida* L. | Gandha-badali | Cl | Common |
| | *Pavetta indica* L. | Banamali | ST | Common |
| | *Prismatomeris tetrandra* (Roxb.) K. Schum. | Chinatita | ST | Common |
| | *Psychotria adenophylla* Wall. | Baro sudma | ST | Common |
| | *P. calocarpa* Kurz | Ranga bhutta | U | Common |
| | *P. monticola* Kurz | Hatichotra | S | Common |
| | *P. symplocifolia* Kurz | - | ST | Common |
| | *Spermacoce articularis* L.f. | Ahtharogia | H | Common |
| | *S. hispida* L. | Ahtharogia | H | Common |
| | *Tarenna companiflora* (Hook. f.) N.P. Balakr. | Kakra | ST | Common |
| | *Tarennoidea wallichii* Tirveng. & Sast. | - | T | Common |
| | *Uncaria scandens* (Smith) Hutch. | Vailful lata | Cl | Rare |
| | *Wendlandia tinctoria* subsp. *orientalis* Cowan | Tulaload | ST | Common |
| Gentianaceae | *Exacum tetragonum* Roxb. | Kuchuri | H | Rare |
| | *Fagraea ceilanica* Thunb. | - | ST | Rare |

**Table 1 Contd.**

| Family | Taxa | Bangla name | Habit | Status of occurrence |
|---|---|---|---|---|
| Apocynaceae | *Alstonia scholaris* (L.) R. Br. | Chaitan | T | Common |
| | *Anodendron paniculatum* A.DC. | Dul | WC | Rare |
| | *Asclepias curassavica* L. | Moricha | H | Rare |
| | *Calotropis gigantea* (L.) Dryand. | Akanda | S | Common |
| | *Chonemorpha fragrans* (Moon) Alston | Gar badero | WC | Endangered |
| | *Gymnema acuminatum* Wall. | Khara lata | WT | Rare |
| | *Hemidesmus indicus* (L.) R.Br. *ex* Schult. | Anantamul | H | Rare |
| | *Holarrhena pubescens* (Roth) Wall. *ex* A. DC. | Kurchi, Kuruj | ST | Rare |
| | *Hoya globulosa* Hook. f. | Pargacha | Ep | Rare |
| | *H. parasitica* (Roxb.) Wall. *ex* Wight | Pargacha | E | Common |
| | *Ichnocarpus frutescens* (L.) W. T. Aiton | Syamalota | CS | Rare |
| | *Melodinus cochinchinensis* (Lour.) Merr. | - | WT | Vulnerable |
| | *Rauvolfia serpentina* (L.) Benth. *ex* Kurz | Sarpagandha | H | Endangered |
| | *Strophanthus wallichii* A. DC. | - | WC | Endangered |
| | *Tabernaemontana divaricata* (L.) R. Br. *ex* Roem. & Schult. | Togor | S | Common |
| | *Tylophora hirsuta* Wight | Anantamul | Tw | Rare |
| | *Vallaris solanacea* (Roth) Kuntze | Hadpur | SS | Critically endangered |
| | *Wrightia arborea* (Dennst.) Mabb. | Dudhkurush | T | Endangered |
| Boraginaceae | *Cordia dichotoma* G. Forst. | Buha | T | Rare |
| | *Heliotropium indicum* L. | Hatisur | H | Common |
| Convolvulaceae | *Argyreia argentea* (Roxb.) Choisy. | Bara rupatola | Cl | Common |
| | *A. capitiformis* (Poir.) Ooststr. | Vogalata | Cl | Rare |
| | *A. splendens* (Hornem.) Sweet | Borogandobaduli | Cl | Rare |
| | *Calystegia hederacea* Wall. | - | Tw | Common |
| | *Cuscuta reflexa* Roxb. | Swornalata | P | Rare |
| | *Evolvulus nummularius* (L.) L. | Bhui-akra | H | Common |
| | *Hewittia malabarica* (L.) Suresh | - | Cl | Common |
| | *Ipomoea alba* L. | Dudhikalmi | Tw | Rare |
| | *I. aculeata* var. *mollissima* (Zoll.) Hallier f. *ex* Oostr. | - | Cl | Rare |
| | *I. aquatica* Forssk. | Kalmishak | AH | Common |
| | *I. carnea* Jacq. | Dholkolmi | H | Common |
| | *I. fistulosa* Mart. *ex* Choisy | Dhol Kolmi | S | Common |
| | *I. hederifolia* L. | Neela-kalmi | Tw | Rare |
| | *I. mauritiana* Jacq. | Bhui-Kumra | Tw | Rare |
| | *I. pes-tigridis* L. | Languli lata | Cl | Common |
| | *I. pileata* Roxb. | - | Cl | Common |
| | *Jacquemontia paniculata* (Burm. f.) Hallier f. | Montilata | Tw | Rare |
| | *Merremia umbellata* (L.) Hallier | Sadakamni | Cl | Common |
| | *M. vitifolia* (Burm. f.) Hallier f. | - | Tw | Common |
| | *Operculina turpethum* (L.) Silva Manso | Dudh kalmi | Tw | Rare |

**Table 1 Contd.**

| Family | Taxa | Bangla name | Habit | Status of occurrence |
|---|---|---|---|---|
| Solanaceae | *Datura metel* L. | Kala datura | H | Common |
| | *Nicotiana plumbaginifolia* Viv. | Bon tamak | H | Common |
| | *Physalis angulata* L. | Fotka | H | Common |
| | *P. micrantha* Link. | Bantipariya | H | Common |
| | *P. minima* L. | Phutka | H | Common |
| | *Solanum ilicifolium* Dunal | Tepari | H | Common |
| | *S. indicum* L. | Bon Begun | S | Common |
| | *S. nigrum* L. | Kakmachi | H | Common |
| | *S. sisymbrifolium* Lam. | Kanta begun | H | Common |
| | *S. torvum* Sw. | Tit begun | S | Common |
| | *S. virginianum* L. | Kantakari | H | Common |
| Oleaceae | *Jasminum scandens* Vahl | - | WC | Rare |
| | *Ligustrum robustum* (Roxb.) Blume | - | LS | Rare |
| | *Myxopyrum smilacifolium* (Wall.) Bl. | - | SS | Rare |
| Scrophulariaceae | *Adenosma indianum* (Lour.) Merr. | Barakesuti | H | Common |
| | *Buddleja asiatica* Lour. | Neemda | S | Rare |
| | *Lindernia antipoda* (L.) Alston | - | H | Common |
| | *Picria fel-terrae* Lour. | - | H | Rare |
| | *Scoparia dulcis* L. | Bandhane | H | Common |
| | *Torenia asiatica* L. | - | H | Common |
| | *T. diffusa* D. Don | - | H | Common |
| | *T. flava* Buch.-Ham. *ex* Benth. | - | H | Common |
| Acanthaceae | *Andrographis laxiflora* (Bl.) Lindau | Algatita | H | Rare |
| | *A. paniculata* (Burm. f.) Nees | Kalmegh | H | Rare |
| | *Barleria strigosa* Willd. | Katapol | H | Rare |
| | *Dicliptera chinensis* (L.) Juss. | - | H | Rare |
| | *Ecbolium ligustrinum* (Vahl) Voll. | - | S | Rare |
| | *Eranthemum pulchellum* Andrew | - | S | Common |
| | *E. strictum* Colebr. *ex* Roxb. | Khara murali | H | Rare |
| | *Justicia adhatoda* L. | Basak | S | Common |
| | *J. gendarussa* Burm. f. | Jugmadan | U | Common |
| | *Hygrophila polysperma* (Roxb.) T. Anders | - | H | Common |
| | *H. ringens* (L.) R. Br. *ex* Spreng. | Kakmasha | H | Common |
| | *Lepidagathis incurva* Buch.-Ham. *ex* D. Don | Karoggthis | H | Common |
| | *Nelsonia canescens* (Lam.) Spreng. | Paramul | H | Common |
| | *Phaulopsis imbricata* (Forssk.) Sweet | Kantasi | H | Common |
| | *Ruellia acuminata* L. | - | H | Common |
| | *R. tuberosa* L. | Chotpoty | H | Common |
| | *Rungia pectinata* (L.) Nees | Pindi | H | Common |
| | *Staurogyne argentea* Wall. | - | H | Rare |
| | *Strobilanthes auriculatus* Nees | Kurinji | S | Rare |
| | *S. scaber* Nees | - | H | Rare |
| | *Thunbergia alata* Bojer *ex* Sims | Ghontolata | Cl | Rare |
| | *T. grandiflora* (Roxb. *ex* Rottl.) Roxb. | Nallata | Cl | Common |
| Bignoniaceae | *Oroxylum indicum* (L.) Kurz | Thona | T | Rare |
| | *Stereospermum chelonoides* (L.f.) DC. | Kam sonalu | T | Rare |
| | *S. tetragonum* DC. | Dharmara | T | Rare |

**Table 1 Contd.**

| Family | Taxa | Bangla name | Habit | Status of occurrence |
|---|---|---|---|---|
| Verbenaceae | *Duranta erecta* L. | Kantamehedi | SS | Common |
| | *Lantana camara* L. | Lantana | S | Common |
| | *Lippia alba* (Mill.) N.E. Br. *ex* Britton & P. Wilson | Motmotia | S | Common |
| | *Phyla nodiflora* (L.) Greene | Bhuiokra | H | Common |
| Lamiaceae | *Anisomeles indica* (L.) O. Kuntze | Gobura | H | Common |
| | *Callicarpa arborea* Roxb. | Bormala | T | Rare |
| | *C. macrophylla* Vahl | Bormala | T | Rare |
| | *Clerodendrum indicum* (L.) Kuntze | Bamunhatti | H | Common |
| | *C. infortunatum* L. | Ghetu | S | Common |
| | *C. laevifolium* Bl. | - | S | Common |
| | *Gmelina arborea* Roxb. | Gamar | H | Common |
| | *Hyptis brevipes* Poit. | - | H | Common |
| | *H. capitata* Jacq. | - | H | Common |
| | *H. suaveolens* (L.) Poit. | Tokma | H | Common |
| | *Leonurus sibiricus* L. | Roktodron | H | Rare |
| | *Leucas aspera* (Willd.) Link | Shetodron | H | Common |
| | *L. zeylanica* (L.) Aiton | Dondokolosh | H | Common |
| | *Ocimum americanum* L. | Bon Tulsi | H | Common |
| | *O. basilicum* L. | Babuitulshi | H | Common |
| | *Pogostemon auricularius* (L.) Hassk. | - | H | Common |
| | *Premna esculenta* Roxb. | Lalana | S | Common |
| | *Vitex altissima* L.f. | Awalya | T | Rare |
| | *V. glabrata* R. Br. | Ashal | T | Rare |
| | *V. negundo* L. | Nishinda | ST | Common |
| | *V. peduncularis* Wall. *ex* Schauer | Awal | ST | Rare |
| Campanulaceae | *Cyclocodon lancifolius* (Roxb.) Kurz | - | H | Rare |
| Asteraceae | *Acmella paniculata* (Wall. *ex* DC.) R.K. Jansen | Kannaghas | H | Common |
| | *Adenostemma lavenia* (L.) Kuntze | Baro-kesuti | H | Common |
| | *A. viscosum* J. R. Forst. & G. Forst. | Boro kesuti | H | Rare |
| | *Ageratum conyzoides* (L.) L. | Fulkuri | H | Common |
| | *Blumea fistulosa* (Roxb.) Kurz | - | H | Common |
| | *B. lacera* (Burm. f.) DC. | Kuksung | H | Common |
| | *B. lanceolaria* (Roxb.) Druce | - | H | Common |
| | *B. virens* DC. | Nilsabus | H | Common |
| | *Blumeopsis flava* (DC.) Gagnep | - | H | Common |
| | *Chromolaena odorata* (L.) R. M. King & H. Rob. | Assam-lata | H | Common |
| | *Crassocephalum crepidioides* S. Moore | - | H | Common |
| | *Cyanthillium cinereum* (L.) H. Rob. | Sahadebi | H | Common |
| | *Eclipta prostrata* (L.) L. | Kesraj | H | Common |
| | *Elephantopus scaber* L. | Gejiashak | H | Common |
| | *Emilia sonchifolia* (L.) DC. *ex* DC. | Mechitra | H | Common |
| | *Grangea maderaspatana* (L.) Poir. | Namuti | H | Rare |
| | *Laphangium luteoalbum* (L.) Tzvelev | Dudh ghas | H | Common |
| | *Mikania scandens* (L.) Willd. | Tufanilata | H | Common |
| | *Pseudelephantopus spicatus* (B.Juss. *ex* Aubl.) Rohr *ex* Baker | Kukurgihba | H | Rare |

**Table 1 Contd.**

| Family | Taxa | Bangla name | Habit | Status of occurrence |
|---|---|---|---|---|
| Asteraceae | *Sonchus wightianus* DC. | - | H | Common |
| | *Sphaeranthus africanus* L. | Gongasag | H | Rare |
| | *S. indicus* L. | Murmuri | H | Rare |
| | *S. senegalensis* DC. | Senigalsag | H | Rare |
| | *Spilanthes acmella* (L.) L. | Nag phul | H | Common |
| | *Synedrella nodiflora* (L.) Gaertn. | Relanodi | H | Common |
| | *Tridax procumbens* L. | Tridhara | H | Common |
| | *Xanthium strumarium* L. | Ghagra | H | Common |
| Araliaceae | *Brassaiopsis glomerulata* (Bl.) Regel | Kurila | ST | Common |
| | *Trevesia palmata* (Roxb. *ex* Lindl.)Vis. | Vombal | LS | Common |
| Apiaceae | *Centella asiatica* (L.) Urb. | Thankuni | H | Common |
| | *Hydrocotyle sibthorpioides* Lam. | - | H | Common |

H= Herb, CS= Climbing shrub, Cr= Creeper, WT= Woody twinner, T= Tree, S= Shrub, Cl= Cimber, WC= Woody climber, TS= Twining shrub, SS= Scandent shrub, Tw= Twiner, U= undershrub, LS= Large shrub, MT= Medium-sized tree, ST= Small tree, CH= Creeping herb, E= Epiphyte, P= Parasite, AH= Aquatic herb.

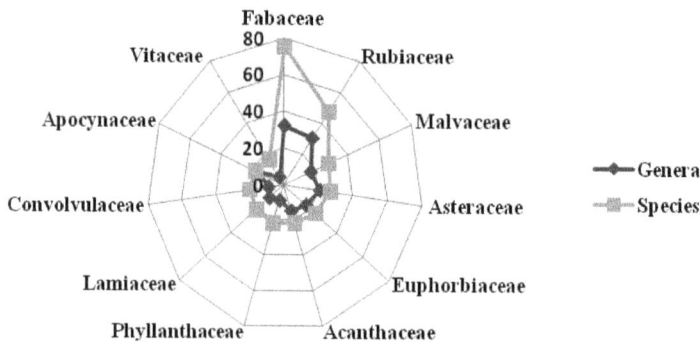

Fig. 2. Radar diagram shows ten dominat plant families in the Baraiyadhala National Park.

critically endangered and 4 species (0.76%) as vulnerable. All species belonging to the families Anacardiaceae (10 spp.), Bignoniaceae (3 spp.), Dipterocarpaceae (4 spp.), and Meliaceae (8 spp.) are trees and under threat in the forest. Rahman (2013) reported *Artabotrys caudatus* of the family Annonaceae as possibly extinct in Bangladesh. Recently this species has been reported from Hazarikhil Wildlife Sanctuary by Rahman (2017) and current investigation also reports its distribution in the Baraiyadhala National Park.

The present study reveals that communities around the area depend more or less on forest resources and some organized encroachers are also noticed. Furthermore, forest fire, livestock and other unsustainable resource utilization practices degrade the habitat in and around the forest. Climate change is an additional threat which may directly affect habitat and biodiversity, as well as indirectly by further increasing human pressure due to migration of peoples from coastal parts of Sitakunda and Mirsharai to the forest area. Development of tourism may enhance a future threat to the park. To address these challenges sustainable management plan for the protected area should be undertaken in the light of National Conservation Strategy, and proper implementation of the action plan is urgently required with collaboration and cooperation of stakeholders and local administrations. More particularly, the rare and threatened species along with their degraded habitats to be protected by taking proper conservation management programmes.

## Acknowledgements

This study was supported by Research Allocation Fund under Revenue Budget, University of Chittagong, Bangladesh. We are thankful to Prof. Dr. M. Atiqur Rahman, Department of Botany, University of Chittagong for his generous help and cooperation during the course of this study.

## References

Ahmed, Z.U., Begum, Z.N.T., Hassan, M.A., Khondker, M., Kabir, S.M.H., Ahmad, M., Ahmed, A.T.A., Rahman, A.K.A. and Haque, E.U. (Eds) 2008. Encyclopedia of Flora and Fauna of Bangladesh, Vol. **6**. Angiosperms: Dicotyledons (Acanthaceae - Asteraceae). Asiatic Society of Bangladesh, Dhaka, pp. 1–408.

Ahmed, Z.U., Hassan, M.A., Begum, Z.N.T., Khondker, M., Kabir, S.M.H., Ahmad, M., Ahmed, A.T.A., Rahman, A.K.A. and Haque, E.U. (Eds) 2009a. Encyclopedia of Flora and Fauna of Bangladesh, Vol. **7**. Angiosperms: Dicotyledons (Balsaminaceae – Euphorbiaceae). Asiatic Society of Bangladesh, Dhaka, pp. 1–546.

Ahmed, Z.U., Hassan, M.A., Begum, Z.N.T., Khondker, M., Kabir, S.M.H., Ahmad, M., Ahmed, A.T.A., Rahman, A.K.A. and Haque, E.U. (Eds) 2009b. Encyclopedia of Flora and Fauna of Bangladesh, Vol. **8**. Angiosperms: Dicotyledons (Fabaceae – Lythraceae). Asiatic Society of Bangladesh, Dhaka, pp. 1–478.

Ahmed, Z.U., Hassan, M.A., Begum, Z.N.T., Khondker, M., Kabir, S.M.H., Ahmad, M. and Ahmed, A.T.A. (Eds) 2009c. Encyclopedia of Flora and Fauna of Bangladesh, Vol. **9**. Angiosperms: Dicotyledons (Magnoliaceae – Punicaceae). Asiatic Society of Bangladesh, Dhaka, pp. 1–488.

Ahmed, Z.U., Hassan, M.A., Begum, Z.N.T., Khondker, M., Kabir, S.M.H., Ahmad, M., and Ahmed, A.T.A. (Eds) 2009d. Encyclopedia of Flora and Fauna of Bangladesh, Vol. **10**. Angiosperms: Dicotyledons (Ranunculaceae – Zygophyllaceae). Asiatic Society of Bangladesh, Dhaka, pp. 1–580.

APG IV. 2016. An update of the Angiosperm Phylogeny Group classification for the orders and families of flowering plants: APG IV. Bot. J. Linn. Soc. **181**: 1–20.

Das, D.K. and Alam, M.K. 2001. Trees of Bangladesh. Bangladesh Forest Research Institute, Chittagong, pp. 1–342.

Dey, C.K., Rahman, M.A. and Wilcock, C.C. 1999. An enumeration of Tree species of Chittagong. Biodiver. Bull. Bangladesh **1**: 1–81.

Heinig, R.L. 1925. List of Plants of Chittagong Collectorate and Hill Tracts. Darjeeling, India, pp. 1–84.

Hossain, M.K. 2015. Protected Area Management Plan for Hazarikhil Wildlife Sanctuary (HWS) and Baraiyadhala National Park (BNP) for 2015-2025. Bangladesh Forest Department, Ministry of Environment and Forests, Government of the People's Republic of Bangladesh, pp. 1–109.

Huq, A.M. 1986. Plant Names of Bangladesh. Bangladesh National Herbarium, Dhaka, pp. 1–289.

Nguyen, H.T. 2010. *Cycas pectinata*. The IUCN Red List of Threatened Species 2010: e.T42062A-10617695. http://dx.doi.org/10.2305/IUCN.UK.20103.RLTS.T42062A10617695.en. Accessed on 20 August 2018.

Prain, D. 1903 (reprint 1963). Bengal Plants. Vols. **1** & **2**. Botanical Survey of India, Bishen Singh Mahendra Pal Singh, Dehra Dun, India, pp. 1–1013.

Rahman, M.A. (Ed.). 2013. IUCN Red List Categories of Plants: Red Data Book of Flowering Plants of Bangladesh. Vol. **1**. Published by Editor, Chittagong, pp. 1-256.

Rahman, M.A. 2017. Plant diversity in Hazarikhil Wildlife Sanctuary of Chittagong and its Conservation Management. J. Biodivers. Conserv. Bioresour. Manage. **3**(2): 43–56.

Rashid, M.E. and Rahman, M.A. 2011. Updated nomenclature and taxonomic status of the plants of Bangladesh included in Hook. f., The Flora of British India: Volume I. Bangladesh J. Plant Taxon. **18**(2): 177–197.

Rashid, M.E. and Rahman, M.A. 2012. Updated nomenclature and taxonomic status of the plants of Bangladesh included in Hook. f., The Flora of British India: Volume II. Bangladesh J. Plant Taxon. **19**(2): 173–190.

The Plant List 2013. Version 1.1. Published on the Internet; http://www.theplantlist.org/ (Accessed on August 2018).

# 19

# *GLOCHIDION TALAKONENSE* SP. NOV. (PHYLLANTHACEAE) FROM SESHACHALAM BIOSPHERE RESERVE, ANDHRA PRADESH, INDIA

M. Sankara Rao[1], J. Swamy[2], S. Nagaraju[2], S.B. Padal[3], M. Tarakeswara Naidu[3], K. Chandramohan[2] and T. Thulasiah[4]

*Botanical Survey of India, Sikkim Himalayan Regional Centre, Gangtok, Sikkim- 737103, India*

*Keywords:* *Glochidion talakonense*; New species; Phyllanthaceae; Andhra Pradesh; India.

**Abstract**

*Glochidion talakonense* M. Sankara Rao, J. Swamy, S. Nagaraju, S.B. Padal, M. Tarakeswara Naidu, K. Chandramohan & T. Thulasiah, a new species of Phyllanthaceae from Talakona hills, Seshachalam Biosphere Reserve, Andhra Pradesh, India, is described and illustrated. It is allied to *G. karnaticum* Chakrab. & M. Gangop., but differs from the latter by stamens, ovary, style and fruit characters.

## Introduction

The Seshachalam hills, spread in Chittoor and Kadapa districts of Andhra Pradesh, were declared as Biosphere Reserve by Government of India in 2010. The Reserve lies between the latitudes of $13^0 38''$ and $13^0 55''$ N and the longitudes of $79^0 07''$ and $79^0 24''$ E with an area of c. 4755.99 sq. km. The vegetation of the reserve is chiefly Southern dry mixed deciduous forests, dry deciduous scrub, dry savannah, Red sander forests and *Hardwickia* forests (Champion and Seth, 1968). A total of 1756 species of flowering plants belonging to 176 families are estimated to occur in this area (Sudhakar, 2012). The hill ranges vary in elevation from 400 to 1370 m with an average altitude of 700 m above sea level.

The genus *Glochidion* J.R. Forst & G. Forst. is represented by c. 320 species distributed in tropical Asia to Northern Australia and Polynesia, a few species in Madagascar and tropical America (Chakrabarty and Gangopadhyay, 1995, 2012; Balakrishnan and Chakrabarty, 2007). In India, the genus is represented by c. 22 species and 8 varieties, of which 3 species and one variety are reported from Andhra Pradesh (Babu, 1997; Chakrabraty and Gangopadhyay, 1995, 2012). Recently, one new species was described from the Seshachalam Hills (Rasingam *et al.,* 2014). While exploring the Talakona area of Seshachalam Biosphere Reserve, the authors collected an interesting species of *Glochidion*, which on critical observations showed conspicuous differences from all other known species of the genus. The same is therefore described and illustrated here as a new species *Glochidion talakonense* sp. nov. and compared with the allied *G. karnaticum* Chakrab. & M. Gangop.

**Glochidion talakonense** M. Sankara Rao, J. Swamy, S. Nagaraju, S.B. Padal, M. Tarakeswara Naidu, K. Chandramohan & T. Thulasiah **sp. nov.**                    (**Figs 1 & 2**).

**Diagnosis:** *Glochidion talakonense* is allied to *G. karnaticum* Chakrab. & M. Gangop., but differ in the asymmetric-rounded leaf-base, 5 stamens, 6-locular ovary and capsules and the longer columnar style, inflexed at apex (Table 1).

[1]Corresponding author. Email: mudadlas@gmail.com
[2]Botanical Survey of India, Deccan Regional Centre, Hyderabad – 500048, Telangana, India.
[3]Department of Botany, Andhra University, Visakhapatnam, Andhra Pradesh, India.
[4]S.L.V.Garden & Landscape developers, Tirupati, Andhra Pradesh, India.

*Type:* India. Andhra Pradesh, Chittoor Dist.,Talakona hills ($13^0$ 48'43.7" N & $79^0$ 13' 05.5" E), 852 m, 16 Apr 2014, *M. Sankara Rao & Party,* 4465 (*Holotype:* CAL; *Isotypes:* BSID); ibid.,9 Sep 2014, *M. Sankara Rao & Party* 5503 (*Paratypes:* BSID).

Trees, up to 7 m high; branches spreading; branchlets terete, puberulous when young, glabrescent at maturity, greenish-yellow. Leaves oblong to oblong-elliptic, 4.0-12.5 x 2.5-6.2 cm, asymmetric-rounded at base, entire, acuminate at apex, thinly coriaceous, glabrous, green above, glaucous beneath; lateral veins 7-8 pairs, prominent on both sides; petioles 4-5 x c. 2 mm, glabrous; stipules subulate, 1-2 mm long, puberulous. Inflorescences axillary, sessile, fascicled, 8-many-flowered, unisexual or bisexual. Male flowers: up to 18 in each fascicle, c. 2.0 x 1.8 mm, brownish-pink; pedicels filiform, up to 4 mm long; tepals 3+3, free, unequal, spreading and recurved, puberulous outside, glabrous inside; outer ones 1.8 x 1.3 mm, ovate, acute at apex; inner ones 1.8-2.0 x c. 1 mm, elliptic-oblong, rounded at apex; stamens 5; anthers connate into an oblong mass, c. 0.8 x 0.6 mm, connectives c. 0.3 x 0.3 mm; anther thecae linear, longitudinally dehiscent. Female flowers: many in each fascicle, 4 -5 x 4.5-4.7 mm, greenish with purple tinge; pedicels 2-4 mm long; tepals 3+3, free or occasionally shortly connate at base, unequal, tawny-puberulous on both surfaces; outer ones 1.8-2.2 x 1.6-1.8 mm, ovate, acute at apex; inner ones 1.4-2.0 x 0.8-1.2 mm, oblong, rounded or acute at apex; ovary subglobose, c. 2.3 x 2.6 mm, glabrous or puberulous, 6-locular, locules biovulate; styles columnar, inflexed at apex, c. 2.3 x 0.9 mm; apical lobes 6, linear or triangular, tawny-puberulous. Fruits capsular, 6-7 x 8-9 mm, shallowly lobed, slightly depressed, glabrous or puberulous. Seeds 12, glabrous.

*Phenology:* Flowering & Fruiting: April - September.

*Etymology:* The species is named after the type locality, "Talakona", a famous water fall in Seshachalam Biosphere Reserve in Chittoor district of Andhra Pradesh.

*Habitat:* Along waterfalls in moist deciduous forests at about 852 m elevation growing in association with *Allophylus cobbe, Syzygium alternifolium* and *Phoenix loureiroi.*

**Table 1. Morphological comparison of *Glochidion talakonense* sp. nov. and its allied *G. karnaticum.***

| Characters | *Glochidion talakonense* **sp. nov.** | *Glochidion karnaticum* |
|---|---|---|
| Leaves | Asymmetric-rounded at base | Acute at base |
| Petioles | 4-5 × c.2.2 mm | 2-3 x 1.0-1.5 mm |
| Male pedicels | Up to 4 mm long | 7-10 mm long |
| Male tepals | Unequal, outer ones ovate, 2.3-2.6 × 1.5-1.7 mm; inner ones elliptic-oblong, 1.8-2.0 × 0.9-1.0 mm | Equal, oblong, elliptic-ovate, 1-2 × 0.5-1.0 mm |
| Anthers | 5 | 3 |
| Female pedicels | 2-4 mm long | c. 2 mm long. |
| Female tepals | Unequal, outer ones ovate, 1.8-2.2 × 1.6-1.85 mm; inner ones oblong, 1.4 -2.0 × 0.8-1.2 mm | Equal, oblong, 1.0-1.5 × c. 0.5 mm |
| Ovary | 6-locular, 2.0-2.6 mm in diameter | 4-locular, c.1 mm in diameter |
| Style | Columnar, inflexed at apex, c. 2.3 × 0.9 mm | Columnar, c.1 × 0.4 mm |
| Fruits | 6 -7 × 8-9 mm, 6-locular | 2-3 × 4-5 mm, 4-locular |

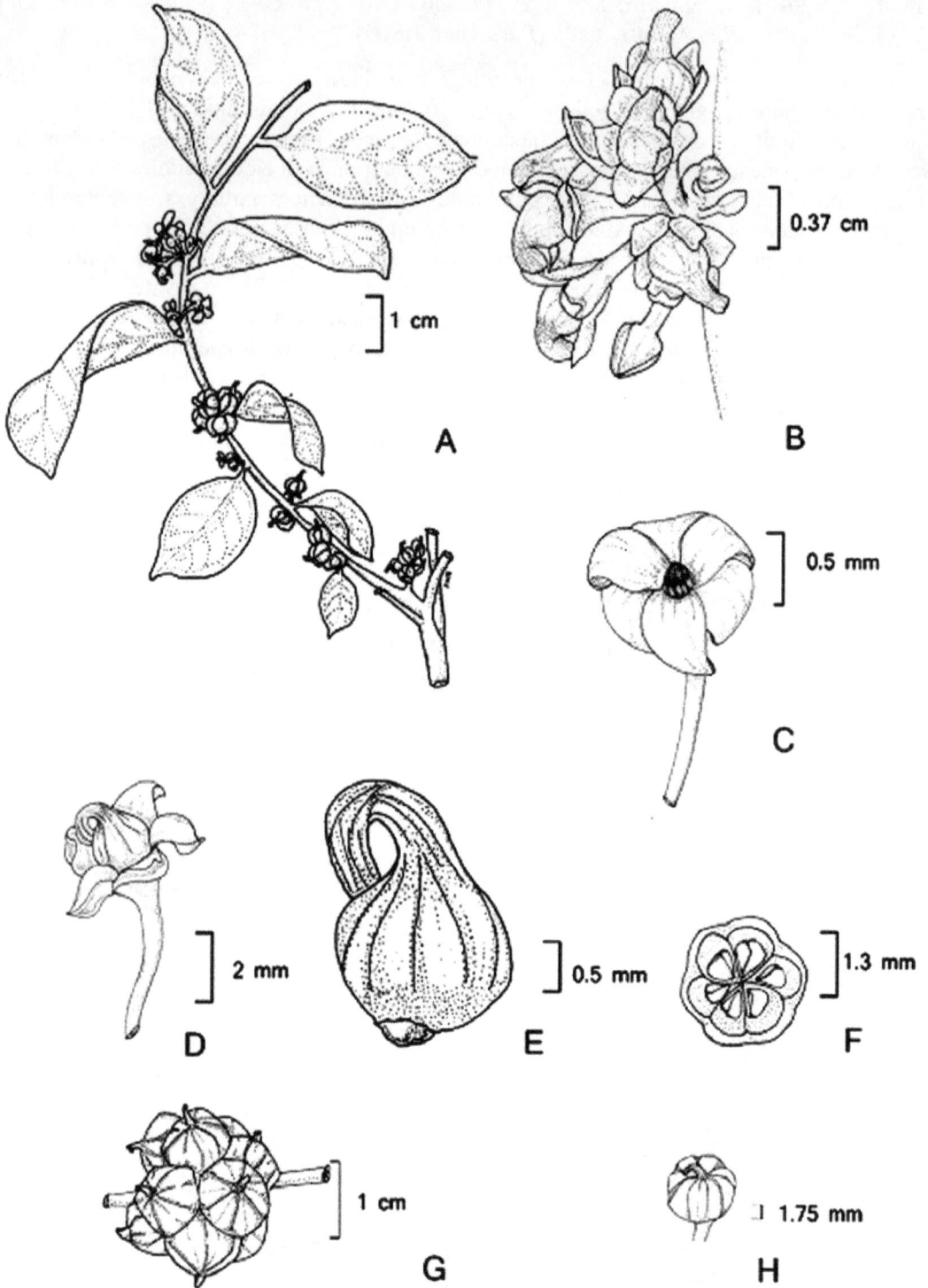

Fig. 1. **Glochidion talakonense, sp. nov**. A. Flowering branch; B. Inflorescence; C. Male flower; D. Female flower; E. Gynoecium; F. T.S. of ovary; G-H. Fruits.

Fig. 2. **Glochidion talakonense, sp. nov.** A. Habit; B. Adaxial and abaxial view of leaves; C. Inflorescence (mixed); D. Male inflorescence; E. Female inflorescence; F. Fruits; G. Male flower; H. Female flower; I. Androecium; J. Gynoecium; K. T. S. of ovary. (from type specimen)

*Conservation status: Glochidion talakonense* is a local endemic and so far known only from its type locality with nearly five [Criteria D] mature individuals. As it is known from a single location [Criteria B(a)], the Extent of Occurrence, EOO [Criteria B1] and Area of Occupancy, AOO [Criteria B2] is calculated as 4 km$^2$ by taking the minimum grid size of 2 km$^2$. The quality of habitat is declining [Criteria B(b-iii)] due to climate change and high anthropogenic pressure through tourism. In view of the above, by applying IUCN Red Listing criteria and guidelines (IUCN, 2012) the threat perspective of this species has been assessed as "Critically Endangered" [CR B1ab (iii)+2ab(iii); D]. Habitat management and propagation studies are suggested conservation measures for this species.

## Acknowledgements

The authors are thankful to the Director, Botanical Survey of India, Kolkata and Scientist In-charge, Botanical Survey of India, Deccan Regional Centre, Hyderabad for providing facilities and to the Forest Department officials, Government of Andhra Pradesh for permission and logistic support during field surveys. Thanks are due to Dr. D.K. Agrawala, Scientist In-charge, Botanical Survey of India, Sikkim Himalayan Regional Centre, Gangtok for encouragements and valuable suggestions and to Dr. T. Chakrabarty, ex Scientist, Botanical Survey of India for suggesting the alliance.

## References

Babu, S.P.1997. Euphorbiaceae *In*: Pullaiah, T. and Ali Moulali, D. (Eds), Flora of Andhra Pradesh (India) **2**: 836–890. Scientific Publishers, Jodhpur, India.

Balakrishnan, N.P. and Chakrabarty, T. 2007. The Family Euphorbiaceae in India - A synopsis of its Profile, Taxonomy and Bibliography. Bishen Singh, Mahendrapal Singh, Dehradun, India, pp. 1–500.

Chakrabarty, T. and Gangopadhyay, M. 1995. The genus *Glochidion* (Euphorbiaceae) in the Indian Subcontinent. J. Econ. Taxon. Bot **19**: 173–234.

Chakrabarty, T. and Gangopadhyay, M. 2012. *Glochidion. In* Balakrishan, N.P, Chakrabarty, T., Sanjappa, M., Lakshminarasimhan, P. and Singh, P. (Eds), Flora of India, **23**:87–511 Botanical Survey of India, Kolkata.

Champion, H.G. and Seth, S.K. 1968. A revised survey of the forest types of India. Manager of publications, New Delhi, pp. 1–404.

IUCN. 2012. IUCN Red List Categories and Criteria: Version 3.1. Second edition. Gland, Switzerland and Cambridge, UK: IUCN. iv + 32 pp.

Rasingam, L., Chorghe, A.R., Prasanna, P.V. and Sankara Rao, M. 2014. *Glochidion tirupathiense* (Phyllanthaceae) - A new species from Seshachalam Biosphere Reserve of Andhra Pradesh, India. Taiwania **59**(1): 9–12.

Sudhakar, G. 2012. Seshachalam Biosphere Reserve – Deccan South, India. *In*: Palni, L.M.S. and Rawal, R.S. (Eds), Compendium on Indian Biosphere Reserve, Progression during two decades of Conservation. Ministry of Environment & Forests, New Delhi. pp. 178–183.

# 20

# LICHEN DIVERSITY IN AMADIYA AND ROWANDUZ DISRICTS IN IRAQ

Zakaria S. Almola[1], Basheer A. Al-Ni'ma and Nadeem A. Ramadan

*Department of Biology, College of Sciences, University of Mosul, Iraq*

*Key words:* Lichens; Iraq; physiogeographic regions; Mountain region.

## Abstract

The lichen biota of the Amadiya and Rowanduz districts in the Mountain physiogeographic region in Iraq was sampled in 2013. The samples provided 47 species belonging to 29 genera and 14 families. Among them 37 species are new records for Iraq. All species except *Lichinella cribellifera* and *Thelidium* sp. were found in Amadiya district whereas only 13 species occurred in Rowanduz district. Most of the species (59.5%) were crustose, while 27.6% were foliose, 12.7 % squamulose and none fruticose. The three most species-rich genera are *Caloplaca* with 7 species, *Collema* with 5 species and *Aspicilia* with 3 species; 6 genera were represented by 2 species and 20 by single species. All saxicolous lichens were calciphilic while the corticolous lichens were acidophilic. The most common and dominant species is *Lecanora muralis*, found in all 17 studied locations.

## Introduction

Iraq is located in the Middle East or South West Asian region above 29° latitude. Up to date the available knowledge concerning the lichen biota of Iraq is very limited in comparison with those of the neighbour countries, especially Turkey and Iran. In Turkey the study of lichens started in the 19th century by visiting foreign scientists. while in the 1980s Turkish researchers started publishing their contributions. In two decades, between 1980 and 2000, they published more than 200 papers (John, 2007). In Iran the first reports on lichens have been published by Göbel in 1830 and Eversmann in 1831 (Seaward *et al.*, 2004). The preliminary check list of Iranian lichens published in 2004 by Seaward *et al.* (2004) consists of 396 species of lichenized fungi in addition to 8 species of lichenicolous or allied fungi. For Iraq, however, so far only three publications dealt with the biodiversity of lichens. The first one was published by Steiner (1921). It contains identification results of the lichen specimens that have been collected by Handel Mazzetti during his expedition to Mesopotamia, Kurdistan, Syria and Prinkipo, posthumously published after revision by Alexander Zahlbruckner. The second one was published by Schubert (1973) who collected lichens in Iraq during 1969. The last one was published by Poelt and Sulzer (1974) about *Buelliaepigaea* in the country. Feuerer (2006) brought the 32 species of these three papers in a single list.

Four physio-geographic regions are distinguished in Iraq viz. Mountain, Upper plains and Foothills, Desert Plateau and Lower Mesopotamian region. Each one is subdivided into a number of districts (Guest, 1966). Our study concerns the Mountain region. This is divided into four districts namely, Amadiya, Rowanduz, Sulaimaniya and Jabel Sinjar. Mountain is considered the richest region by lichen biota due to the suitability of its climate and substrate for lichen growth. Therefore it was selected for the present study that is considered as the first step in preparing a complete check list which will cover all Iraqi physio-geographic regions and districts.

---

[1] Corresponding author. Email: zakaria_sami@yahoo.com

## Materials and Method

### The study area

The Mountain region of Iraq extends from West Zakho to southeast of Halabja, and is delimited by the 500 m contour (the approximate lower limit of the mountain forest). This contour runs irregularly following the margin of the mountain in a general south-easterly direction, very roughly along a line passing near FaishKhabur-Zakho-Simel-Dohuk-Alqosh-Erbil-Kirkuk, then some way north of Tuz and Kifri to Halabja. In the North and Northeast the region is bordered by the Iraqi frontiers with Turkey and Iran, respectively (Guest,1966). According to Guest (1966), the Mountain region contains four physio-geographic districts, two of which are the subject of our study, viz. Amadiya and Rowanduze (Figure 1). Amadiya district is bordered in the east by the Greater Zab tributary, and in the North by the Iraqi frontier with Turkey, while Rowanduz district is located between the Greater and Lesser Zab tributaries of the Tigris River.

FIGURE No. I

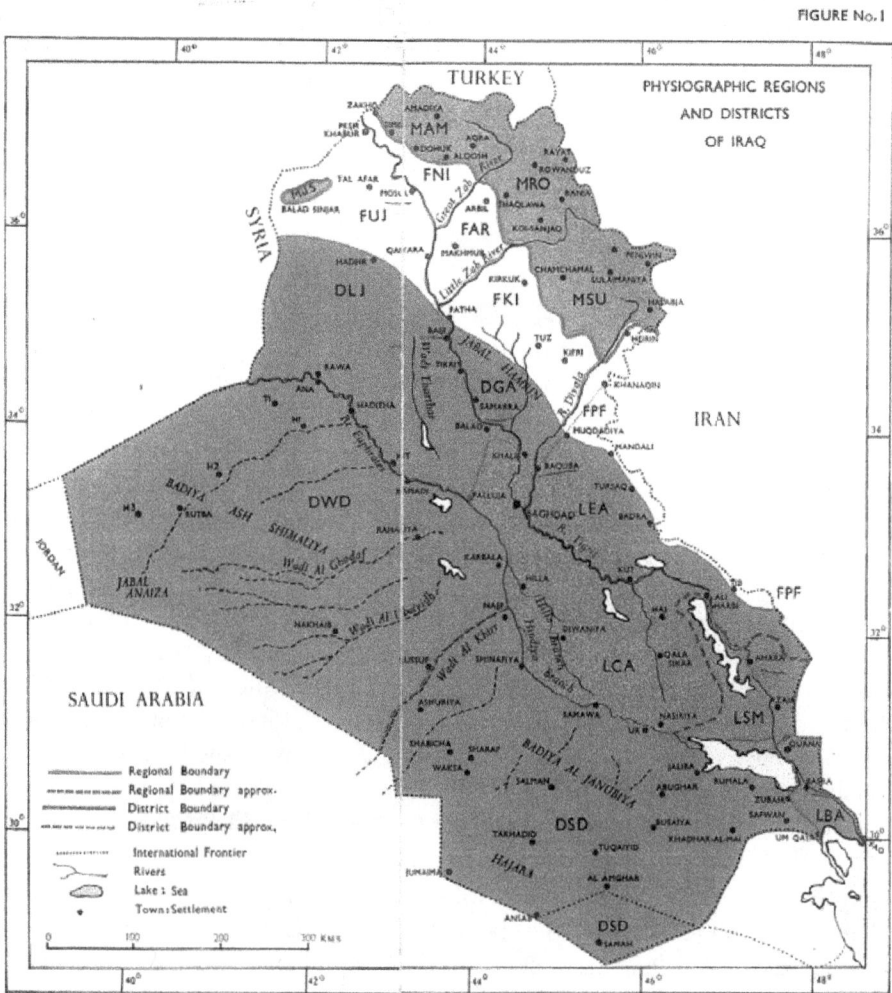

Drawn by E. R. Guest, 1958                                        *frontispiece*

Fig. 1. Physiogeographic regions and districts of Iraq, after Guest (1966). The upper north eastern part represents the mountain region; MAM= Amadiya district, MRO= Rowanduz district, MSU= Sulaimaniya district and MJS= Jabal Sinjar district.

Geologically, the Iraqi mountains consist mainly of Cretaceous and Eocene limestone and shale, with Paleozoic strata exposed here and there in the cores of eroded anticlines, followed towards the south-west by Miocene Lower Fars gypsum with red shale and thin limestone bands, overlying a basal conglomerate; and ending with Miocene Upper Fars and red shale and the Pliocene Bakhtiari conglomerates.

A belt of confused topography stretching along the north-east frontier from Rowanduz to Halabja is made up of a variety of basic and ultrabasic igneous and metamorphic rocks, with radiolarian chert and a little limestone (Macfadyen, 1966).

The climate of the area is generally warm-temperate. July is the hottest month where air temperature may surpass $40^0$ C and January is the coldest month where air temperature approaches zero $^0$ C. The rainy period extends, generally, from October to May and the amount of rain is variable during the season and from year to year, for example the minimum and maximum amount of rain that have been fallen on Dohuk city centre between (1980-2005) were 284.3 and 909.7 mm/year during 1994 and 1999, respectively (Al-Rijabo and Bleej, 2010).

**Table 1. Average climatological data for two city centers located within the study area.**

|                              | Unit      | Dohuk     | Erbil     |
|------------------------------|-----------|-----------|-----------|
| Range of maximum air temp.   | $^0$C     | 11.0-42.0 | 12.4-42.0 |
| Range of minimum air temp.   | $^0$C     | 3.0-27.0  | 2.4-24.9  |
| Annual rain fall             | mm/year   | 616       | 543       |

The data represent an average of the last two decades and are obtained from Wikipedia.

*Collection, preservation and identification of specimens*

Lichens from different substrates (soil, rocks and tree trunks) have been collected between February and May 2013 from 17 locations, of which 12 located in Amadiya district and 5 in Rowanduz district (Table 2).

**Table 2. Names and coordinates of the study locations.**

|                      | Sl. No. | Locations                                      | Coordinates                        |
|----------------------|---------|------------------------------------------------|------------------------------------|
| Amadiya District     | 1.      | Sharanish                                      | 37° 13′ 52” N<br>42° 50′ 45” E     |
|                      | 2.      | Sarsang                                        | 37° 01′ 04” N<br>43° 13′ 44” E     |
|                      | 3.      | Zawita                                         | 36° 51′ 50” N<br>43° 05′ 34” E     |
|                      | 4.      | By the road, about 2 km north Gali Balkaif     | 36° 52′ 37” N<br>43° 20′10” E      |
|                      | 5.      | By the road about 1 km south east GaliBalkaif  | 36° 52′ 22” N<br>43° 20′ 45” E     |
|                      | 6.      | By the road of ShaikhAdi-Zawita                | 36° 50′ 01” N<br>43° 14′ 00” E     |
|                      | 7.      | Atrush                                         | 36° 49′ 59” N<br>43° 19′ 38” E     |

(Contd.).

|  | Sl. No. | Locations | Coordinates |
|---|---|---|---|
|  | 8. | Dinarta | 36° 48′ 50” N 43° 59′49” E |
|  | 9. | GaliBalkaif | 36° 48′ 12” N 43° 18′ 07” E |
|  | 10. | ShaikhAdi | 36° 46′ 30” N 43° 18′ 59” E |
|  | 11. | Shaikh Aid road fork | 36° 45′ 05” N 43° 19′ 48” E |
|  | 12. | GaliZanta | 36° 45′ 03” N 43° 58′ 32” E |
| Rowanduz District | 13. | Gali Ali Beg | 36° 37′ 55” N 44° 26′ 49” E |
|  | 14. | Khalifan | 36° 37′ 07” N 44° 25′ 04” E |
|  | 15. | Heran | 36° 26′ 07” N 44° 23′ 33” E |
|  | 16. | Shaqlawa | 36° 24′ 30” N 44° 18′ 06” E |
|  | 17. | Kori-Kori sheer | 36° 23′ 02” N 44° 15′ 56” E |

In the field, each specimen was preserved in a numbered paper bag and the related essential data were recorded. In the laboratory, the specimens were air dried and preserved permanently in labeled paper bags and deposited in the herbarium of the Biology Department, College of Sciences- Mosul University (MOS).

Using Olympus compound and stereomicroscopes the lichen specimens were identified after the keys following Brodo *et al.* (2001), Dobson (2011) and Goward *et al.* (1994).

The identifications were confirmed by Harrie Sipman from the Freie University of Berlin (Germany) on the basis of notes and photos of whole specimens, cross sections and spores.

## Results

List of the taxa with life form, substrate, and distribution in the two districts # = new record for the district; * = new record for Iraq.

### Family: Acarosporaceae
### Genus: *Acarospora*

#1. *Acarospora cervina* A. Massal. (1852)
Squamulose; on limestone, in Amadiya district only: locations 1, 2, 9, 11 & 12.

### Genus: *Sarcogyne*

*2. *Sarcogyne regularis* Körb. (1855)
Crustose; on limestone, in Amadiya district only: location 7.

## Family: Candelariaceae
## Genus: *Candelariella*

#3. *Candelariella aurella* (Hoffm.) Zahlbr. (1928)
Crustose; on exposed calcareous rocks, in Amadiya district only: locations 1 & 3.

*4. *Candelariella xanthostigma* (Pers. *ex* Ach.) Lettau (1912)
Crustose; on bark of oak trees, in Amadiya district only: location 2.

## Family: Collemataceae
## Genus: *Collema*

*5. *Collema cristatum* (L.) Weber *ex* F. H. Wigg. (1780)
Foliose; on hard limestone and soil, in Amadiya district only: locations 1, 5, 7, 10 & 11.

*6.*Collema fuscovirens* (With.) J.R. Laundon (1984)
Foliose; on exposed, hard, calcareous rocks and dolomitic limestone, In Amadiya and Rowanduz district: locations 2 & 16.

*7. *Collema ligerinum* (Hy) Harm. (1905)
Foliose; on bark of oak trees, in Amadiya district only: location 5

*8. *Collema polycarpon* Hoffm. (1796)
Foliose; on exposed, hard, calcareous rocks, in Amadiya district only: location 11.

*9. *Collema tenax* (Sw.) Ach. (1810)
Foliose; on soil containing calcium, in Amadiya district only: location 9.

## Family: Graphidaceae
## Genus: *Diploschistes*
#10. *Diploschistes ocellatus* (Fr.) Norman (1853)
Crustose; on calcareous and shale rocks, in Amadiya and Rowanduz districts: locations 4, 5, 6, 10, 12, 13 & 14.

## Family: Lecanoraceae
## Genus: *Lecanora*
#11. *Lecanora muralis* (Schreber) Rabenh. (1845)
Crustose; on dolomite, limestone, shale and siliceous rocks, in Amadiya and Rowanduz district: all 17 locations.

## Family: Lichinaceae
## Genus: *Lichinella*

*12. *Lichinella cribellifera* (Nyl.) P. P. Moreno &Egea (1992)
Foliose; on limestone and siliceous rocks, in Rowanduz district only: location 14.

## Family: Megasporaceae
## Genus: *Aspicilia*

#13. ***Aspicilia calcarea*** (L.) Mudd (1861)
Crustose; on limestone, dolomitic limestone and shale rocks, in Amadiya and Rowanduz districts: locations 1, 3, 5, 7, 8, 11, 12, 13, 14 & 16.

*14. ***Aspicilia contorta*** (Hoffm.) Kremp. (1861)
Crustose; on limestone and non-calcareous rocks, in Amadiya district only: locations 10, 11 & 12.

*15.  ***Aspicilia ferruginea*** (J. Steiner) Szatala
Crustose; on dolomitic limestone, in Amadiya district only: locations 1, 5 & 9.

### Genus: *Lobothallia*

*16. ***Lobothallia praeradiosa*** (Nyl.) Hafellner (1991)
Crustose; on limestone and siliceous rocks, in Amadiya district only: locations 6, 7 & 10.

*17.  ***Lobotha lliaradiosa*** (Hoffm.) Hafellner (1991)
Crustose; on calcareous rocks, in Amadiya district only: location 12.

### Genus: *Megaspora*

*18. ***Megaspora verrucosa*** (Ach.) Hafellner and V. Wirth (1987)
Crustose; on bark of oak trees, in Amadiya district only: locations 2 & 12.

## Family: Parmeliaceae
## Genus: *Melanelia*

*19. ***Melanelia glabra*** (Schaer.) Essl. (1987)
Foliose; on bark of oak trees, in Amadiya district only: location 5.

### Genus: *Parmelina*

*20.  ***Parmelina tiliacea*** (Hoffm.) Hale (1974)
Foliose; on bark of oak trees, in Amadiya district only: location 6.

## Family: Physciaceae
## Genus: *Anaptychia*

*21.  ***Anaptychia desertorum*** (Rupr.) Poelt (1969)
Foliose; on bark of oak trees, in Amadiya district only: locations 2, 9 & 10.

### Genus: *Diplotomma*

*22.  ***Diplotomma hedinii*** (H. Magn.) P. Clerc& Cl. Roux (2004)
Crustose; on calcareous rocks, in Amadiya district only: location 1.

### Genus: *Phaeophyscia*

*23.  *Phaeophyscia orbicularis* (Neck.) Moberg (1977)
Foliose; on bark of oak and pine trees, in Amadiya district only: locations 2, 10 & 12.

### Genus:*Physcia*

*24. *Physcia biziana* (A. Massal.) Zahlbr. (1901)
Foliose; on bark of oak and hawthorn trees, in Amadiya district only: locations 2, 6, 10 & 12.

### Genus: *Physconia*

*25.  *Physconia distorta* (With.) J. R. Laundon (1984)
Foliose; on bark of oak trees, in Amadiya district only: locations 5, 6 & 7.

### Genus: *Rinodina*

#26. *Rinodina bischoffii* Hepp (1855)
Crustose; on limestone, in Amadiya and Rowanduz districts: locations 3 & 17.

### Family: Psoraceae
### Genus: *Psora*

#27. *Psora decipiens* (Hedwig) Hoffm. (1794)
Squamulose; on calcareous soil, sandy stones and limestone, in Amadiya district only: locations 4, 5, 6, 9, 10 & 11.

*28.  *Psora vallesiaca* (Schaer.) Timdal (1984)
Squamulose; on calciferous soil, limestone and shale rocks, in Amadiya and Rowanduz districts: locations 1,8 & 13.

### Family: Ramalinaceae
### Genus: *Lecania*

*29.  *Lecania koerberiana*J. Lahm (1859)
Crustose; on bark of oak trees, in Amadiya district only: location 2.

### Genus: *Squamarina*

*30.  *Squamarina cartilaginea* (With.) P. James (1980)
Squamulose; on calcareous rocks and stony calcareous soil, sometimes on mosses that grow on soil, in Amadiya and Rowanduz districts: locations 1, 4, 5, 6, 10, 11, 12 & 14.

#31.  *Squamarina lentigera* (Weber) Poelt (1958)
Squamulose; on sandy stones and on calcareous, especially gypsiferous soils, in Amadiya district only: location 9.

**Genus:** *Toninia*

**#32. *Toninia sedifolia*** (Scop.) Timdal (1991)
Squamulose; on sandy stones and soil, in Amadiya and Rowanduz districts: locations 1, 8, 9 & 14.

**Family: Stereocaulaceae**
**Genus:** *Lepraria*

***33.  Lepraria vouauxii*** (Hue) R. C. Harris (1987)
Crustoseleprose (powdery), on limestone and siliceous rocks, in Amadiya district only: location 1.

**Family: Teloschistaceae**
**Genus:** *Caloplaca*

***34. Caloplaca aegyptiaca*** (Müll. Arg.) J. Steiner (1893)
Crustose; on limestone, in Amadiya district only: location 4.

***35.  Caloplaca aurantia*** (Hoffm.) Hafellner (1991)
Crustose; on limestone, In Amadiya and Rowanduz district: locations 4, 5, 10, 11, 12 & 14.

***36. Caloplaca biatorina*** (A. Massal.) J. Steiner (1910)
Crustose; on dolomitic limestone, in Amadiya and Rowanduz districts: locations 4, 16 & 17.

***37. Caloplaca chalybaea*** (Fr.) Mull. Arg. (1862)
Crustose; on hard calcareous rocks, in Amadiya district only: location 7.

***38. Caloplaca erythrocarpa*** (Pers.) Zwackh (1862)
Crustose; on calcareous rocks, in Amadiya district only: location 9.

***39. Caloplaca polycarpoides*** (J. Steiner) M. Steiner & Poelt (1982)
Crustose; on bark of oak and pine trees, in Amadiya district only: location 2.

***40. Caloplaca variabilis*** (Pers.) Mull. Arg. (1862)
Crustose; on calcareous rocks, in Amadiya district only: location 1.

**Genus:*Fulgensia***

***41. Fulgensia schistidii*** (Anzi) Poelt (1965)
Crustose; on moss *Schistidiumapocarpum*, in Amadiya district only: location 1.

***42. Fulgensia subbracteata*** (Nyl.) Poelt (1961)
Crustose; on limestone, dolomitic limestone, shale rocks, sandstones and soil, in Amadiya and Rowanduz district: locations 4, 6, 7, 8, 9, 10, 11, 13, 14 & 16.

**Family: Verrucariaceae**
**Genus:** *Dermatocarpon*

***43. Dermatocarpon miniatum*** (L.)W. Mann (1825)
Foliose; on limestone, in Amadiya district only: locations 3 & 12.

## Genus: Placocarpus

*44. *Placocarpus schaereri* (Fr.) Breuss (1985)
Crustose; on limestone, dolomitic limestone and shale rocks, in Amadiya and Rowanduz districts; locations 1, 4, 5, 6, 8, 13, 14, 15 & 16.

## Genus: *Thelidium*

*45. *Thelidium sp.* A. Massal. (1855)
Crustose; on limestone, in Rowanduz district only: location 14.

## Genus: *Verrucaria*

*46. *Verrucaria macrostoma* Dufour *ex* DC. (1805)
Crustose; on siliceous rocks, in Amadiya district only: location 4.

#47. *Verrucaria nigrescens* Pers. (1795)
Crustose; on dolomitic limestone and shale rocks, in Amadiya and Rowanduz districts: locations 1, 6, 13, 15 & 17.

## Discussion

Lichen specimens collected in Amadiya and Rowanduz districts revealed 47 species in 29 genera and 14 families. All these species, except *Lichinella cribellifera* and *Thelidium* sp., were found in Amadiya district whereas, 13 species were found in Rowanduz district. Out of 47 species recorded, 37 species are new records for Iraq (marked by an asterisk (*) in the list) and 10 were known before from the country but are new records for the two districts (marked by "#"). Of the recorded species 59.5 % were crustose, 27.6% foliose and 12.7 % squamulose. No fruticose species were found.

Considering the substrates, most species grow only on one type, either soil, rock or trees, with few exceptions such as *Aspicilia calcarea, Fulgensia schistidii* and *Psora decipiens* which were collected from a variety of substrates as rock, soil and sandstone. The epiphyte *Physcia biziana* was found on bark of oak and hawthorn trees.

Like Iraqi higher plants, the epilithic (saxicolous) lichens are lime lovers (calciphiles) since the rocks of the studied area are all calcareous, with available lime, which is normally present as calcium carbonate giving aqueous extracts of higher pH on the alkaline side of the neutral point (Guest,1966), however the epiphytic species, according to Oran (2011) can be called calcifuges or acidophilic as their results showed that the highest average pH values of the bark of 18 oak species was 6.16 and the lowest one was 4.76. The epiphytic (corticolous) lichens were found on four species only, viz. *Quercus aegilops*, *Q. infectoria*, *Crataegus azarolus* and *Pinus brutia*. The largest genera were *Caloplaca* represented by 7 species, *Collema* with 5 species and *Aspicilia* with 3 species. Six genera were represented by two species and 20 by one species only. The distribution of the recorded species was also widely variable, some species as *Caloplaca chalybaea, Collema tenax, Sarcogyne regularis* and another 18 species were found in one location only, i.e showed narrow distribution. On the other hand, *Lecanora muralis* was found in all 17 locations, hence it is considered the most dominant species. Comparatively less dominant species are *Fulgensia subbracteata* and *Aspicilia calcarea,* both found in 10 locations, then *Placocarpus schaereri,* which appeared in 9 locations. The most dominant species, *Lecanora muralis* can be described as an ubiquitous crustose lichen, it grows all over the world including Europe, Asia, North America,

South Africa, Africa, Macronesia, Oceania and Australasia (Nash *et al.*, 2001). This wide spreading is possibly due to the ability of the species to grow on a wide variety of rocks, basalt, pumice, rhyolite, granite, sandstone and limestone (*op. cit.*). Moreover it is a desiccation tolerant species and it can overcome dry spells of considerable duration.

## Acknowledgements

The authors are grateful to Dr. Harrie Sipman of Botanical Garden, Berlin of the Freie University, Germany for his helpful collaboration in this work.

## References

Al-Rijabo, W.I. and Bleej, D.A. 2010. Variation of Rainfall with Space and Time in Duhok. Education Science **23**(1): 32-43.

Brodo, I.M., Sharnoff, S.D. and Sharnoff, S. 2001. Lichens of North America. New Haven, Yale University Press, London.

Dobson, F.S. 2011. Lichens an Illustrated Guide to the British & Irish Species. Sixth Revised & Enlarged Edition. The Richmond Publishing Co. Ltd., England.

Eversmann, E. 1831. In: Lichen emesculentum Pall. et species consimilesadversaria. Nova Acta Acad. Caes. Leopold. Carol. **15**(2): 349-362.

Feuerer, T. 2006. Check List of Lichens and Lichenicolous Fungi, Version 1.http://www.checklist.de.

Goward, T., McCune, B. and Meidinger, D. 1994. The Lichens of British Columbia. Ministry of Forests Research Program.

Göbel, F. 1830. Chemische Untersuchungeiner in Persienherabgeregneten Substanz, der *Parmelia esculenta* .J. Chem. Phys. **60**: 393-399.

Guest, E. 1966. Flora of Iraq. Vol. **1**, Ministry of Agriculture, Republic of Iraq.

John, V. 2007. Lichenological studies in Turkey and their relevance to environmental interpretation. Bocconea **21**: 85-93.

Macfadyen, W.A. 1966. The Geology of Iraq. *In*: Guest, E. (Ed.). Flora of Iraq. Vol. 1, Ministry of Agriculture, Republic of Iraq.

Nash, T.H., Ryan, B.D., Gries, C. and Bungartz, F. 2001. Lichen Flora of the Greater Sonoran Desert Region. Vol. 2. Arizona state university, Tempe, Arizona, USA.

Oran S. 2011. Investigations on the bark pH and epiphytic lichen diversity of *Quercus* taxa found in Marmara Region. J. Appl. Biol. Sci. **5**(1): 27-33.

Poelt, J. and Sulzer, M. 1974. Die Erdflechte *Buelliaepigaea*, eine Sammelart. Nova Hedwigia **25**(1+2): 173-194.

Schubert, R. 1973. Notizenzur Flechtenflora des nördlichen Mesopotamien(Irak). Feddes Rept. **83**: 585-589.

Seaward, M.R.D., Sipman, H.J.M., Schultz, M., Maassoumi, A.A., Haji MoniriAnbaran, M. & Sohrabi, M. 2004. A preliminary Lichen Checklist for Iran. Willdenowia **34**: 543-576.

Steiner, J. 1921. Lichenesaus Mesopotamien und Kurdistan sowie Syrien und Prinkipo. Wissenschaftliche Ergebnisse der Expedition nach Mesopotamien 1910. Ann. Naturhist. Staatsmus. Wien **34**:1-68.

# 21

# PLANT DIVERSITY OF SONADIA ISLAND – AN ECOLOGICALLY CRITICAL AREA OF SOUTH-EAST BANGLADESH

M.S. Arefin, M.K. Hossain[1] and M. Akhter Hossain

*Institute of Forestry and Environmental Sciences, University of Chittagong, Chittagong 4331, Bangladesh*

*Keywords:* Plant Diversity; Ecologically Critical Area; Sonadia Island; Mangroves.

## Abstract

The study focuses the plant diversity in different habitats, status and percentage distribution of plants in Sonadia Island, Moheshkhali, Cox's Bazar of Bangladesh. A total of 138 species belonging to 121 genera and 52 families were recorded and the species were categorised to tree (56 species), shrub (17), herb (48) and climber (17). Poaceae represents the largest family containing 8 species belonging to 8 genera. Homestead vegetation consists of 78% species followed by roadside (23%) and cultivated land (10%), mangroves (9%), sandy beaches (4%) and wetland (1%). The major traditional use categories were timber, food and fodder, fuel, medicine and fencing where maximum plant species (33% of recorded) were traditionally being used for food and fodder.

## Introduction

Sonadia Island at Moheshkhali of Cox's Bazar is situated in the southern-eastern coastal region of Bangladesh with partial regular inundations of saline water. The island covers an area of 10,298 hectares including coastal and mangrove plantations, salt production fields, shrimp culture firms, plain agriculture lands, human settlements etc. Ecosystem of this island was adversely affected due to increasing rate of anthropogenic disturbances. To protect the ecosystem of this island, it was declared as Ecologically Critical Area (ECA) in 1999 under section of the Bangladesh Environment Conservation Act, 1995 (MoEF, 2015). ECAs are ecologically defined areas or ecosystems affected adversely by the changes brought through human activities. This island is floristically composed of a number of mangrove and terrestrial plant species. The island is important not only as renewable resources but also as an essential in conservation of nature, wildlife, fish and environment of the island and the surrounding areas. The ECA needs special attention for environmental conservation in terms of both flora and fauna aspects. For this, a comprehensive list of the flora and fauna existing in Sonadia Island is essential. Moloney (2006) reported 60 vascular plants from Sonadia in the draft Sonadia Island ECA Conservation and Management Plan. There had been gradual changes in the ecological conditions due to increased anthropogenic interference. Since, no complete study was carried out throughout the period, it is completely unknown if any changes in the floristic composition of the critically endangered ecosystem has occurred in the last decade. Therefore, the present study was undertaken with the aim of assessing the plant resources of Sonadia Island, an ECA based on extensive field observations.

## Materials and Methods

*Study area*

Sonadia Island is located in the far south-eastern corner of Bangladesh at 21°N and 91°E, the site lies a few kilometers north of Teknaf Peninsula, north-west of Cox's Bazar town and is bounded by the Bay of Bengal on the West and East (Fig. 1).

[1]Corresponding author: Email: mkhossain2009@gmail.com

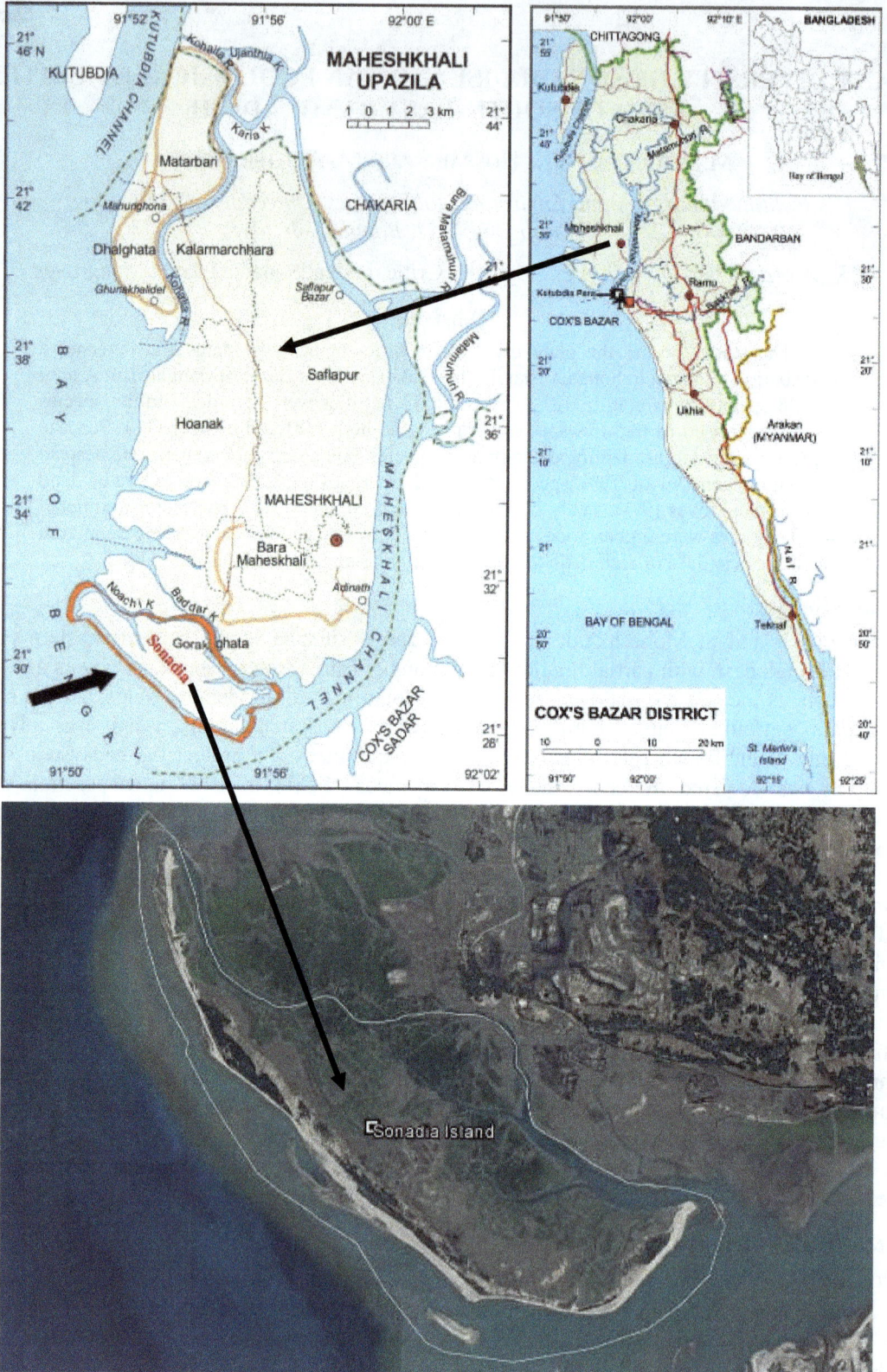

Fig. 1. Location of Sonadia island in Moheshkhali upazila of Cox's Bazar district, Bangladesh.

The Island is separated from the mainland by the Moheshkhali channel and from Moheshkhali Island by the Bara Canal. The soil of this area is the admixture of sand and clay in varying proportion. The soil of the North part is clay and is inundated by sea water. The entire soil condition of the South part is almost sandy (DoE, 1999). The whole island has a mild temperature and high humidity. The summer begins from March and continues till the beginning of June. The annual average temperature in Cox's Bazar is 34.8°C and a minimum of 16.1°C. Sonadia Island is a gently sloping low-lying barrier island with an altitude range of 0-4 metres (DoE, 1999).

*Field visit, data collection and analysis*

A reconnaissance survey was conducted in the Sonadia island ECA prior to the field work to have a general idea of the site, topography, species composition, habitat condition and socio-economic status of the local people. The flora study methods include key informant interview, reconnaissance survey, and field data collection through whole area survey, homestead plant survey, and focused group discussion from October, 2015 to August 2016. Field work was scheduled in such a way that enable plant observation and specimen collections of unknown plant species during the flowering and fruiting time of maximum number of species.

A total 9 foot trails of different length (3-7 km each) in two villages (*Purbo para* and *Passchim para*) and 9 boat journey through the surrounding canals, rivers and sea shore (total 33 km (approx.) were made to record the flora of homesteads and mangrove forests of Sonadia island. Survey was continued until occurrence of new species. The observed plant species were identified and recorded in the field. Habitat and habit form were also recorded. Herbarium specimen of rare and unidentified plant samples with fertile material (flower, fruit and seed) were collected and prepared for identification after necessary processing. Plant specimens with only vegetative part were also collected for herbarium preparation in case of unavailability of fertile materials. Photographs of the characteristic plant species from suitable projection were taken to keep a digital record of morphological features of the plants. Along with verification of the local names, local use of the recorded plants was explored through focused group discussion in the two villages of Sonadia Island.

Herbarium specimens were identified by consultation with voucher specimens and taxonomists of Bangladesh Forest Research Institute as well as recognized references, viz. Prain (1903); Heinig (1925); Siddiqui *et al.* (2007) and Ahmed *et al.* (2008). The identified taxa were arranged alphabetically with species names.

## Results

*Floristic composition*

A total of 138 plant species belonging to 121 genera and 52 families were identified from the Sonadia island (Table 1). Among the recorded 138 species, Poaceae appeared as the largest family with 8 species under 8 genera followed by Cucurbitaceae (7 genera and 8 species), and Mimosaceae (8 species and 6 genera) (Fig. 2). Most of the families (28 nos.) were represented by only 1 species each (Table 1).

*Growth (habit) forms of the plants*

The recorded flora of Sonadia Island is grouped under tree, shrubs, herbs and climbers growth (habit) forms. Trees constitute the major category (56 species) of plant species followed by herbs (48 species), shrubs (17 species), and climbers (17 species) (Fig. 3). Number of tree species in Mimosaceae was maximum (5 genera and 7 species), whereas shrubs were maximum in Verbenaceae (3 genera and 4 species). In case of herbs and climbers Amaranthaceae (4 genera and 7 species) and Cucurbitaceae (7 genera and 8 species) were represented by maximum species respectively.

**Table 1. List of plant species recorded from Sonadia Island of Bangladesh.**

| SN | Scientific name | Local name | Family name | Habit | Habitat |
|----|----------------|-----------|-------------|-------|---------|
| 1 | *Acacia auriculiformis* A. Cunn. *ex* Benth. & Hook. | Akashmoni | Mimosaceae | T* | Homestead, Roadside |
| 2 | *Abelmoschus esculentus* (L.) Moench | Vandi | Malvaceae | H | Cultivated |
| 3 | *Acacia farnesiana* (L.) Willd. | Bilati Babla | Mimosaceae | T | Homestead |
| 4 | *Acanthus ilicifolius* L. | Hargoza | Acanthaceae | S | Mangrove |
| 5 | *Aegialitis rotundifolia* Roxb. | Nunia gach | Plumbaginaceae | S | Mangrove |
| 6 | *Albizia lebbeck* (L.) Benth. | Kala koroi | Mimosaceae | T | Homestead |
| 7 | *Albizia procera* (Roxb.) Benth. | Sada koroi | Mimosaceae | T | Homestead |
| 8 | *Alocasia macrorrhizos* (L.) G. Don | Mankachu | Amaranthaceae | H | Homestead |
| 9 | *Alternanthera philoxeroides* (Mart.) Griseb. | Helencha | Amaranthaceae | H | Cultivated, Roadside |
| 10 | *Alternanthera sessilis* (L.) R. Br. *ex* Roem. & Schult. | Saci Shak | Amaranthaceae | H | Cultivated |
| 11 | *Amaranthus spinosus* L. | Katashak | Amaranthaceae | H | Homestead |
| 12 | *Amaranthus tricolor* L. | Lalshak | Amaranthaceae | H | Homestead, Cultivated |
| 13 | *Amaranthus viridis* L. | Datashak | Amaranthaceae | H | Homestead, Roadside |
| 14 | *Anacardium occidentale* L. | Kajubadam | Anacardiaceae | T | Homestead |
| 15 | *Areca catechu* L. | Supari | Arecaceae | T | Homestead |
| 16 | *Argyreia capitiformis* (poir.) Oostr. | Voga Lata | Convolvulaceae | C | Roadside |
| 17 | *Artocarpus heterophyllus* Lamak. | Kathal | Moraceae | T | Homestead |
| 18 | *Asystasia gangetica* (L.) T. Anders. | | Acanthaceae | H | Roadside |
| 19 | *Averrhoa carambola* L. | Kamranga | Averrhoaceae | T | Homestead |
| 20 | *Avicennia alba* Blume. | Sada Baen | Avicenniaceae | T | Mangrove |
| 21 | *Avicennia marina* (Forsk.) Vierh. | Moriccha Baen | Avicenniaceae | T | Mangrove |
| 22 | *Avicennia officinalis* L. | Kalo Baen | Avicenniaceae | T | Mangrove |
| 23 | *Azadirachta indica* A.Juss. | Neem | Meliaceae | T | Homestead |
| 24 | *Bambusa vulgaris* Schrad.*ex* Wendl. | Baijja Bans | Poaceae | T | Homestead |
| 25 | *Basella rubra* L. | Poi Shak | Basellaceae | C | Homestead |
| 26 | *Benincasa hispida* (Thunb.) Cogn. | Chal Kumra | Cucurbitaceae | C | Homestead |
| 27 | *Blumea lacera* (Burm.f.) | Kukur Muta | Asteraceae | H | Roadside |
| 28 | *Brassica juncea* (L.) Czern. | Rai Sorisa | Brassicaceae | H | Homestead |
| 29 | *Calotropis procera* (Ait.) R. Br. | Akanda | Asclepiadaceae | T | Homestead |
| 30 | *Canavalia virosa* (Roxb.) Wight & Arn. | Kalo Shim | Fabaceae | H | Homestead |
| 31 | *Capsicum frutescens* L. | Morich | Solanaceae | H | Homestead, Cultivated |
| 32 | *Carica papaya* L. | Pepe | Caricaceae | S | Homestead |
| 33 | *Carissa carandas* L. | Koromcha | Apocynaceae | S | Homestead |
| 34 | *Cassia fistula* L. | Sonalu | Caesalpiniaceae | T | Homestead |
| 35 | *Casuarina equisetifolia* Forst. | Jau | Casuarinaceae | T | Sandy beach, Roadside |

(Contd.

| SN | Scientific name | Local name | Family name | Habit | Habitat |
|---|---|---|---|---|---|
| 36 | *Catharanthus roseus* (L.) G.Don | Nayantara | Apocynaceae | H | Homestead |
| 37 | *Ceiba pentandra* (L.) Gaertn. | Burma Simul | Bombacaceae | T | Homestead |
| 38 | *Cicca acida* (L.) Merr. | Orboroi | Euphorbiaceae | T | Homestead |
| 39 | *Citrus aurantifolia* (Christm. & Panzer) Swingle | Lebu | Rutaceae | T | Homestead |
| 40 | *Citrus grandis* (L.) Osbeck | Jambura | Rutaceae | T | Homestead |
| 41 | *Citrullus lanatus* (Thunb.) Matsum. & Nakai | Tormuj | Cucurbitaceae | C | Cultivated |
| 42 | *Clerodendrum inerme* (L.) Gaertn. | Bonjui | Verbenaceae | S | Roadside |
| 43 | *Cocos nucifera* L. | Narikel | Arecaceae | T | Homestead |
| 44 | *Colocasia esculenta* (L.) Schott | Kachu | Araceae | H | Homestead |
| 45 | *Commelina benghalensis* L. | Kanchira | Commelinaceae | H | Cultivated |
| 46 | *Corypha umbraculifera* L. | | Arecaceae | T | Homestead |
| 47 | *Crotalaria juncea* L. | Junjuni | Fabaceae | H | Roadside |
| 48 | *Cucumis melo* L. | Bangi | Cucurbitaceae | C | Cultivated |
| 49 | *Cucumis sativus* L. | Khira | Cucurbitaceae | C | Cultivated |
| 50 | *Cucurbita maxima* Duch. *ex* Lamk. | Misti Kumra | Cucurbitaceae | C | Homestead, Cultivated |
| 51 | *Curcuma longa* L. | Halud | Zingiberaceae | H | Homestead |
| 52 | *Cynodon dactylon* (L.) Pers. | Durbagass | Poaceae | H | Roadside |
| 53 | *Cyperus javanicus* Houtt. | Kucha | Cyperaceae | H | Roadside |
| 54 | *Dalbergia spinosa* Roxb. | Churilla kanta | Fabaceae | C | Mangrove |
| 55 | *Delonix regia* Rafin. | Krishnachura | Caesalpiniaceae | T | Homestead |
| 56 | *Dendrocalamus giganteus* Wall. *ex* Munro | Budhum bans | Poaceae | T | Homestead |
| 57 | *Dioscorea bulbifera* L. | Pagla alu | Dioscoreaceae | S | Homestead |
| 58 | *Eclipta alba* (L.) Hassk. | Kesaraj | Asteraceae | H | Roadside |
| 59 | *Elaeis guineensis* Jacq. | Oil Palm | Arecaceae | T | Homestead |
| 60 | *Erythrina fusca* Lour. | Kata Mandar | Fabaceae | T | Homestead |
| 61 | *Eucalyptus camaldulensis* Dehnh. | Euclyptus | Myrtaceae | T | Homestead |
| 62 | *Eupatorium odoratum* L. | Assam Gach | Asteraceae | H | Cultivated |
| 63 | *Excoecaria agallocha* L. | Gewa | Euphorbiaceae | T | Mangrove |
| 64 | *Ficus benghalensis* L. | Bot | Moraceae | T | Homestead |
| 65 | *Garuga pinnata* Roxb. | Bhadi | Burseraceae | T | Homestead |
| 66 | *Gmelina arborea* Roxb. | Gamar | Verbenaceae | T | Homestead |
| 67 | *Hedyotis corymbosa* (L.) Lam. | Khetpapra | Rubiaceae | H | Roadside, Cultivated |
| 68 | *Heliotropium curassavicum* L. | Hatisur | Boraginaceae | S | Mangrove |
| 69 | *Heliotropium indicum* L. | Hatisur | Boraginaceae | H | Roadside |
| 70 | *Hibiscus rosa-sinensis* L. | Joba | Malvaceae | S | Homestead |
| 71 | *Hopea odorata* Roxb. | Telsur | Dipterocarpaceae | T | Homestead |
| 72 | *Hyptis suaveolens* (L.) Poit. | Tokma | Lamiaceae | S | Roadside |
| 73 | *Imperata cylindrica* (L.) P. Beauv. | Chan | Poaceae | H | Roadside |

(Contd.)

| SN | Scientific name | Local name | Family name | Habit | Habitat |
|---|---|---|---|---|---|
| 74 | *Ipomea batatas* (L.) Lam. | Misti alu | Convolvulaceae | C | Homestead, Cultivated |
| 75 | *Ipomea pes-caprae* (L.) R. Br. | Sagorlata | Convolvulaceae | C | Sandy beach |
| 76 | *Ipomoea aquatica* Forsk. | Kolmi Shak | Convolvulaceae | C | Homestead |
| 77 | *Ipomoea fistulosa* Mart. *ex* Choisy | Dolkolomi | Convolvulaceae | S | Roadside |
| 78 | *Jatropha curcas* L. | Baghverenda | Euphorbiaceae | S | Roadside |
| 79 | *Justicia gendarussa* Burm. f. | Jagmodon | Acanthaceae | H | Roadside |
| 80 | *Lablab purpureus* (L.) Sweet | Sheem | Fabaceae | C | Homestead |
| 81 | *Lagenaria vulgaris* Seringe | Lao | Cucurbitaceae | C | Homestead |
| 82 | *Lagerstroemia speciosa* (L.) Pers. | Jarul | Lythraceae | T | Roadside |
| 83 | *Lannea coromandelica* (Houtt.) Merr. | Bhadi | Anacardiaceae | T | Homestead |
| 84 | *Lantana camara* L. | Moggula | Verbenaceae | S | Homestead, Roadside |
| 85 | *Launaea sarmentosa* (Wild.) Sch. Bip. *ex* Kantze | | Asteraceae | H | Roadside |
| 86 | *Lawsonia inermis* L. | Mendi | Lythraceae | S | Homestead |
| 87 | *Leucaena leucocephala* (Lam.) de Wit. | Ipil-Ipil | Mimosaceae | T | Homestead |
| 88 | *Leucas aspera* (willd.) Link. | Shetodhrona | Lamiaceae | H | Roadside |
| 89 | *Leucas cephalotes* (Roth) Spreng. | Bara-halkus | Lamiaceae | H | Roadside |
| 90 | *Lindernia ciliata* (Colsm.) Pennell | Bhui | Scrophulariaceae | T | Roadside |
| 91 | *Ludwigia adscendens* (L.) Hara | Kesra-dum | Onagraceae | H | Roadside |
| 92 | *Luffa cylindrica* M. Roem. | Dundul | Cucurbitaceae | C | Roadside |
| 93 | *Lumnitzera racemosa* Willd. | Kirpa | Combretaceae | T | Mangrove |
| 94 | *Lycopersicon esculentum* Mill. | Tomato | Solanaceae | H | Homestead, Cultivated |
| 95 | *Mangifera indica* L. | Aam | Anacardiaceae | T | Homestead |
| 96 | *Mimosa pudica* L. | Lojjaboti | Mimosaceae | H | Roadside |
| 97 | *Moringa oleifera* Lamk. | Shajna | Moringaceae | T | Homestead |
| 98 | *Musa paradisiaca* L. | Kola | Musaceae | H | Homestead |
| 99 | *Neolamarckia cadamba* (Roxb.) Bosser. | Kadam | Rubiaceae | T | Homestead |
| 100 | *Opuntia dillenii* Haw. | Foni Monsha | Cactaceae | C | Homestead |
| 101 | *Oryza sativa* L. | Dhan | Poaceae | H | Cultivated |
| 102 | *Oxystelma secamone* (L.) Karst. | Dudhia kata | Asclepiadaceae | H | Roadside |
| 103 | *Pandanus fascicularis* Lamk. | Keyakata | Pandanaceae | T | Sandy beach |
| 104 | *Pandanus foetidus* Roxb. | Keyakata | Pandanaceae | S | Sandy beach |
| 105 | *Paspalum vaginatum* Sw. | | Poaceae | H | Cultivated |
| 106 | *Passiflora foetida* L. | Jumka lata | Passifloraceae | C | Homestead |
| 107 | *Phoenix sylvestris* (L.) Roxb. | Deshi Khejur | Arecaceae | T | Homestead |
| 108 | *Pithecellobium dulce* (Roxb.) Benth. | Jilapi | Mimosaceae | T | Homestead |
| 109 | *Porteresia coarctata* (Roxb.) Tateoka | Urigrass | Poaceae | H | Mangrove meadow |
| 110 | *Portulaca oleracea* L. | Nuinnashak | Portulacaceae | H | Mangrove meadow |
| 111 | *Psidium guajava* L. | Payara | Myrtaceae | T | Homestead |

(Contd.)

| SN | Scientific name | Local name | Family name | Habit | Habitat |
|---|---|---|---|---|---|
| 112 | *Psilotrichum ferrugineum* (Roxb.) Moq.-Tand. | Khetapada Shak | Amaranthaceae | H | Homestead, Roadside |
| 113 | *Raphanus sativus* L. | Mula | Brassicaceae | H | Cultivated |
| 114 | *Ricinus communis* L. | Varenda | Euphorbiaceae | T | Homestead |
| 115 | *Samanea saman* (Jacq.) Merr. | Raintree | Mimosaceae | T | Homestead |
| 116 | *Senna tora* (L.) Roxb. | Terasena | Caesalpiniaceae | H | Roadside |
| 117 | *Sida cordifolia* L. | Berela | Malvaceae | H | Homestead |
| 118 | *Solanum melongena* L. | Begun | Solanaceae | H | Homestead |
| 119 | *Sonneratia apetala* Buch.-Ham. | Keora | Sonneratiaceae | T | Mangrove |
| 120 | *Spinacia oleracea* L. | Palon Shak | Chenopodiaceae | H | Homestead |
| 121 | *Spondias pinnata* (L. f.) Kurz. | Amra | Anacardiaceae | T | Homestead |
| 122 | *Streblus asper* Lour. | Sheora | Moraceae | T | Homestead |
| 123 | *Suaeda maritima* (L.) Dumort. | | Chenopodiaceae | H | Roadside |
| 124 | *Swietenia mahagoni* Jacq. | Mahogoni | Meliaceae | T | Homestead |
| 125 | *Synedrella nodiflora* (L.) Gaertn. | Not known | Asteraceae | H | Roadside |
| 126 | *Syzygium fruticosum* DC. | Putijam | Myrtaceae | T | Homestead |
| 127 | *Tamarindus indica* L. | Tentul | Caesalpiniaceae | T | Homestead |
| 128 | *Tamarix gallica* L. | Nona jau | Tamaricaceae | S | Mangrove |
| 129 | *Tephrosia purpurea* (L.) Pers. | Bon-neel | Fabaceae | H | Cultivated |
| 130 | *Terminalia arjuna* (Roxb. Ex DC.) | Arjun | Combretaceae | T | Roadside |
| 131 | *Terminalia catappa* L. | Kat Badam | Combretaceae | T | Homestead |
| 132 | *Thevetia peruviana* (Pers.) K. Schum. | Halde Karabi | Apocynaceae | T | Homestead |
| 133 | *Thysanolaena maxima* (Roxb.) O. Kuntze | Jahruful | Poaceae | H | Homestead |
| 134 | *Trichosanthes anguina* L. | Chichinga | Cucurbitaceae | C | Cultivated, Homestead |
| 135 | *Typha domingensis* (Pars.) *ex* Steud. | Hogla | Typhaceae | H | Wetland |
| 136 | *Vitex negundo* L. | Nil Nishinda | Verbenaceae | S | Sandy beach, Roadside |
| 137 | *Vitex trifolia* L. f. | Nishinda | Verbenaceae | S | Sandy beach, roadside |
| 138 | *Ziziphus mauritiana* Lamk. | Boroi | Rhamnaceae | T | Homestead |

[* T- Tree, S-Shrub, H-Herb, C-Climber]

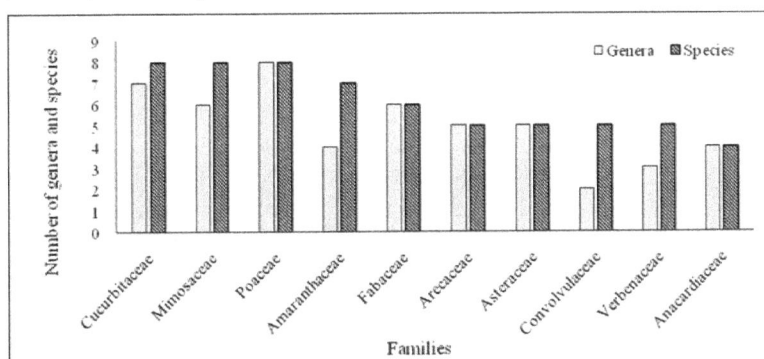

Fig. 2. Number of species belonging to dominant Family in Sonadia Island.

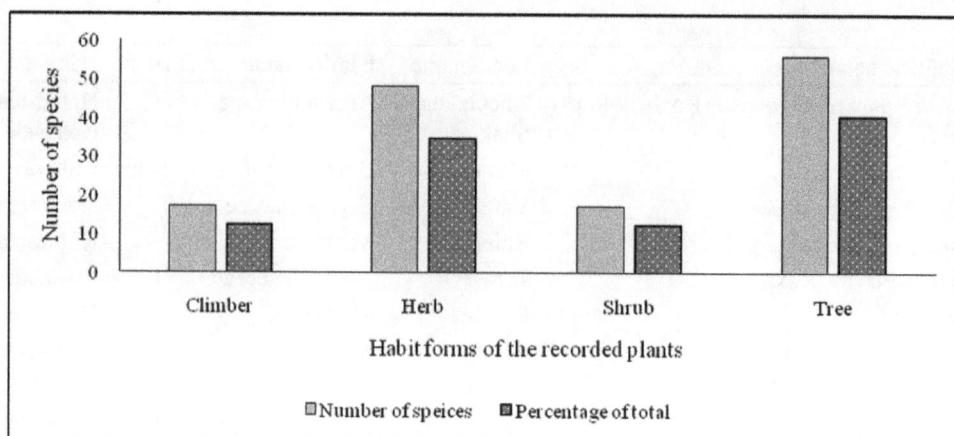

Fig. 3. Number of species belonging to habit form in Sonadia island.

## Major plant habitats in Sonadia Island

The Sonadia Island supports vegetation growing in 6 broad categories of habitats including sand dunes or sandy beach area, homestead, mangrove, mangrove meadow, bounds or foot trail or roadside and cultivated land. Homestead represented 78 species constituting 53% of total species followed by 23% in roadside, 10% in cultivation firms, 9% in 10% in cultivation firms, 9% in mangrove, and 1% in wetland. Plants common in the sand dunes constitute 4% of total are species, in particular *Ipomea pes-caprae*, *Vitex trifolia*, *Pandanus foetidus* and *Casuarina equisetifolia*. Plants commonly occurring in the homesteads are *Acacia auriculiformis*, *Cocos nucifera* and *Eucalyptus camaldulensis* etc. *Avicennia officinalis*, *Avicennia alba* and *Acanthus illicifolius* appeared as very common in the natural mangrove forest, whereas in the plantations *Casuarina equisetifolia*, *Eucalyptus camaldulensis*, *Acacia auriculiformis* and *Sonneratia apetala* were commonly found.

## Traditional uses of the recorded plants

Knowledge about the various uses of the available plants was gained through conversations made with the local peoples living within the island. Traditional use of the recorded plants indicate that most of the plants (33%) have food value as fruit, flower, seed and different parts of those plants are edible in raw or after processing. Plants also used substantially as fuel wood (18%), timber (11%), biological fence (9%), medicine (7%) etc. It is found that many medicinal plants, their medicinal values and uses are not known to local people. Plants that provide fodder, oils, weeds etc. are grouped under miscellaneous category which constitutes 11% of all recorded plant species.

## Discussion

The study reveals that Sonadia island currently harbours 138 plant species (tree 56, shrubs 17, herbs 48, climbers 17) that belong to 111 genera and 55 families which is higher in comparison to Moloney (2006) that recorded only 60 vascular plants from Sonadia island (14 trees, 8 shrubs, 27 herbs and 11 climbers species). In the first report on the angiospermic flora of this land (Khan *et al.*, 1977) the number of species was for less. According to the people living in the island, vegetation coverage in mangrove forest was dense. The findings conform with the reports of Thompson and Islam (2010) who indicated 144 angiospermic plants from Saint Martin's Island of Cox's Bazar. Sandwip, another island of Bangladesh harbours much higher plants (438 vascular plants) due to its comparatively larger area coverage and varied households with diversified domestic flora (Sajib *et al.*, 2016). The floristic records of different island's of Bangladesh also reported 149 species from Moheshkhali (Huq and Khan, 1984), 151 species from the same island (Rashid *et al.*, 2000), 91 species from Kutubdia island (Huq 1986), 98 plant species from Hatiya island (Huq 1988) and 37 species from Nijhum Dwip (Khan *et al.*, 1985) and 152 species from Nijhum Dwip (Uddin *et al.*, 2015).

The presence of some exotic tree species, i.e., *Acacia auriculiformis, Swietenia mahagoni* and *Eucalyptus camaldulensis* species was due to the plantations conducted by Bangladesh Forest Department and the local people. Major area of the island is occupied by natural mangroves, but encroachment is becoming a serious concern because of the conversion of forest lands to salt bed and shrimp cultivation. Jhau, the successful species in the sandy beaches of Cox's Bazar (Hossain, 2010)) is also promising in Sonadia Island but illegal felling is a common threat in the island).

## Acknowledgements

The authors are grateful to the Research Cell authority of University of Chittagong for providing funds for field works. We are also grateful to the officers and field staffs of Chittagong Coastal Forest Division, Bangladesh Forest Department for helping in the field work. Thanks are due to Taxonomists of Forest Botany Division, BFRI and Dr. Mohammed Yusuf, Ex-Director of BCSIR for their supports in identification of the plant samples.

## References

Ahmed, Z.U., Begum, Z.N.T., Hassan, M.A., Khondker, M., Kabir, S.M.H., Ahmad, M., Ahmed, A.T.A., Rahman, A.K.A. and Haque, E.U. (Eds). 2008. Encyclopedia of Flora and Fauna of Bangladesh, vol. 5-12. Asiatic Society of Bangladesh, Dhaka.

DoE (Department of Environment). 1999. GIS and Cartographic Services – Final Report, Pre-Investment Facility Study: Coastal and Wetland Biodiversity Management Project (Project BGD/94/G41), Dhaka, Bangladesh.

Heinig, R.L. 1925. List of Plants of Chittagong Collectorate and Hill Tracts. Darjeeling, India, 84 pp.

Hossain, M.K. 2010. *Casuarina equisetifolia*- a promising species for green belt project of coastal and off-shore islands of Bangladesh. *In:* Zhong, C., Pinyopusarerk, K., Kalinganire, A. and Franche C. (eds.), Improving Smallholder Livelihoods Through Improved *Casuarina* Productivity: Proceedings of the 4th International *Casuarina* Workshop, Haikou, China 21-25 March 2010. pp. 200–206.

Huq, A.M. 1986. Preliminary studies on the anthropogenic flora of Kutubdia Island in Bangladesh. J Asiatic Soc. Bangladesh (Sci.) 12: 59-70.

Huq, A.M. 1988. A Preliminary taxonomic report on the Angiospermic flora of Hatia Islan (Noakhali district) (Dicotyledons). Bull. Bangladesh Nat. Herb., Dhaka **1**: 1–10.

Huq, A.M. and Khan, M.S. 1984. A preliminary taxonomic report on the angiospermic flora of Maheshkhali Island-1 (Dicotyledons). Dhaka Univ. Stud. B **32**: 19–31.

Khan, M.S., Huq, M.A. Rahman, M.M. and Hassan, M.A. 1977. A preliminary report on the angiospermic flora of Sonadia island, Bangladesh. J. Asit. Sco. Bangladesh **3**(1): 125–126.

Khan, M.S., Huq, A.M. and Rahman, M.M. 1985. Studies on the angiospermic flora of Nijhum Dwip (Char Osman) in the Bay of Bengal. Dhaka Univ. Stud. B **33**: 145–151.

MoEF (Ministry of Environment and Forests) 2015. Fifth National Report to the Convention on Biological Diversity. Government of the People's Republic of Bangladesh, Dhaka, 164 pp.

Moloney, L. 2006. Coastal and Wetland Biodiversity Management Plan BGD/ 99/ G31 Sonadia Island ECA Draft Conservation Management Plan.

Prain, D. 1903 (Reprinted.1981). Bengal Plants. Calcutta, **1**: 120 pp.

Rashid, M.H., Rahman, E. and Rahman, M.A. 2000. Additions to the angiospermic flora of the Moheskhali island. Cox's Bazar. Bangladesh J Plant Taxon 7: 43–63.

Sajib, N.H., Uddin, S.B. and Islam, M.S. 2016. Vascular plant diversity and their distribution pattern in Sandwip Island, Chittagong, Bangladesh. J Biodivers Manage Forestry **5**: 2

Siddiqui, K.U., Islam, M.A., Ahmed, Z.U., Begum, Z.N.T., Hassan, M.A., Khondker, M., Rahman, M.M., Kabir, S.M.H., Ahmed, M., Ahmed, A.T.A., Rahman, A.K.A. and Haque, E.U. (eds). 2007. Encyclopedia of Flora and Fauna of Bangladesh, Vol. 11. Angiosperms: (Agavaceae - Najadaceae). Asiatic Society of Bangladesh, pp. 399.

Thompson, P.M. and Islam, M.A. (Eds.). 2010. Environmental Profile of St. Martin's Island,        United Nations Development Programme, Dhaka. Washington, DC: Island Press, pp. 112–117.

Uddin, M.Z., Kibria, M.G. and Hassan, M.A. 2015. Assessment of angiosperm plant diversity of Nijhum Dweep. Bangladesh J. Asiat. Soc. Bangladesh, Sci. **41**(1): 19–52.

# Permissions

# List of Contributors

**N. M. George and A. Ghareeb**
Department of Botany, Faculty of Science, Zagazig University, Egypt

**N. M. Fawzi and S. Saad**
Flora and Phytotaxonomy Research Department, Horticultural Research Institute, Agriculture Research Center, Cairo-Egypt

**S. Mesut Pinar**
Yüzüncü Yıl University, Faculty of Science, Department of Biology, 65080, Van/Turkey

**Lütfi Behçet**
Bingöl University, Science and Art Faculty, Department of Biology, 12000, Bingöl/Turkey

**Maksuda Khatun, Md. Abul Hassan and Shaikh Nazrul Islam**
Department of Botany, University of Dhaka, Dhaka 1000, Bangladesh

**M. Oliur Rahman**
Institute of Nutrition and Food Science, University of Dhaka, Dhaka 1000, Bangladesh

**M. Oliur Rahman, Momtaz Begum and Md. Wajib Ullah**
Department of Botany, University of Dhaka, Dhaka 1000, Bangladesh

**M. Sivadasan**
Department of Botany and Microbiology, College of Science, King Saud University, Riyadh 11451, Kingdom of Saudi Arabia

**V. Abdul Jaleel and Ahmed H. Alfarhan**
Department of Post-graduate Studies and Research in Botany, Sir Syed College, Taliparamba, Kannur 670 142, Kerala, India

**P. Lakshminarasimhan**
Central National Herbarium, Botanical Survey of India, Howrah 711 103, India

**M. Ajmal Ali and Fahad M. A. Al-Hemaid**
Department of Botany and Microbiology, College of Science, King Saud University, Riyadh-11451, Saudi Arabia

**Arun K. Pandey**
Department of Botany, University of Delhi, Delhi-110007, India

**Joongku Lee**
International Biological Material Research Center, Korea Research Institute of Bioscience and Biotechnology, Daejeon- 305806, South Korea

**M. S. Kiran Raj**
Department of Botany, Sree Narayana College, Cherthala 688 582, Alappuzha, Kerala, India

**A. H. Alfarhan**
Department of Botany, University of Calicut, Kerala, India

**M. Sivadasan**
Department of Botany, University of Calicut, Kerala, India
Department of Biology, College of Science, Princess Nora bint Abdulrahman University, Riyadh 11652, Kingdom of Saudi Arabia

**J. F. Veldkamp**
Department of Botany & Microbiology, College of Science, King Saud University, Riyadh 11451, Kingdom of Saudi Arabia

**A. S. M. Amal Tamimi**
Naturalis Biodiversity Center, 2300 RA Leiden, The Netherlands

**Tulika Talukdar and Sobhan Kumar Mukherjee**
Department of Botany, University of Kalyani, Kalyani, Nadia 741235, West Bengal, India

**Rajeev Kumar Singh and Arti Garg**
Botanical Survey of India (BSI), Central Regional Centre (CRC), 10-Chatham Lines, Allahabad 211 002, Uttar Pradesh, India

**Paramjit Singh**
Botanical Survey of India, Head Quarter, CGO Complex, Salt Lake City, Kolkata 700 064, West Bengal, India

**Lyubov A. Kameneva and Inna M. Koksheeva**
Botanical Garden-Institute, Far Eastern Branch of Russian Academy of Sciences (BGI FEB RAS), Vladivostok, Russia

**N. Sarojini Devi, Y. Padma, C. L. Narasimhudu and R. R. Venkata Raju**
Biosystematics and Phytomedicine Division, Department of Botany, Sri Krishnadevaraya University, Anantapur - 515 055, Andhra Pradesh, India

**Pravin Patil**
National Bureau of Plant Genetic Resources, Pusa Campus, New Delhi 110012, India

**Shrikant Sutar, Surendra Kumar Malik, Shrirang Yadav and Kangila Venkataraman Bhat**
Botany Department, Shivaji University, Kolhapur 416004, India

**Joseph John**
National Bureau of Plant Genetic Resources Regional Station, KAU PO, Thrissur 680656, India

**A. K. M. Golam Sarwar**
Laboratory of Systematic Botany, Graduate School of Agriculture, Hokkaido University, Japan

**Hideki Takahashi**
Department of Crop Botany, Bangladesh Agricultural University, Mymensingh 2202, Bangladesh

**M. Oliur Rahman, Md. Zahidur Rahman and Ayesa Begum**
Department of Botany, University of Dhaka, Dhaka 1000, Bangladesh

**V. Abdul Jaleel**
Department of Botany, University of Calicut, 673 635, Kerala, India Materials and Methods

**M. Sivadasan**
Department of Botany, Sir Syed College, Taliparamba, Kannur-670 142, Kerala, India Department of Botany & Microbiology, College of Science, King Saud University, Riyadh-11451, Kingdom of Saudi Arabia

**Ahmed H. Alfarhan, Jacob Thomas and A. A. Alatar**
Department of Botany & Microbiology, College of Science, King Saud University, Riyadh-11451, Kingdom of Saudi Arabia

**Huan-Fang Liu, Yun-Fei Deng and Jing-Ping Liao**
Key Laboratory of Plant Resources Conservation and Sustainable Utilization, Chinese Academy of Sciences, CN-510650 Guangzhou, PR China

**M. M. Islam and M. K. Huda**
Department of Botany, University of Chittagong, Chittagong 4331, Bangladesh

**M. Halim**
Chittagong Education Board, Chittagong, Bangladesh

**Mohammad Harun-Ur-Rashid, Saiful Islam and Sadia Binte Kashem**
Department of Botany, University of Chittagong, Chittagong 4331, Bangladesh

**M. Sankara Rao**
Botanical Survey of India, Sikkim Himalayan Regional Centre, Gangtok, Sikkim- 737103, India

**J. Swamy, K. Chandramohan and S. Nagaraju**
Botanical Survey of India, Deccan Regional Centre, Hyderabad – 500048, Telangana, India

**S. B. Padal and M. Tarakeswara Naidu**
Department of Botany, Andhra University, Visakhapatnam, Andhra Pradesh, India

**T. Thulasiah**
S.L.V.Garden & Landscape developers, Tirupati, Andhra Pradesh, India

**Zakaria S. Almola, Basheer A. Al-Ni'ma and Nadeem A. Ramadan**
Department of Biology, College of Sciences, University of Mosul, Iraq

**M. S. Arefin, M. K. Hossain and M. Akhter Hossain**
Institute of Forestry and Environmental Sciences, University of Chittagong, Chittagong 4331, Bangladesh

# Index